Springer Series in
ADVANCED MICROELECTRONICS   2

Springer

*Berlin
Heidelberg
New York
Barcelona
Hong Kong
London
Milan
Paris
Singapore
Tokyo*

Springer Series in
ADVANCED MICROELECTRONICS

The Springer Series in Advanced Microelectronics provides systematic information on all the topics relevant for the design, processing, and manufacturing of microelectronic devices. The books, each prepared by leading researchers or engineers in their fields, cover the basic and advanced aspects of topics such as wafer processing, materials, device design, device technologies, circuit design, VLSI implementation, and subsystem technology. The series forms a bridge between physics and engineering and the volumes will appeal to practicing engineers as well as research scientists.

1 **Cellular Neural Networks**
   Chaos, Complexity, and VLSI Processing
   By G. Manganaro, P. Arena, and L. Fortuna
2 **Technology of Integrated Circuits**
   By D. Widmann, H. Mader, and H. Friedrich
3 **Ferroelectric Memories**
   By J.F. Scott

D. Widmann   H. Mader   H. Friedrich

Technology of Integrated Circuits

With 300 Figures

Springer

Dr.-Ing. Dietrich Widmann †
semiconductor technology development
Infineon Technologies
Balanstraße 73
D-81609 Munich
Germany

Prof. Dr.-Ing. Hermann Mader
Department of Electrical Engineering
and Information Technologies
Munich University of Applied Sciences
Lothstraße 34
D-80335 Munich
Germany

Dr.-Ing. Hans Friedrich
TELA Versicherungs AG
Pranner Straße 8
D-80333 Munich
Germany

Prof. Dr. rer.nat. Walter Heywang
Technical University Munich
Arcisstraße 21
D-80333 Munich
Germany

Prof. Dr. techn. Rudolf Müller
Technical University Munich
Arcisstraße 21
D-80333 Munich
Germany

Series Editor:
Dr. Kiyoo Itoh
Hitachi Ltd., Central Research Laboratory
1-280 Higashi-Koigakubo
Kokubunji-shi
Tokyo 185-8601
Japan

Professor Takayasu Sakurai
Center for Collaborative Research
University of Tokyo
7-22-1 Roppongi, Minato-ku,
Tokyo 106-8558
Japan

Library of Congress Catalog Card Number 61642

ISSN 1437-0387
ISBN 3-540-66199-9 Springer-Verlag Berlin Heidelberg New York

This work is subject to copyright. All rights are reserved, whether the whole or part of the material is concerned, specifically the rights of translation, reprinting, re-use of illustrations, recitation, broadcasting, reproduction on microfilms or in other ways, and storage in data banks. Duplication of this publication or parts thereof is only permitted under the provisions of the German Copyright Law of September 9, 1965, in its current version, and permission for use must always be obtained from Springer-Verlag. Violations are liable for prosecution under the German Copyright Law.

Springer-Verlag is a company in the BertelsmannSpringer publishing group
© Springer-Verlag Berlin Heidelberg 2000
Printed in Germany

The use of registered names, trademarks, etc. in this publication does not imply, even in the absence of a specific statement, that such names are exempt from the relevant protective laws and regulations and therefore free for general use.

Typesetting: MEDIO GmbH, Berlin
Cover design: *design & production* GmbH, Heidelberg
production: Produserv
Printed on acid-free paper   SPIN: 10729795   62/3020 Pt - 5 4 3 2 1 0

Preface to the english edition

The first edition of this book was very well received. Because of its proximity to the application of the most advanced technologies it became a standard for industrial process engineers as well as for scientists at universities and research labs.

The fundamental revision for the second edition sustained the modern profile of the book and its role as a technology guide. Many of our english speaking colleagues continously encouraged and finally convinced us to prepare an english edition. The authors hope that this edition will also be well received and they wish that it can contribute to the exciting progress in microelectronics technology. The focus on practical applications should provide the book with a firm position within technology literature.

Sadly we remember Dr. D. Widmann, who died in August 1999. Dr. Widmann was the spirit of this book. His profound knowledge and experience in semiconductor technology and industrial applications were highly respected internationally and giving a substantial formative impact to this book. It will be his scientific legacy and so he will continue his outstanding contribution to technology development.

We want to express our thanks to Dr. Middlehurst, who did the English translation of this complicated subject. Our thanks also go to Infineon technologies and especially to Dr. W. Beinvogl for promotion and generous support of this english edition.

Finally, we wish to thank Dr. D. Merkle, Ms. G. Maas, Ms. R. Peters and Mr. J. Wandtke for editing of the manuscript.

Munich, June 2000   H. Mader, H. Friedrich

Preface by the editors

Semiconductor devices dominate a large sector of today's electrical industry. The evidence lies in both the huge range of new components available and in the average growth in manufactured output, running at a rate of about 20% per year over the last 20 years. It is the special physical and functional properties of semiconductor devices that have enabled the development of such complex electronic systems as are found today in data processing and information technology. This progress was only possible because of combined efforts in the research of fundamental physical properties and the development of engineering solutions.

Success in working with this range of components, and the ability to keep pace with future requirements, demands mastery of a broad spectrum of knowledge, from understanding basic physical principles to recognizing device applications according to their performance characteristics. This applies not just to the device developer but also to the circuit design engineer.

The "Semiconductor Electronics" series of books recognizes this close relationship between physical operation and the objectives of electrical design. It covers the physical principles behind the operation of semiconductor devices (diodes, transistors, thyristors etc.), their method of manufacture and their electrical data.

The series has been designed on a "modular" basis. This is the best means of keeping abreast of continued developments, and of providing readers with a useful tool both in their studies and professional work.

The first two volumes are intended as introductions to the field. Volume 1 presents the basic physical principles of semiconductors, and defines and explains the relevant terminology. Volume 2 deals as simply as possible with those semiconductor devices and integrated circuits that are of importance today. These two volumes are accompanied by volumes 3 to 5 and volume 19, which deal in more depth with semiconductor band structure and transport phenomena, and also provide an introduction to the basic technological principles of semiconductor manufacture. All these volumes are based on single-term lecture courses for core and specialized options at technical universities.

W. Heywang and R. Müller

Preface to the 2nd german edition

Since the first edition of this book originally appeared eight years ago, there have been considerable developments in many areas of integrated circuit technology. The second edition takes these advances into account, focusing on those technologies and processes that are used today, or will be introduced in the near future in progressive industrial production lines.

We have completely revised chapter 8 "Process Integration". It now presents a comprehensive summary of all important technologies, starting from the basic CMOS process. Each processing module within a technology is dealt with in detail, in particular the isolation of transistors, the transistors themselves, memory cells, planarization and contacts and interconnections. The chapter is completed by detailed descriptions of the processing sequence for four selected technologies.

Our thanks go to all those who have helped us to write this second edition: Mrs Vogs for typing some of the manuscript; Professor Higelin for his critique of part of chapter 8; and finally Dr. Arden, Dipl.-Ing. Bitto, Dipl.-Phys. Enders, Dr. Erb, Dr. Frank, Dipl.-Ing. Hiller, Dr. Mathuni, Dipl.-Phys. Melzner and Dipl.-Phys. Pöhle for their valuable advice on individual chapters.

Once again we give our heartfelt thanks to our wives for their understanding over the (free) time we have devoted to writing this book.

Munich, April 1996   D. Widmann, H. Mader, H. Friedrich

Contents

List of Symbols XIII
Key to the different layers in integrated circuits XVIII

1   Introduction 1

2   Basic principles of integrated circuits technology 3

3   Film technology 13
    3.1   Film production processes 13
          3.1.1   The CVD process 13
          3.1.2   Thermal oxidation 20
          3.1.3   Vapour phase deposition 27
          3.1.4   Sputtering 29
          3.1.5   Spin coating 33
          3.1.6   Film production by ion implantation 34
          3.1.7   Film production using wafer-bonding and
                  back-etching 34
          3.1.8   Annealing techniques 35
    3.2   The monocrystalline silicon wafer 38
          3.2.1   Geometry and crystallography of silicon wafers . 38
          3.2.2   Doping of silicon wafers 39
          3.2.3   Monocrystalline silicon growing techniques 39
    3.3   Epitaxial layers 41
          3.3.1   Uses for epitaxial layers 41
          3.3.2   Diffusion of doping atoms from the substrate
                  into the epitaxial layer 43
    3.4   Thermal SiO_2 layers 46
          3.4.1   Uses of thermal SiO_2 layers 46
          3.4.2   The LOCOS technique 47
          3.4.3   Properties of thin thermal SiO_2 films 53
    3.5   Deposited SiO_2 films 59
          3.5.1   Creating deposited SiO_2 films 59
          3.5.2   Applications of deposited SiO_2 films 60
          3.5.3   Spacer technology 60
          3.5.4   Trench isolation 62
          3.5.5 SiO_2 isolation films for multi-level metallization 62
    3.6   Phosphorus glass films 63
          3.6.1   Producing phosphorus glass films 64

		3.6.2 Flow-glass	66
		3.6.3 Thermal phosphorus glass	67
	3.7	Silicon nitride films	67
		3.7.1 Producing silicon nitride films	68
		3.7.2 Nitride films as an oxidation barrier	68
		3.7.3 Nitride films as a capacitor dielectric	69
		3.7.4 Using nitride films for passivation	70
	3.8	Polysilicon films	70
		3.8.1 Producing polysilicon films	70
		3.8.2 Grain structure of polysilicon films	71
		3.8.3 Conductivity of polysilicon films	72
		3.8.4 Uses of polysilicon films	74
	3.9	Silicide films	78
		3.9.1 Producing silicide Films	79
		3.9.2 Polycide films	81
		3.9.3 Silication of source/drain regions	83
	3.10	Refractory metal films	83
	3.11	Aluminium films	85
		3.11.1 Producing aluminium films	85
		3.11.2 Crystal structure of aluminium films	86
		3.11.3 Electromigration in aluminium interconnections	87
		3.11.4 Aluminium-silicon contacts	88
		3.11.5 Aluminium-aluminium contacts	90
	3.12	Organic films	91
		3.12.1 Spin-on glass films	91
		3.12.2 Polyimide films	92
4	**Lithography**		95
	4.1	Linewidth dimension, placement errors and defects	96
	4.2	Photolithography	98
		4.2.1 Photoresist films	98
		4.2.2 Formation of photoresist patterns	102
		4.2.3 Light intensity variation in the photoresist	105
		4.2.4 Special photoresist techniques	110
		4.2.5 Optical exposure techniques	116
		4.2.6 Resolution capability of optical exposure techniques	119
		4.2.7 Alignment accuracy of optical exposure equipment	130
		4.2.8 Defects occurring in optical lithography	133
	4.3	X-ray lithography	134
		4.3.1 Wavelength region for X-ray lithography	135
		4.3.2 X-ray resists	136
		4.3.3 X-ray sources	137
		4.3.4 X-ray masks	142
		4.3.5 Alignment procedure for X-ray lithography	144

		4.3.6	Radiation damage in X-ray lithography	144
		4.3.7	Opportunities for Y-ray lithography	144
	4.4	Electron lithography		145
		4.4.1	Electron resists	145
		4.4.2	Resolution capability of electron lithography	146
		4.4.3	Electron beam pattern generators	148
		4.4.4	Electron projection equipment	153
		4.4.5	Alignment techniques in electron lithography	154
		4.4.6	Radiation damage in electron lithography	154
	4.5	Ion lithography		156
		4.5.1	Ion resists	156
		4.5.2	Ion beam writing	158
		4.5.3	Ion beam projection	160
		4.5.4	Resolution capability of ion lithography	162
	4.6	Pattern generation without using lithography		166
5	**Etching technology**			169
	5.1	Wet etching		170
		5.1.1	Wet chemical etching	170
		5.1.2	Chemical-mechanical polishing	171
	5.2	Dry etching		174
		5.2.1	Physical dry etching	174
		5.2.2	Chemical dry etching	176
		5.2.3	Physical-chemical dry etching	178
		5.2.4	Chemical etching reactions	186
		5.2.5	Etching gases	188
		5.2.6	Process optimization	188
		5.2.7	Endpoint detection	193
	5.3	Dry etch processes		196
		5.3.1	Dry etching of silicon nitride	197
		5.3.2	Dry etching of polysilicon	197
		5.3.3	Dry etching of monocrystalline silicon	199
		5.3.4	Dry etching of metal silicides and refractory metals . .	200
		5.3.5	Dry etching of silicon dioxide	201
		5.3.6	Dry etching of aluminium	203
		5.3.7	Dry etching of polymers	205
6	**Doping technology**			207
	6.1	Thermal doping		208
	6.2	Doping by ion implantation		209
		6.2.1	Ion implantation machines	209
		6.2.2	Implanted doping profiles	211
	6.3	Activation and diffusion of dopant atoms		219
		6.3.1	Activating implanted dopant atoms	219

	6.3.2	Intrinsic diffusion of dopant atoms	220
	6.3.3	Diffusion for high concentrations of dopant atoms ...	223
	6.3.4	Oxidation enhanced diffusion	224
	6.3.5	Diffusion of dopant atoms at interfaces	225
	6.3.6	Diffusion of dopant atoms in films	227
	6.3.7	Sheet resistance of doped layers	229
	6.3.8	Diffusion at the edge of doped regions	230
6.4	Diffusion of non-doping materials		231

7 Cleaning technology 235
- 7.1 Contaminants and their effect 235
- 7.2 Clean rooms, clean materials and clean processes 239
  - 7.2.1 Clean rooms 239
  - 7.2.2 Clean materials 242
  - 7.2.3 Clean processing 244
- 7.3 Wafer cleaning 244

8 Process integration 249
- 8.1 The various MOS and bipolar technologies 249
  - 8.1.1 Active components in integrated circuits 249
  - 8.1.2 Comparsion of MOS and bipolar technologies 249
  - 8.1.3 Passive components in integrated circuits 252
- 8.2 Technology architecture 252
  - 8.2.1 Architecture of MOS technology 252
  - 8.2.2 Architecture of bipolar and BICMOS technologies ... 254
- 8.3 Transistors in integrated circuits 256
  - 8.3.1 Design of MOS transistors and their isolation 256
  - 8.3.2 Design of DMOS transistors 262
  - 8.3.3 Design of bipolar transistors and their isolation 264
- 8.4 Memory cells 267
  - 8.4.1 Design of static memory cells 267
  - 8.4.2 Design of dynamic memory cells 269
  - 8.4.3 Design of non-volatile memory cells 272
- 8.5 Multilayer metallization 276
  - 8.5.1 Planarization of surfaces in integrated circuits 277
  - 8.5.2 Contacts in integrated circuits 281
  - 8.5.3 Metallization in integrated ciruits 284
  - 8.5.4 Passivation of integrated circuits 285
- 8.6 Detailed process sequence of selected technologies 286
  - 8.6.1 Digital CMOS process 286
  - 8.6.2 BICMOS process 286
  - 8.6.3 Microwave bipolar process 286
  - 8.6.4 DRAM process 295

References 323

Subject Index 329

List of Symbols

Parameter	Meaning	Unit
A	proportionality constant for the oxidation rate	m
A	area	m²
a	minimum resist pattern width	m
B	magnetic flux density	T
B	parabolic oxidation constant	m²s⁻¹
B	image site dimension	m
B/A	linear oxidation constant	ms⁻¹
b	pattern width	m
b_{min}	minimum pattern width	m
b	contact hole size	m
C	capacitance	F
C_D	drain-substrate capacitance	F
C_{ox}	oxide capacitance (per unit area)	Fm⁻²
c	speed of light	ms⁻¹
c	particle concentration	m⁻³
c_0	maximum particle concentration	m⁻³
c_v	specific heat of resist	Jm⁻³
D	diffusion constant	m²s⁻¹
D	exposure dose	Jm⁻²; Cm⁻²
D	image site dimension	m
D	defect density	m⁻²
D_0	resist sensitivity	Jm⁻²
D_{Si}	thickness of Si wafer	m
d	layer thickness	m
d	focal spot size	m
d_m	maximum diameter of an aberration-free feature	m
d_f	diameter of image site in ion projection	m
d_{ox}	thickness of SiO$_2$ layer	m
d_p	particle diameter	m
d_r	resist thickness	m
d_{Si}	thickness of silicon layer	m
$d_{Si_3N_4}$	thickness of silicon nitride layer	m
E	electric field strength	Vm⁻¹
E_c	critical field strength	Vm⁻¹
E	electron energy	J

List of Symbols

Symbol	Description	Unit
E_0	rest-mass energy of electrons	J
E_{BD}	breakdown field strength	Vm^{-2}
E_{Si}	modulus of elasticity of silicon	Nm^{-2}
e	elementary charge = $1.6 \cdot 10^{-19}$ C	C
F	force	N
f	frequency	s^{-1}
f	anisotropy factor	–
G	generation rate	m^{-3}s^{-1}
H	magnetic field strength	Am^{-1}
I	electric current	A
I	radiation density, radiation intensity	Wm^{-2}
I_0	radiation density incident on resist	Wm^{-2}
J	ion current density	Am^{-2}
J	gas flow	kgs^{-1}
j	current density	Am^{-2}
j	particle flow density	m^{-2}s^{-1}
K	scaling factor	–
k	diffraction order	–
k	Boltzmann's constant $k = 1.380 \cdot 10^{-23}$ JK^{-1}	JK^{-1}
kT	thermal energy	J
L	channel length of a MOS transistor	m
L_G	gate length, geometrical channel length	m
L_{eff}	effective channel length	m
l	length	m
l_{min}	minimum controllable grid dimension	m
l'	length of contact hole through which current flows	m
M	ion mass	kg
MTF	electromigration lifetime (MTF - Mean Time to Failure)	s
	gas flow rate	kgs^{-1}
m_0	rest mass of electrons	kg
N	ion density	m^{-3}
NA	numerical aperture	–
N_A	acceptor density	m^{-3}
N_D	donor density	m^{-3}
N_D	doping dose	m^{-2}
N_{D0}	total doping dose	m^{-2}
n	refractive index	–
n	number of critical lithography planes	–
n	density of electrons, donor density	m^{-3}
n_r	refractive index of resist	–
n_s	refractive index of substrate	–
n$^+$	high donor density	–
n$^-$	low donor density	–

List of Symbols

P	electrical power	W
p	pressure	Pa
p	density of holes, acceptor density	m^{-3}
p^+	high acceptor density	–
p^-	low acceptor density	–
ppm	parts per million (=10^{-6})	–
ppb	parts per billion (=10^{-9})	–
Q	electric charge	As
Q	charge density	Cm^{-2}
Q_{bd}	defining charge for the breakdown of silicon	As
Q_f	fixed interface charge	As
Q_{it}	interface trapped charge	As
Q_m	mobile charge	As
Q_{ot}	oxide trapped charges	As
R	radius of the electron orbit	m
R	electron range	m
R	electrical resistance	Ω
R	gas constant	JK^{-1}mol^{-1}
R_k	contact hole resistance	Ω
R_{KK}	resistance of a contact hole chain	Ω
R_p	range of ions for ion implantation	m
R_s	sheet resistance	Ω/
r	etch rate	ms^{-1}
r_h	horizontal etch rate	ms^{-1}
r_v	vertical etch rate	ms^{-1}
S	selectivity of an etch process	–
S	distance between focal point and mask in X-ray lithography	m
s	minimum resist pattern separation	m
s	proximity separation: distance between semiconductor wafer and mask	m
T	temperature	K, °C
T	period	s
t	time	s
t_d	delay time	s
U	electrical voltage	V
U_D	drain voltage	V
U_{DD}	supply voltage for MOS circuits	V
U_{DS}	drain-source voltage	V
U_{Diff}	diffusion voltage	V
U_T	threshold voltage of a MOS transistor	V
U_G	gate voltage	V
V	electric potential	V
V	volume	m^3

List of Symbols

Symbol	Description	Units
v	velocity	ms^{-1}
W	channel width of a MOS transistor	m
W	energy	J, eV
W_A	activation energy	J, eV
W_c	energy at bottom of conduction band	J, eV
W_F	Fermi energy	J, eV
W_v	energy at top of valence band	J, eV
w	width	m
w	probability	–
x	spatial co-ordinate parallel to the wafer surface	m
x_{iK}	spatial co-ordinate for the edge of a pattern	m
x_m	spatial co-ordinate for the centre point of a pattern	m
Y	yield	–
Y_A	area yield	–
Y_M	mounting yield	–
Y_T	test yield	–
Y_{TS}	test station yield	–
Y_{chip}	chip yield	–
y	spatial co-ordinate parallel to the wafer surface	m
z	spatial co-ordinate perpendicular to the wafer surface	m
z_j	depth of pn junction	m
α	lens aperture angle	degrees
α	edge angle	degrees
α	convergence angle of an electron beam machine	degrees
α	absorption coefficient	m^{-1}
β	reaction coefficient	$s^{-1}m^{-2}$
γ	resist contrast	–
ΔB	change in size of the image site during X-ray lithography	m
Δb	width of the transition region in X-ray lithography	m
Δf	distance between the image and focal plane	m
Δf_R	Rayleigh depth	m
Δl	path difference for light beams	m
ΔR_p	standard deviation for the doping density in ion implantation	m
δ	line width variance	m
ε	dielectric constant	$AsV^{-1}m^{-1}$
ε_0	dielectric constant of a vacuum $\varepsilon_0 = 8.854 \cdot 10^{-12}\ AsV^{-1}m^{-1}$	$AsV^{-1}m^{-1}$
ε_{Si}	dielectric constant of silicon	$AsV^{-1}m^{-1}$
ε_{SiO_2}	dielectric constant of SiO_2	$AsV^{-1}m^{-1}$
$\varepsilon_{Si_3N_4}$	dielectric constant of Si_3N_4	$AsV^{-1}m^{-1}$
λ	wave length	m

λ_p	wavelength of synchrotron radiation	m
μ	mobility of charge carriers	cm^{2}V^{-1}s^{-1}
$\mu 0$	mobility at low field strengths	cm^{2}V^{-1}s^{-1}
ρ	electrical resistivity	Ωm
ρc	specific contact resistance	Ωm2
ζ	spatial co-ordinate	m
σ	diffusion length	m
σ	parameter of the standard deviation formula	m
τ	time constant	s
Θ	diffraction angle	degrees
ϑ	angle between each diffraction order	degrees
φ	difference in work functions of metal and semiconductor with respect to the elementary charge	V
ψ	divergence angle	degrees
ω	angular frequency	s^{-1}

Key to the different layers in integrated circuits

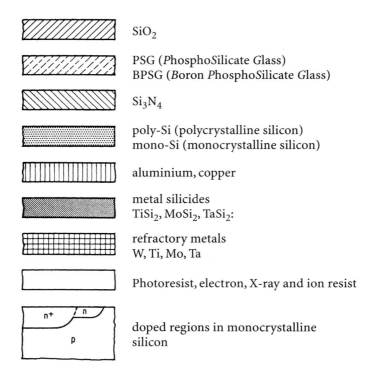

n ≙ donor doping
n⁺≙ high donor density
n⁻≙ low donor density
p ≙ acceptor doping
p⁺≙ high acceptor density
p⁻≙ low acceptor density

1
Introduction

There has probably never been another technology that has developed so rapidly as microelectronics. The global market for integrated circuits is growing at an annual rate of about 15 % to 20 %. One can assume that there will be no slow-down in this growth for many years to come (Fig. 1.1 a). By the year 2010 the world market for integrated circuits could outstrip that for cars.

The main stimulus for this growth is the demand for a particular electronic function at an ever reduced cost. Engineers have managed to continuously reduce the manufacturing costs, for instance of memory cells, and thus to achieve constant rises in productivity (Fig. 1.1 b). Productivity is likely to increase still further in the future, doubling every two to three years. The key factors providing this rise in productivity are CMOS technology and feature size reduction, both enabling very high packing densities to be achieved (Fig. 1.1 c, d). At this point in time there is no sign of any ceiling to this trend.

The fact that the requirements for faster circuits and lower power consumption can both best be met by smaller CMOS circuits can be considered a lucky circumstance. CMOS technology is thus by far the most important technology. Nevertheless, in many applications CMOS cannot replace bipolar, BICMOS or power technologies.

Chapter 8 describes all the major technologies used in integrated circuits. The level of detail depends on the importance of the technology being described. In Chaps. 3-7, which deal with processing technology, the emphasis is again on those processes and techniques which have proved themselves commercially, or are likely to be used in future in the industry. The reader should therefore gain an impression of the tremendous world-wide effort being devoted to ensure the future progress of integrated circuits.

It is remarkable that the basic design of the two fundamental elements of the integrated circuit – the MOS transistor and the bipolar transistor – has not changed since they were first developed in the 1950s, and will remain unchanged for many years to come. The efforts of engineers and physicists are thus directed less at new components, but more at how to control the individual processes and at the art of process integration, in order to achieve high yields (and thus low costs) and high reliability of integrated circuits.

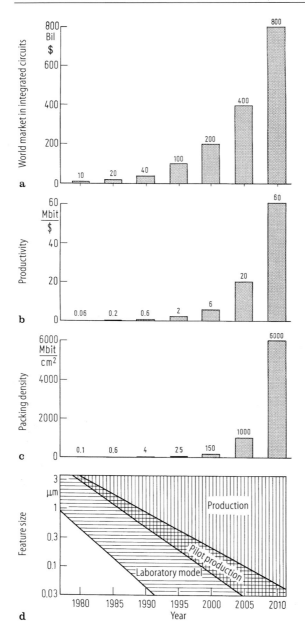

Fig. 1.1 a–d. Major trends in integrated circuits. The main driving force behind the rapid development of integrated circuits is the rise in productivity, which was mostly achieved by increases in packing density. The increase in packing density is in turn achieved predominantly by feature size reduction. **b** and **c** refer to DRAM (Dynamic Random Access Memory), which can store 1 bit of information in one memory cell

2
Basic principles of integrated circuit technology

Integrated circuits are manufactured using planar technology. This refers to the successive technological processing steps which are performed on monocrystalline semiconductor wafers. The processing steps can be classified into the four categories below.
– Thin film technology
– Lithography
– Etching technology
– Doping technology

Figure 2.1 illustrates the basic process sequence in the manufacture of integrated circuits, using the example of a transistor in a MOS integrated circuit.

The first two processing steps performed on the silicon wafer belong to the thin film technology category. In these steps an insulating SiO_2 film and a polycrystalline silicon film are deposited on the silicon surface. Polycrystalline means that the layer is made up of silicon crystal grains of different orientation. A light-sensitive coating (photoresist) is then deposited on the polycrystalline silicon, and exposed to light through a suitable mask. For a positive acting photoresist, those areas exposed to light are dissolved away when immersed in developer, whilst the unexposed areas remain intact. Transferring the mask pattern into the photoresist pattern is called lithography. The photoresist pattern now acts as the mask for the subsequent etching process. By immersing in an etching solution, or by attacking with reactive particles, those areas of polycrystalline silicon not covered by resist are etched off, so that the photoresist pattern is transferred to the layer below. The remaining photoresist is then chemically removed, and the semiconductor is doped with foreign atoms. Doping means introducing foreign atoms (e.g. phosphorus, arsenic or boron atoms) in order to selectively change the conductivity of the silicon.

Finished integrated circuits containing e.g. MOS transistors and other electronic components are produced by repeatedly applying the techniques shown in Fig. 2.1 under the categories thin film technology, lithography, etching technology and doping technology.

Table 2.1 shows the key processing steps in the manufacture of CMOS integrated circuits (CMOS; Complementary *MOS*). Each diagram shows the cross-section of the silicon wafer after the processing step just described.

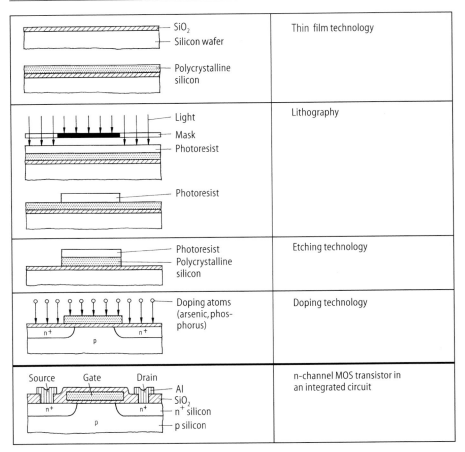

Fig. 2.1. Basic process sequence to manufacture a MOS transistor in an integrated circuit

The processing steps specified in Table 2.1 are described in detail in Chaps. 3 (thin film technology), 4 (lithography), 5 (etching technology) and 6 (doping technology).

By the time the semiconductor wafer is finished, it will have undergone up to 600 process steps. If one of these goes wrong, all this effort is wasted. Particular attention needs to be paid to providing a dust-free environment and clean processing systems. A single dust particle left on the semiconductor wafer during one of the several hundred processing steps may ruin the operation of the integrated circuit. Cleaning technology, which is dealt with in Chap. 7, is therefore particularly important.

Chapter 8 describes how the individual processing steps are integrated into complete processes. It covers those processes used to manufacture CMOS, bipo-

Table 2.1. Possible process sequence in the manufacture of CMOS circuits using polysilicon gate technology [2.11, 2.12]

No	• Process step • Cross-section through the silicon wafer after the processing step just described	• Description of each process step	Process module
1		• Initial material: n$^+$ doped silicon ($\varrho \approx 0.01\ \Omega$ cm) with n-doped single-crystal silicon layer ($\varrho \approx 5\ \Omega$ cm) produced by epitaxy	
2		• Oxidation of silicon wafer • Si$_3$N$_4$ deposition for localized oxidation	Wells

3		• Photolithography with mask 1
		  – for defining n-type wells
		  – processing steps: applying photoresist, exposure with mask 1, developing photoresist
		• Si$_3$N$_4$ etching
		  – with photoresist mask
		• Ion implantation of phosphorus
		  – for doping the n-type well
		  – mask: photoresist
			Wells

4		• Photoresist etching
		  – for removing the photoresist mask ("resist stripping")
		• Annealing of implanted zones (Sect. 6.3)
		• Localized oxidation
		  – oxidation only of those areas not covered by Si$_3$N$_4$, which acts as a diffusion barrier to the oxygen (Sect. 3.4.2)
		  – oxidation is self-aligned over the n-type well
		  – the thick oxide acts as a mask for subsequent p-type well implantation
			Wells

5		• Si$_3$N$_4$ etching
		  – for removing the Si$_3$N$_4$ mask
		• Boron ion implantation
		  – to create the p-type well
		  – mask: thick oxide
			Wells

Table 2.1. continued

6	p-type well / n-type well / n⁻ / n⁺	• Diffusion (1st well drive)
		  – diffusion of implanted doping atoms into the silicon crystal ("drive in")
		  – to create the wells
		• SiO₂ etching
		  – to remove thick and thin oxide
			Wells

7	Si₃N₄ / Poly-Si / SiO₂ over p-type well, n-type well, n⁻, n⁺	• Oxidation
		  – to isolate mono-Si and poly-Si
		• Deposition of poly-Si (polycrystalline silicon)
		  – to shorten the LOCOS beak (Sect. 3.4.2)
		• Si₃N₄ deposition
		  – for subsequent localized oxidation
			Insulation

8	Mask 2 / Photoresist / Si₃N₄ / Poly-Si / SiO₂ over p, n, n⁻, n⁺	• Photolithography with mask 2
		  – to define the active areas
		  – processing steps: applying photoresist, exposure with mask 2, developing photoresist
		• Si₃N₄ etching
		  – with photoresist mask
			Insulation

9	Si₃N₄ / Poly-Si / SiO₂ over p, n, n⁻, n⁺	• Photoresist stripping
		  – to remove photoresist mask
		• Localized oxidation plus diffusion
		  – to create field oxide regions (thick oxide areas)
		  – during oxidation, diffusion of implanted doping atoms
			Insulation

10	SiO₂ over p, n, n⁻, n⁺	• Si₃N₄ etching
		  – to remove the Si₃N₄ mask
		• Poly-Si etching
		  – to remove the Poly-Si layer
		• Diffusion (2nd well drive)
		  – to deepen the p-type and n-type wells
		• SiO₂ etching
		  – to remove the contaminated thin oxide
			Insulation

Table 2.1. continued

11	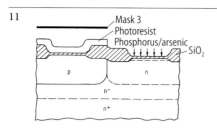	• Oxidation
		  – to create the scattering oxide for ion implantation
		• Ion implantation of boron
		  – to set the threshold voltage of the n-channel transistors
		• Photolithography with mask 3
		  – for ion implantation of active n-type well regions
		• Phosphorus/arsenic ion implantation
		  – to avoid the punch-through effect between source and drain of p-channel transistors, and to set the threshold voltage of the p-channel transistors
			CMOS transistors

12	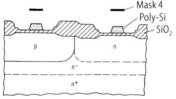	• Photoresist stripping
		  – to remove photoresist mask
		• SiO$_2$ etching
		  – to remove contaminated thin oxide
		• Oxidation
		  – to create gate oxide
		• Poly-Si deposition
		  – to create gate electrodes
		• Photolithography with mask 4
		  – to define the gate
		• Poly-Si etching
		  – with photoresist mask
		• Photoresist stripping
		  – to remove photoresist
			CMOS transistors

13	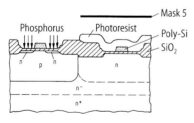	• Oxidation
		  – to create the scattering oxide for subsequent ion implantation
		• Photolithography with mask 5
		  – for source/drain ion implantation of n-channel transistors
		• Phosphorus ion implantation
		  – to produce the lightly doped n-type regions in the source and drain of the LDD transistors (Lightly Doped Drain, Sect. 8.3)
		  – photoresist as implantation mask
			CMOS transistors

Table 2.1. continued

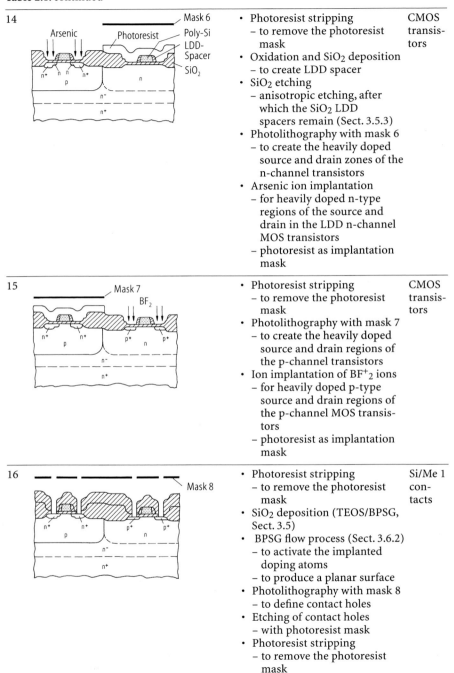

14	• Photoresist stripping
	  – to remove the photoresist mask
	• Oxidation and SiO$_2$ deposition
	  – to create LDD spacer
	• SiO$_2$ etching
	  – anisotropic etching, after which the SiO$_2$ LDD spacers remain (Sect. 3.5.3)
	• Photolithography with mask 6
	  – to create the heavily doped source and drain zones of the n-channel transistors
	• Arsenic ion implantation
	  – for heavily doped n-type regions of the source and drain in the LDD n-channel MOS transistors
	  – photoresist as implantation mask
		CMOS transistors

15	• Photoresist stripping
	  – to remove the photoresist mask
	• Photolithography with mask 7
	  – to create the heavily doped source and drain regions of the p-channel transistors
	• Ion implantation of BF$^+_2$ ions
	  – for heavily doped p-type source and drain regions of the p-channel MOS transistors
	  – photoresist as implantation mask
		CMOS transistors

16	• Photoresist stripping
	  – to remove the photoresist mask
	• SiO$_2$ deposition (TEOS/BPSG, Sect. 3.5)
	• BPSG flow process (Sect. 3.6.2)
	  – to activate the implanted doping atoms
	  – to produce a planar surface
	• Photolithography with mask 8
	  – to define contact holes
	• Etching of contact holes
	  – with photoresist mask
	• Photoresist stripping
	  – to remove the photoresist mask
		Si/Me 1 contacts

Table 2.1. continued

17	*figure: Mask 9, Metal 1, Poly-Si, SiO₂*	• Metal deposition (barrier layer, Sect. 3.10, + Al alloy, Sect. 3.11.1)
		  – to create the 1st metallization layer
		• Photolithography with mask 9
		  – to define the 1st metallization level
		• Metal etching
		  – to create the interconnects of the 1st metallization level
		• Photoresist stripping
		  – to remove the photoresist mask
			Metal 1

18	*figure: Mask 10, Spin-on glass, Metal 1, Poly-Si, SiO₂*	• SiO₂ deposition (plasma oxide, Sect. 3.5.1)
		  – to isolate the 1st and 2nd metallization layers
		• Spin-on glass deposition (Sect. 3.5.5)
		  – to create a planar surface (Sect. 8.5)
		• Etch-back of spin-on glass and SiO₂
		  – for further planarization (Sect. 8.5)
		• SiO₂ deposition (plasma oxide, Sect. 3.5.1)
		  – to enclose the spin-on glass
		• Photolithography with mask 10
		  – to define the contact holes (vias) between 1st and 2nd metallization levels
		• Etching of vias
		  – using photoresist mask
		• Photoresist stripping
		  – to remove the photoresist mask
			Me1/
			Me2
			contacts

19	*figure: Mask 11, Spin-on glass, Metal 1, Metal 2, Poly-Si, SiO₂*	• Metal deposition
		  – to create the 2nd metallization layer
		• Photolithography with mask 11
		  – to define the 2nd metallization level
		• Metal etching
		  – to create the tracks in the 2nd metallization level
		• Photoresist stripping
		  – to remove the photoresist mask
			Metal 2

Table 2.1. continued

20

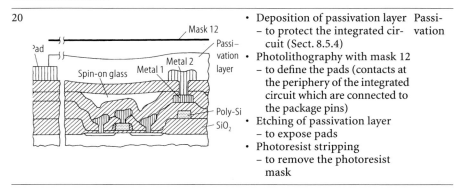

- Deposition of passivation layer — to protect the integrated circuit (Sect. 8.5.4)   Passivation
- Photolithography with mask 12 — to define the pads (contacts at the periphery of the integrated circuit which are connected to the package pins)
- Etching of passivation layer — to expose pads
- Photoresist stripping — to remove the photoresist mask

lar, and BICMOS circuits (combination of bipolar and CMOS technology) as well as smart power circuits (combination of DMOS, bipolar and CMOS technology). The architecture and a detailed processing sequence is given for each complete process. Chapter 8 also includes a description of the electronic components used in integrated circuits, the electrical connections between components, the cell designs of semiconductor memories and the principle of scaling in integrated circuits.

This book is focussed on the technology of integrated circuits, neglecting circuit engineering or device physics, which are dealt with in detail in [2.1–2.4]. The reader can find the fundamental principles of semiconductor technology in [2.5], and further technical data in the comprehensive information source [2.6]. The books [2.7–2.9] provide useful extra material on the subject of VLSI circuit technology.

Up to several hundred identical integrated circuits are produced on the semiconductor wafer by the processes described in Table 2.1 (wafer technology). Figure 2.2 shows the final processing steps that follow wafer processing.

The integrated circuits are electrically tested whilst still coherent on the semiconductor wafer using test programs. The wafer is then diced into the individual circuits (chips), and the good chips are mounted in suitable packages, wire-

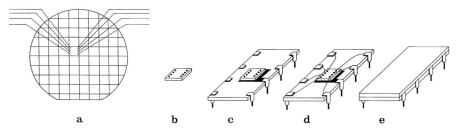

Fig. 2.2 a–e. Final processing steps in the manufacture of integrated circuits. **a** Test; **b** dicing; **c** chip assembly; **d** wire bonding; **e** encapsulation

bonded and encapsulated [2.10]. The circuit then undergoes final testing to ensure that it is still working.

The yield has particular economic significance in the manufacture of integrated circuits. It is defined as the ratio of good products to total number of products, and is made up of the following contributors:

1. Process yield Y_P,

$$Y_P = \frac{\text{number of wafers reaching wafer test}}{\text{number of wafers started for wafer processing}}$$

2. Wafer test yield Y_T,

$$Y_T = \frac{\text{number of chips that pass the wafer test}}{\text{number of chips submitted for wafer test}}$$

3. Assembly yield Y_A,

$$Y_A = \frac{\text{number of mounted chips without defects}}{\text{number of chips submitted for mounting}}$$

4. Final test yield, Y_P,

$$Y_F = \frac{\text{number of good products}}{\text{number of products submitted for final testing}}$$

The total yield Y is obtained from:

$$Y = Y_P \cdot Y_T \cdot Y_A \cdot Y_F$$

Semiconductor manufacturers aim to achieve a yield Y of more than 70 % to remain internationally competitive.

3
Film technology

The starting point for the manufacture of an integrated circuit is a monocrystalline silicon wafer. In the course of manufacture, various layers of material are added to the silicon wafer. A pattern is normally produced in these layers, or films, by retaining selected areas, and these patterned films act as electrical interconnects or as isolating and passivating layers. They may also perform a masking, doping or gettering function during integrated circuit manufacture.

The following sections describe the most important film production techniques used in silicon technology, as well as the films themselves.

3.1
Film production processes

The most important film production techniques used in silicon technology are the CVD process (CVD = Chemical Vapour Deposition), thermal oxidation, sputtering and spin-coating. The vapour phase deposition method has become less important as sputtering techniques have advanced. Ion implantation, however, being a new technique for film production may prove useful in some specific applications in the future. The various film production techniques are discussed in more detail below.

3.1.1
The CVD process

Gas phase deposition, also called CVD, is one of the most important techniques in silicon technology [3.1, 3.2]. The CVD process is used to create silicon epitaxial layers, polycrystalline silicon layers, SiO_2 films, plus films of boron and phosphorus glass, silicon nitride, and more recently of metals and metal silicides.

The basic principle behind CVD involves passing selected gases over the heated substrates onto which the required layer is to be deposited. The process gases react on the hot substrate surface (400–1250 °C, depending on reaction gases) to produce the required layer, the other reaction products being gases which are extracted from the reactor.

Figure 3.1.1 uses the example of silicon epitaxy to illustrate the three key physical-chemical processes occurring in gas phase deposition. $SiCl_4$ and H_2 are used as the reaction gases here. The $SiCl_4$ breaks down at high temperatures

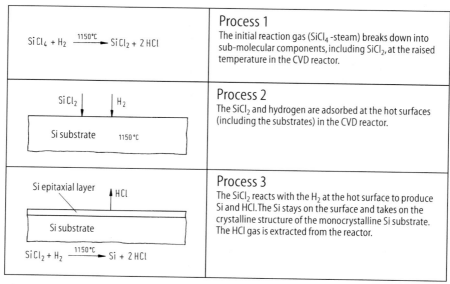

Fig. 3.1.1. Schematic diagram using the example of SiCl₄ epitaxy to illustrate the physical-chemical processes involved in gas phase deposition (CVD)

(1150 °C in this case) into submolecular components (process 1), which are adsorbed on the hot silicon wafers (as well as on the other hot surfaces of solids in the reactor) (process 2). Finally, the SiCl$_2$ and H$_2$ react on these surfaces to produce Si and HCl (process 3). The Si remains as a surface layer, whilst the HCl gas is extracted from the reactor.

The reaction gases move towards the wafers by convection until close to the silicon wafer surface, where diffusion takes over as the gas transport mechanism. This is because the gas flow close to the wafer surface is parallel to the surface, reducing to zero at the surface itself. According to the diffusion law:

$$j \sim D \frac{dc}{dz}$$

where j is the gas flow needed for the surface reaction, D is the diffusion constant and dc/dz is the concentration gradient perpendicular to the surface. If the ratio j/D is small (i.e. for a slow reaction rate and/or large diffusion constant) then a small concentration gradient is sufficient to maintain the required gas flow. The reaction gas concentration at the wafer surface is then only slightly lower than the concentration in the gas mixture fed into the reactor. This case is referred to as a reaction-controlled process, because the reaction rate at the surface determines the layer growth rate.

The other extreme case is the diffusion-controlled process. Here the surface reaction rate is so high that the reaction gas concentration falls off sharply close to the surface. This depletion means that the reaction rate also slows, so that the reaction rate settles at a level corresponding to the maximum possible diffusion flow rate.

In CVD processes for silicon technology, the usual aim is to achieve a reaction-controlled process, because it has several advantages over the diffusion-controlled process.

- The uniform deposition rate enables so-called conformal film deposition to be achieved. This means that sharp steps on the wafer surface are covered with the same film thickness as horizontal surface areas (Fig. 3.1.2 b). Thus features such as narrow trenches can be filled (Fig. 3.1.2 d), since a sufficient supply of reaction gases is available even in poorly accessible places. This is not the situation in a diffusion-controlled process (Fig. 3.1.2 a, c).
- The orientation of the silicon wafers in the CVD reactor is relatively uncritical. For example, the silicon wafers can be positioned close together in the CVD reactor, even across the direction of gas flow, because gas transport to the silicon surface is still guaranteed even in these adverse flow conditions.

The key parameter that can tip the balance of the gas phase reaction towards a reaction-controlled process, is the gas pressure in the CVD reactor. If one reduces the pressure by a factor a, the reaction rate and thus the necessary gas flow j in the above diffusion equation falls by the factor a, whilst the diffusion constant D increases by the factor a in accordance with the kinetic gas theory [3.48]. Thus according to the diffusion equation, the concentration gradient dc/dz reduces by the factor a^2. This means that a concentration gradient reduced by factor a^2 is large enough to maintain the required gas flow to the wafer surface. In the extreme case, the concentration gradient is so low that at each point on the wafer surface almost the same concentration of reaction gases is present as in the specified gas mixture. This is precisely the characteristic of a reaction-controlled process. The pressure range for low-pressure CVD processes (LPCVD = *Low Pressure Chemical Vapour Deposition*) lies between 20 and 100 Pa.

There is a second important parameter apart from gas pressure that can control the layer growth in CVD processes. This is the adsorption of the reaction components at the wafer surface (process 2 in Fig. 3.1.1). The electronegativity of the substrate surface is one factor affecting the adsorption, and this property can be used to produce selective layer deposition by coating certain areas of the substrate surface with a material of a different electronegativity. Figure 3.1.3 shows two important applications: selective epitaxy of Si [3.24] and selective SiO$_2$ deposition for filling trenches between aluminium interconnects [3.2]. Both processes operate at pressures in the range 0.1–1 bar (SACVD = *Sub-Atmospheric Pressure CVD*). In both cases the layer grows faster in the lower-lying areas, so that the required surface planarization is achieved, but by a completely different mechanism than in Fig. 3.1.2 d.

Figure 3.1.4 shows sketches of the most widely used designs of CVD reactors. They differ in the type of heating, the gas feed arrangement and the positioning

16   3 Film technology

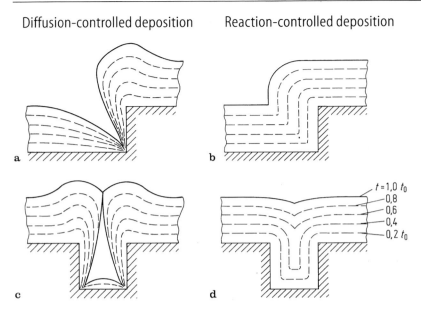

Fig. 3.1.2 a–d. Covering a step (**a** and **b**) or a trench (**c** and **d**) when the CVD reaction is diffusion-controlled (**a** and **c**) or reaction-controlled (**b** and **d**). The dashed lines indicate the layer outline after 20, 40, 60 and 80 % of the coating time t_0.

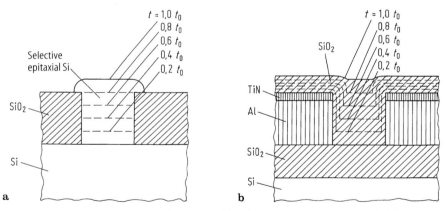

Fig. 3.1.3 a, b. Two applications of selective film deposition using SACVD (=Sub-*A*tmospheric *P*ressure CVD). The dashed lines indicate the layer outline after 20, 40, 60, and 80 % of the coating time t_0. **a** Selective Si epitaxy. In this case the SiH_4 molecules are only adsorbed on the Si surface areas, but not on SiO_2 surfaces. **b** Selective SiO_2 deposition. In this example, four times as many CVD reaction components are adsorbed over the SiO_2 surface regions than over the aluminium or TiN surfaces. This means that SiO_2 growth is four times more rapid in the gaps

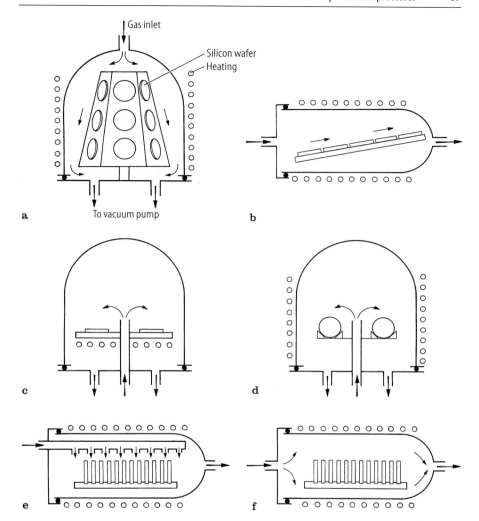

Fig. 3.1.4 a–f. Various designs of CVD reactors. The arrows indicate the direction of gas flow. Reactor type **c** is a cool wall reactor, all other types are hot wall reactors. Types **a** and **b** can also be designed as cool wall reactors.

of the silicon wafers (vertical or horizontal). All reactors in Fig. 3.1.4 are multi-wafer systems. The reactor type shown in Fig. 3.1.4 c is being used increasingly, however, for single-wafer fabrication of large-diameter wafers [3.3].

The silicon wafers can be heated inductively, by resistive heating elements or by radiant heat. In inductive heating systems, the silicon wafer support (called the susceptor) must be made of a material of suitable resistivity (usually graphite), so that the eddy currents induced by the high frequency magnetic field can heat up the susceptor as much as possible. The graphite susceptor is usually

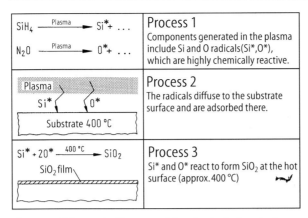

Fig. 3.1.5. Schematic diagram of the physical-chemical processes that occur during PECVD deposition (Plasma Enhanced CVD) using SiO_2 deposition as an example

coated with a silicon carbide layer to protect the silicon wafer from wear on the graphite base. In inductive heating the silicon wafers are heated by thermal conduction and/or radiation from the susceptor, so they must lie flat against the susceptor. Thus only those reactor types in Fig. 3.1.4 a–c can be considered for inductive heating. Of course direct heating of the susceptor can also be achieved here using a heating coil or by thermal radiation from heating lamps (e.g. quartz halogen lamps)[1], and this is employed in some commercial CVD reactors. All CVD reactors with direct susceptor heating have the common advantage that the walls of the reactor are not heated (cool wall reactors). This means that the layer is only deposited on the silicon wafers and the exposed parts of the susceptor.

In the reactor types shown in Fig. 3.1.4 d–f, a heating coil or heating lamps are the only heating option. Unless the walls are cooled, they reach the same temperature as the silicon wafers (hot-wall reactors). Thus the film is deposited on the walls as well. Silicon epitaxy is today solely performed in cool wall reactors, because high temperatures of up to 1250 °C are required, and because of the large thickness of the layer to be deposited. Hot wall reactors, however, are still the main means of producing polycrystalline silicon (polysilicon), Si_3N_4 and SiO_2 (at 400–1000 °C, less than 1 µm layer thickness).

If the silicon wafers stand exposed in the reactor, as in the reactor types of Fig. 3.1.4 d–f, then film deposition occurs both on the front and back of the wafers. In reactor type (Fig. 3.1.4 f), the reactor gases may become depleted as they pass along the tube axis. A temperature gradient can be used to compensate for

---

1 The heating lamps can also be located outside the reactor chamber. The reactor walls must be cooled to prevent them heating up (e.g. by passing air between the quartz walls of a dual wall reactor).

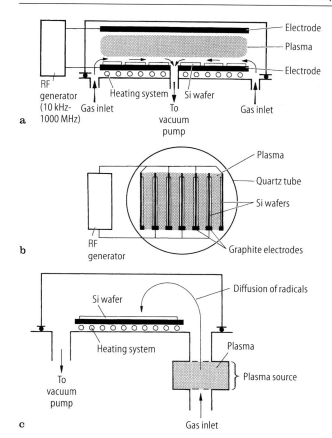

Figure 3.1.6 a–c. Three types of plasma CVD reactors. **a** Parallel plate reactor; **b** tube reactor with graphite electrodes. Several of the twelve-wafer configurations shown can be accommodated down the length of the tube. Heating is by heating coil. **c** Reactor with separate plasma source (remote plasma, downstream-reactor) [3.6]

this depletion, which exploits the fact that the CVD reaction rate increases as the temperature rises.

In the CVD processes described above, the gas phase reactions were triggered by a raised temperature. If the CVD reaction components are excited in a plasma, then this is a PECVD process (*P*lasma *E*nhanced CVD) [3.4]. Figure 3.1.5 shows the physical-chemical processes that occur in PECVD deposition, using SiO_2 deposition as an example.

An essential feature of PECVD deposition is that the temperature can be reduced to below 500 °C. This means that PECVD can be used, for instance, to produce the isolation layers needed in multilayer aluminium metallization. PECVD deposition is best performed in a parallel plate reactor (Fig. 3.1.6 a). For larger wafer diameters this holds only a single wafer [3.5].

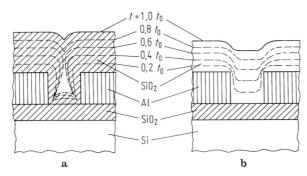

Fig. 3.1.7 a, b. Film growth in a gap (**a**) in PECVD deposition (Plasma enhanced CVD) and (**b**) in the Dep./Etch process, where ion etching occurs at the same time as PECVD deposition. The dashed lines indicate the layer outline after 20, 40, 60, and 80 % of the coating time t_0. The hollow space (void, keyhole) in case (**a**) is unwanted because of possible reliability problems.

Although PECVD processes are usually performed in the same pressure range as LPCVD processes, at 20-100 Pa, film deposition is less conformal, i.e. less material is deposited in narrow gaps than on planar surfaces. This situation - which is not normally desirable - can be improved by making that electrode carrying the wafers in the parallel plate reactor the cathode (cf. Fig. 3.1.6). This causes part of the growing layer to be sputtered off again (Dep./Etch process), by the same mechanism as in bias sputtering (see Sect. 3.1.4). Thanks to redeposition and intensified sputter thinning at non-horizontal surfaces, planar film deposition is achieved without voids being formed (Fig. 3.1.7 b).

Instead of a plasma, a CVD reaction can also be excited by high-energy radiation (RECVD = Radiation Enhanced CVD) [3.7]. Such CVD techniques are not in widespread use as yet, but may be important in the future. In particular, the possibility of focusing the radiation from e.g. an Excimer laser (wavelength approx. 250 nm) onto the surface of the silicon wafers, would enable deposition of the CVD layers solely on the silicon wafers themselves, and nowhere else. This would avoid the creation of particles associated with layer deposition on the reactor walls, which can have a very detrimental effect.

A special case of laser activated CVD deposition is localized film deposition (a few μm^2), which is achieved by using a suitably fine-focused laser beam. This type of localized deposition can be used to repair defective chromium masks for instance (see Sect. 4.2.8).

3.1.2
Thermal oxidation

Looked at from the perspective of the processes that take place, thermal oxidation is very similar to CVD deposition. The essential difference is that one of the

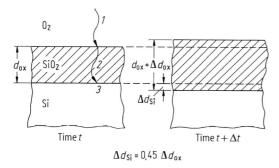

Fig. 3.1.8. Schematic diagram of the processes occurring in the thermal oxidation of silicon. The three mechanisms in operation are O_2 diffusion to the surface 1, O_2 diffusion through the SiO_2 layer 2, and the reaction $Si+O_2 = SiO_2$ at the Si-SiO_2 interface 3. The right half of the diagram shows how the SiO_2 surface and Si-SiO_2 interface move as the oxidation time increases

two reaction partners, namely silicon, is not introduced as a component of a gas, but already exists as a silicon substrate or silicon layer. Therefore unlike the CVD process, where a layer is deposited by a gas phase reaction onto an essentially unchanged substrate, in thermal oxidation the surface of a silicon substrate is converted into SiO_2 in an oxidizing atmosphere.

Figure 3.1.8 illustrates the physical-chemical processes of thermal oxidation [3.8]. It involves three consecutive mechanisms, of which the first and third are the same as in CVD deposition.
- Diffusion of oxygen out of the gas space onto the SiO_2 surface.
- Diffusion of oxygen through the SiO_2 layer
- Reaction of oxygen with silicon to produce SiO_2 at the Si-SiO_2 interface

In this process, the thickness of the Si layer converted to SiO_2 equals 45 % of the SiO_2 layer thickness.

The slowest of the three processes determines the SiO_2 growth rate. In all practical cases, the diffusion of the reaction gas (oxygen) onto the SiO_2 surface is the fastest of the three processes of thermal oxidation. As already discussed in Sect. 3.1.1, one consequence is that the SiO_2 layer grows on steps, overhangs and even in narrow trenches at the same rate as on planar surfaces, given the same conditions. This conformal growth, or topographically-independent growth, is extremely useful in practice.

As long as the SiO_2 layer thickness is small (less than about 0.1 μm), the oxidation process $Si + O_2 = SiO_2$ controls the rate. In such a reaction-controlled process, the SiO_2 thickness d_{ox} increases linearly over time with a proportionality constant B/A:

$$d_{ox} = Bt_1/A \quad \text{(for small } d_{ox}\text{)}$$

where B/A is the linear oxidation constant.

As the SiO$_2$ thickness increases, the diffusion through the SiO$_2$ layer starts to control the growth rate (diffusion-controlled process).[2] If one assumes a linear fall off in the O$_2$ concentration in the SiO$_2$, then for twice the SiO$_2$ thickness only half so much oxygen diffuses to the Si interface. Thus the growth rate reduces linearly with increasing SiO$_2$ thickness, and the SiO$_2$ thickness increases according to the law:

$$d_{ox}^2 = Bt_2 \quad \text{(for large } d_{ox}\text{)}$$

where B is referred to as the parabolic oxidation constant.

$$d_{ox}^2 + Ad_{ox} = B(t_2 + t_1) = Bt.$$

In practice, the two processes – O$_2$ diffusion through the SiO$_2$ layer and oxidation at the Si-SiO$_2$ interface – occur consecutively (see Sect. 3.1.8). Thus the times t_2 and t_1 can be added together to obtain the well-known oxidation formula:

$$d^2{}_{ox} + Ad_{ox} = B(t_2 + t_1) = Bt$$

In order to include the "naturally grown oxide" that is already present before thermal oxidation (0.5–3 nm), the oxidation formula is usually expressed as follows:

$$d_{ox}^2 + Ad_{ox} = B(t + \tau)$$

This oxidation formula describes the observed oxide growth very closely, except for the finding that the first 10 nm or so of the SiO$_2$ layer grows faster than predicted by the linear equation for initial growth, under certain oxidation conditions(cf. Fig. 3.1.13). This phenomenon has not yet been fully explained.

Figures 3.1.9 and 3.1.10 respectively show how the constants B/A and B vary with respect to temperature in pure oxygen and in pure steam at atmospheric pressure. There is no appreciable difference in the constant B for different crystallographic orientations of the Si surface, whereas there is a difference for the constant B/A (Fig. 3.1.9). The curves in Fig. 3.1.9 only apply to silicon substrates that are lightly or moderately doped. For heavily doped silicon substrates with concentrations above 10^{19} cm^{-3}, the linear SiO$_2$ growth rate B/A increases sharply, in particular for heavily n-doped substrates. Figure 3.1.11 shows an example of the different levels of oxide growth for silicon in a moist atmosphere at 900 °C. The differences in thickness vary by up to a factor of 5. This effect, which is accentuated as the oxidation temperature falls, is often exploited in modern silicon technology (see Sect. 3.4.1). An analysis of the effect shows that it is essentially the linear oxidation constant that is involved, but not the parabolic. This suggests that oxidation at the silicon interface is accelerated at high doping concentrations.[3]

---

2 The advantage of conformal layer growth is not lost, however, since the oxygen is still diffusing out of the gas space onto the SiO$_2$ surface at a fast enough rate. Thus there is no need to reduce the gas pressure, as is required in the CVD process.
3 The accelerated reaction occurs at vacancies in the silicon lattice.

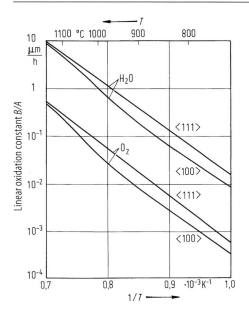

Figure 3.1.9. The linear oxidation constant B/A with respect to temperature for lightly doped $\langle 100 \rangle$ or $\langle 111 \rangle$ silicon surfaces in pure oxygen or pure steam at atmospheric pressure

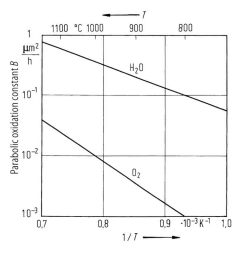

Fig. 3.1.10 The parabolic oxidation constant B with respect to temperature in pure oxygen or pure steam at atmospheric pressure. The curves give a good approximation for $\langle 100 \rangle$ and $\langle 111 \rangle$ silicon surfaces

The different energy states of the doping atoms at the Si-SiO$_2$ interface lead to segregation of the doping atoms between the SiO$_2$ and Si. Section 6.3.5 looks more closely at the segregation of boron, phosphorus and arsenic in the SiO$_2$/Si system.

The breakdown characteristic and the electrical stability of SiO$_2$ layers that are produced with dry oxygen, can be significantly improved by adding a few

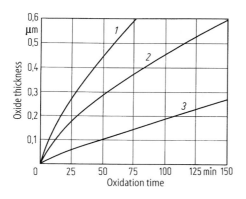

Fig. 3.1.11. Example of the different oxide growths for different high levels of doping of silicon (measured curves) [3.11]. Thermal oxidation was performed in a steam atmosphere at 900 °C for all three cases. Curve 1: $\langle 100 \rangle$ Si, arsenic implant dose $8 \cdot 10^{15}$ cm^{-2}, 100 keV; curve 2: poly-Si, thermal phosphorus doping $3 \cdot 10^{20}$ cm^{-3}; curve 3: $\langle 100 \rangle$ Si, boron doping $7 \cdot 10^{14}$ cm^{-3}

percent by volume of HCl (or other gases containing chlorine, such as trichloroethane or trichloroethylene).[4] The addition of HCl appears to suppress both the formation of stacking faults and the inclusion of metal atoms. Adding HCl also reduces the percentage of H$_2$O in the oxidation atmosphere, which in turn increases the oxide growth rate (Fig. 3.1.12). The SiO$_2$ layers produced with the added HCl contain chlorine. The chlorine is not uniformly distributed through the SiO$_2$ layer, however, but is concentrated near the Si–SiO$_2$ interface in a region 10–20 nm thick. For a percentage volume of 5 % HCl in oxygen, the maximum chlorine concentration in the SiO$_2$ equals about $2 \cdot 10^{20}$ atoms per cm^3 (at an oxidation temperature of 900 °C) and $5 \cdot 10^{20}$ atoms per cm^3 (at 1000 °C).

When HCl is added the oxide growth of very thin SiO$_2$ layers differs from the linear-parabolic growth law. Figure 3.1.13 shows the growth curves for various HCl concentrations at 800 and 900 °C.

If thermal oxidation is performed at increased pressure (10–25 bar), the oxide thickness is found to increase proportionally with pressure if other conditions remain constant. This finding is reasonable, because the oxidation rate is linearly related to the oxygen concentration at the surface, both in the reaction-controlled and diffusion-controlled region of oxide growth. Thus high-pressure oxidation [3.10] enables SiO$_2$ layers to be produced in a shorter time or at a lower temperature. At lower oxidation temperatures, fewer stacking faults are created, and the segregation of doping atoms at the Si–SiO$_2$ interface is less pronounced.

Moving to the opposite extreme, thermal oxidation at reduced partial pressure of oxygen leads to a reduced oxidation rate. This effect is important for very thin gate oxides.

Thermal oxidation is still performed mainly in horizontal tube furnaces, with wafers[5] standing upright in quartz boats (Fig. 3.1.14). For larger wafer diame-

---

4 HCl is frequently used in wet oxidation as well (steam atmosphere), although only for "tube cleaning" (sodium, heavy metals) before the actual wet oxidation process, or as an added component in the drying phase in a dry/wet/dry cycle.
5 Thermal oxidation occurs equally on both front and back of the silicon wafers.

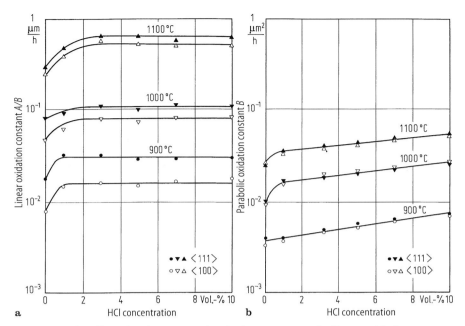

Fig. 3.1.12 a, b. Effect of HCl concentration in dry oxygen on the linear oxidation constant B/A (**a**) and the parabolic oxidation constant B (**b**) in the thermal oxidation of $\langle 100 \rangle$ and $\langle 111 \rangle$ oriented silicon surfaces [3.9]

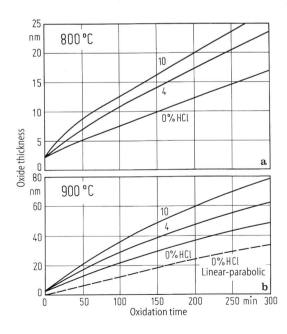

Fig. 3.1.13 a, b. Measured oxide growth [3.11] on a $\langle 100 \rangle$ silicon surface in oxygen for a substrate lightly doped with boron ($7 \cdot 10^{14}$ cm^{-3}) with different amounts of added HCl at 800 °C (**a**) and 900 °C (**b**). In the 900 °C diagram, the curve for 0 % added HCl, which follows the linear-parabolic oxidation law, is shown dashed for comparison

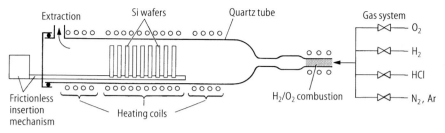

Fig. 3.1.14. Schematic diagram of a horizontal tube furnace for thermal oxidation at atmospheric pressure

ters, vertical furnaces are used, with wafers lying horizontally, spaced one above the other. The advantage of vertical furnaces is that the temperature and gas flow gradients, which are affected by gravity, act at right angles to the wafer plane. Vertical furnaces therefore significantly reduce the risk of temperature gradients along the wafer surface, which can otherwise cause mechanical stress and thus dislocations in the monocrystalline silicon wafer as well as permanent warpage. This design also needs only a simple jig for holding the wafer (three-point support).

Rapid processing techniques (see Sect. 3.1.8) are also an option for thin gate oxides. For instance, RTO (= Rapid Thermal Oxidation) can be used to produce a 10 nm thick SiO_2 layer in 1 min at 1100 °C (cf. Fig. 3.1.12 a).

In horizontal tube furnaces (Fig. 3.1.14), air diffusing from the loading end of the tube to the tube centre can have a detrimental effect. This can be avoided by high gas flow rates or by providing a sufficiently good seal (cap) at the tube end (to be maintained even when the wafers are being taken in and out). The moisture content in the filtered gases is held below 0.5 ppm, and the gas flow rates are controlled using mass flow regulators. Instead of using steam in wet oxidation, better control is achieved by producing H_2O from the combustion of H_2 and O_2 in a burner at the tube inlet. In order to avoid particles being produced as the quartz boat rubs against the tube when it is being moved in and out, cantilevers are used, which either hold the boat suspended during oxidation (Fig. 3.1.14), or deposit it in the tube centre and are then withdrawn. A final point to mention is that in modern oxidation furnaces all oxidation processes and control functions are computer controlled. This is particularly significant for what is known as ramping, i.e. the slow heating and cooling of the silicon wafers, intended to avoid radial temperature gradients across the silicon wafers in the critical temperature range of 800–1000 °C.[6]

---

6 In this temperature range, dislocations form in the silicon even for relatively small mechanical stresses. Charge carriers associated with heavy metal atoms are generated at the dislocations, which can cause leakage currents.

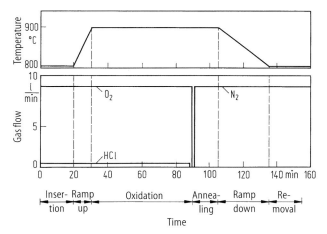

Fig. 3.1.15. Example of an oxidation cycle for creating a 20 nm thick gate oxide in a horizontal tube furnace

Figure 3.1.15 shows a typical oxidation cycle in the creation of a 20 nm thick gate oxide. Annealing in nitrogen (or also in argon) after oxidation is used to eliminate or reduce the fixed interface charges and traps (see Sect. 3.4.3) in the SiO_2 layer.

3.1.3
Vapour phase deposition

Vapour phase deposition has been superseded in almost all areas of modern silicon technology by the sputtering process. It shall therefore only be touched on briefly here.

In vapour phase deposition, the material to be deposited is heated to a point at which the vapour pressure is sufficient to produce significant vaporization. The pressure in the recipient vessel is kept so low (approx. 10^{-4} Pa), that the evaporated atoms travel in a straight line to the substrate some 50 cm away without colliding with any residual gas molecules.

Resistive and inductive heating is used to heat the vaporization source, as well as electron beam bombardment. In electron beam vaporization, the crucible can be cooled in order to suppress metallurgical reactions between the vapour deposition substance and the crucible material. The X-rays produced by bombardment with high-energy electrons (typically 10 keV) can lead to "radiation damage", for instance of MOS transistors on the silicon wafers undergoing vapour deposition.

Conformal coating of surfaces containing steps is not possible in the vapour deposition process, because the atoms impinge from one direction. Even if

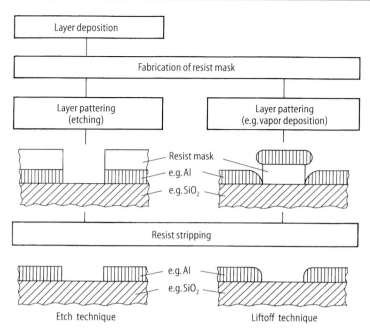

Figure 3.1.16. Processing sequence for creating a pattern using the liftoff technique compared with the standard etching technique. The key difference between the two methods is that in the liftoff technique layer deposition occurs after the resist mask has been laid down. This is also the basic principle behind the Mandrel and the Damascene process (see Sect. 8.5.3)

one ensures that the incident angle of the vaporization atoms varies continuously during deposition, e.g. by planetary motion of the silicon wafers, shadowing effects mean that film deposition is still reduced in steps and trenches (e.g. in contact holes), producing a result comparable to the diffusion-controlled CVD process (cf. Fig. 3.1.2 a, c). Raising the temperature of the substrate (up to 400 °C) during vapour deposition helps to produce a degree of mobility in the atoms on the substrate surface, so that at least cracks can be avoided at sharp edges. If the substrate is not heated its temperature remains below 100 °C. Unlike sputtering, the vapour deposition technique is therefore suitable for liftoff techniques using photoresist masks (Fig. 3.1.16). In this application the poor edge coverage in vapour deposition is actually an advantage [3.12].

One disadvantage of the vapour deposition process is the problem of reproducibility in a specific layer composition (e.g. 99 % Al and 1 % Si). For a single source the problem is caused by the different vapour pressures of each component. For two or more sources, however, there is the problem of simultaneously controlling the evaporation rate of each source, and of achieving a uniform layer composition over all wafers in the vapour deposition facility.

Molecular *B*eam *E*pitaxy (MBE) is a special version of the vapour deposition process [3.13, 3.14]. Although this has yet to make its mark in silicon technology, it could become more important in the future. In molecular beam epitaxy, silicon and the required doping material are vaporized from suitable vaporization sources onto the monocrystalline silicon substrate at e.g. 850 °C. The pressure in the recipient vessel must be reduced to between 10^{-7} and 10^{-8} Pa, in order to reduce the effect of residual gas atoms in the epitaxial layer. So-called Delta doping films, with a defined thickness of about 1 nm, can be produced using MBE, as well as SiGe films on Si (see Fig. 8.3.9).

3.1.4
Sputtering

The drawbacks associated with vapour deposition discussed above (poor edge coverage, difficult control of layer composition when several components used), established the sputtering technique as the most important PVD process (PVD = *P*hysical *V*apour *D*eposition) [3.15].

Figure 3.1.17 illustrates the sputtering principle. A plasma is excited between two electrodes in a recipient vessel at a gas pressure of approximately 1 Pa (mostly argon) by a dc or rf voltage. A voltage drop of typically 1 kV is set up in front of the cathode holding the sputter target. The positively charged argon ions are accelerated along this cathode fall path. When they hit the target they have enough energy to knock out individual atoms or molecules from the target ("sputtering"). Although the argon ions hit the target perpendicularly with high energy (approx. 1 keV) in the vertical electric field, the energy of the sputtered atoms or molecules is low (a few eV) and they leave the target surface in all directions, where a cosine distribution[7] can be assumed for the angles of emission. The substrates, which lie on the anode, then become covered with a layer that has the same composition as the target.

It is not immediately evident why one can actually refer to cathodes and anodes when rf excitation is used – which is the most common option for sputtering equipment. As Fig. 3.1.18 explains, a negative charge actually builds up on the electrodes because of the different mobilities of the argon ions and electrons. This charge build-up can only occur if there is a dc block in the circuit, i.e. if there is a capacitor in the circuit. If the target is a conductor, a capacitor must be inserted in the external circuit (see Fig. 3.1.17). Where an insulator (e.g. quartz) is used for the target, the target plate itself, bonded onto an electrode, constitutes the capacitor.

In a symmetrical configuration of the two electrodes, the negative charge, and the resultant voltage drop in front of the two electrodes would be the same for

---

7 If a grid is inserted between the two electrodes, those atoms sputtered off at larger angles are eliminated (collimated sputtering). This helps to prevent the build up of trenches such as those in Fig. 3.1.19 a, although the layer grows more slowly.

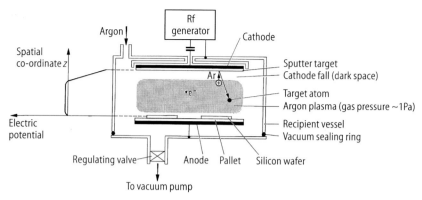

Fig. 3.1.17. Diagram of an *rf* sputtering system illustrating the sputtering principle. Outside, on the left, the electric potential acting on the argon ions is shown

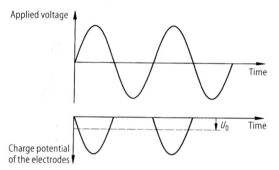

Fig. 3.1.18. Explanation of the negative charge of the electrodes in rf sputtering. Whilst the electrons in the argon plasma can follow the rapid changes in potential of the rf voltage (usually 13.56 MHz), the argon ions are too slow. They "see" the mean negative charge of the electrodes (bias voltage U_0, dashed line). Thus there is a constant flow of argon ions to the electrodes

each. If there is just a slight difference in the electrode surface areas, however, the smaller electrode becomes more negatively charged. Using the Schottky-Langmuir space charge equation, one can deduce that the voltage drops U_1 and U_2 in front of the two electrodes are inversely proportional to the fourth power of the electrode surface areas A_1 and A_2 [3.49]. This is expressed as:

$$\frac{U_1}{U_2} = \left(\frac{A_2}{A_1}\right)^4$$

In standard sputtering equipment one of the two electrodes is usually in electrical contact with the recipient vessel. This increases the electrode surface area to such an extent that the voltage drop in front of this electrode is negligible in comparison with the voltage drop in front of the other electrode (see Fig. 3.1.17). Both electrodes are actually acting as cathodes, but because of the smaller negative charge one refers to the electrode connected to the recipient vessel as the anode, from which almost nothing can be sputtered off.

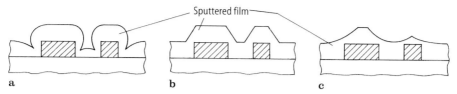

Fig. 3.1.19a–c. Edge coverage of sputtered layers: **a** no bias; **b** moderate bias; **c** heavily biased

In a technique referred to as bias sputtering[8], a specific rf voltage is applied to each electrode, e.g. by using two separate rf generators and associated matching networks. This means that the voltage drops U_1 and U_2 can be varied independently of each other. In this technique, the cathode-type behaviour of the substrate electrode is deliberately enhanced so that some of the layer sputtered onto the silicon substrate is constantly removed by back-sputtering. In order to ensure that a layer is actually grown on the silicon substrate, the rf voltage, and thus the negative charge of the target, must still be higher than that of the substrate electrode. This back-sputter effect is used to produce bevelled edges when coating surfaces containing steps, and thus to achieve better planarization [3.16] (Fig. 3.1.19). The edge bevelling occurs because those parts of the surface that are not horizontal are removed more effectively by back sputtering (ion etching) than horizontal areas of the surface (maximum effect at about 60 °C). Narrow trenches can therefore be completely or partially smoothed off in bias sputtering, because the majority of back-sputtered atoms are redeposited on the trench walls (redeposition).

Moderate back-sputtering in bias sputtering systems can be used before actual sputter coating to clean the substrate surface or to remove unwanted surface layers (sputter cleaning). This is useful before sputtering of the metallization layer, because it removes the naturally occurring oxide on the silicon or aluminium in the contact holes, and any other residues that might be there. During bombardment of the silicon surface, the argon ions penetrate to a depth of some 10 nm into the silicon, and cause massive crystal damage in this thin surface layer. This is not a problem, however, as long as the pn junction lies much deeper.

Accelerated argon ions may not be the only source of "radiation damage" during sputtering. High-energy secondary electrons[9] and shortwave ultraviolet radiation in the plasma can also cause damage. Both radiation effects cause traps in SiO_2 layers, the majority of which can be removed again by annealing in hydrogen or forming gas (90 % N_2, 10 % H_2) at about 450 °C. The secondary electrons heat the silicon wafers during sputtering to 100–350 °C.

---

8 Bias refers to the electrical bias voltage.
9 The secondary electrons are generated at the target during ion bombardment. Initially they only have a low energy of a few eV, but are then accelerated away from the cathode along the cathode fall path (see Fig. 3.1.17), finally reaching an energy of about 1 keV. This is avoided, however, in magnetron sputtering (see Sect. 3.1.21).

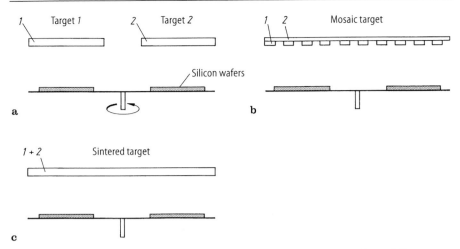

Fig.3.1.20 a–c. Sputtering composite layers by means of co-sputtering of two different targets (**a**), sputtering with a mosaic target (**b**) and sputtering with a sintered target (**c**). The two different materials making up the target are identified by numbers 1 and 2

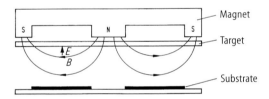

Fig. 3.1.21. Diagram of a magnetron sputtering system. The secondary electrons emitted from the target follow spiral paths in front of the target surface because of the effect of the cathode electric field E and the perpendicular magnetic field B

If composite layers need to be produced, there are three different options, all of which are used. These are sputtering with several targets (also called co-sputtering), sputtering with a mosaic target and sputtering with a sintered target (Fig. 3.1.20). In co-sputtering the silicon wafers are moved under the target in such a way that numerous alternating layers are deposited in a sandwich-like arrangement, rather than creating a mixed layer.

In order to increase the sputter rate for the same rf voltage, sputtering systems are often fitted with permanent or electromagnets [3.17] (magnetron sputtering), positioned behind the (non-ferromagnetic) target (Fig. 3.1.21). Without the additional influence of the magnetic field, the secondary electrons emitted from the target would be accelerated along the cathode fall path towards the anode, rapidly leaving the dark space. The presence of the magnetic field, however, means that they move in spiral paths in front of the target. This process repeats itself many times along the target surface, thus increasing the probability of impacting with the argon to generate extra ion/electron pairs. This technique can produce a 10–100 fold increase in the argon ion current density of up to

100 mA/cm². Since the sputter rate is determined by the current density, increased film growth rates of 1 mm per minute are then possible at a pressure of 0.5 Pa.

In modern sputtering systems the silicon wafers are moved into the recipient vessel through a vacuum interlock. This ensures that specified conditions in the sputtering chamber remain constant because air cannot enter the recipient vessel, preventing any consequent adsorption e.g. of water vapour by the internal walls of the vessel.

3.1.5
Spin coating

Spin coating is the cheapest film production method in silicon technology. It is used to apply photosensitive layers, films containing doping material as well as planarized coatings. The most important coatings produced by this technique are photoresist, spin-on glass and polyimide films.

Figure 3.1.22 illustrates the principle of spin coating. The layer material is dissolved in a solvent, and drops of this are placed on the silicon wafers which are held firmly centred on a turntable by vacuum suction. At typical spin rates of 5000 rpm, the liquid is spun radially outwards by the centrifugal force until it becomes a thin film. The turntable continues to rotate until most of the solvent has evaporated out of the residual film. Normally this is followed by an additional drying stage at 100–200 °C, on a hot plate or in a convection oven.

For planar surfaces, spin coating provides a very uniform layer thickness over the whole silicon wafer (better than 1 %). Good planarization of surfaces containing steps is also achieved with spin-on films (Fig. 3.1.23). This is of fundamental importance for most applications, in particular for photoresists.

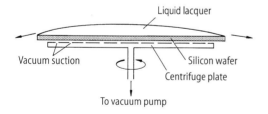

Fig.3.1.22. Schematic diagram of spin coating

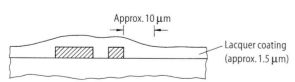

Figure 3.1.23. Step coverage in spin coating. For typical spin conditions, the resist coating at a distance of some 10 μm from a step attains the same thickness as would be formed on a substrate without steps

When producing photoresist films, a strip several mm wide around the edge of the wafer is usually kept free of resist by directing a jet of solvent at the edge of the rotating wafer. This helps to prevent resist particles flaking off under mechanical strain (e.g. during wafer transport), which may be redeposited in the centre of the wafer surface and possibly cause the integrated circuit to fail.

3.1.6
Film production by ion implantation

Ion implantation is primarily used in the field of silicon technology for doping the silicon with acceptors (boron) or donors (arsenic, phosphorus). A separate chapter is devoted to these doped layers (Chap. 6).

This section focuses on the SIMOX technique (SIMOX = *S*eparation by *Im*plantation of *O*xygen), where a high concentration of oxygen (approx. 10^{18} cm^{-2}) is implanted in a monocrystalline silicon substrate to a depth of 0.1–1 mm. This creates a buried SiO$_2$ layer with a 0.1–1 mm thick monocrystalline SOI film above (SOI = *S*ilicon *O*n *I*nsulator) [3.18]. The damage caused to the silicon crystal lattice as the oxygen ions decelerate is annealed in-situ by using a high implantation temperature.

If nitrogen ions are implanted in silicon at a concentration of approx. $5 \cdot 10^{15}$ cm^{-2}, the nitride-like layer which is produced can be used to inhibit thermal oxidation (see Sect. 3.4.2).

Another application of ion implantation for film production is "ion beam induced mixing". In this technique two layers lying one above the other (e.g. molybdenum on silicon) are mixed together in the interface region by the recoil effect, as implanted ions (e.g. arsenic) bombard the atoms in the film. This is used to create an electrical contact or to facilitate silicide formation (see Fig. 3.9.1 c).

3.1.7
Film production using wafer-bonding and back-etching

There is another technique for creating a monocrystalline SOI film, which is completely different from the SIMOX process (Sect. 3.1.6). This is the BESOI technique (BESOI = *B*onded *E*tched-Back *S*ilicon *O*n *I*nsulator) [3.19]. Figure 3.1.24 shows the main processing steps involved.

BESOI offers several advantages over SIMOX.
- A wide variation in the thickness of the monocrystalline SOI layer can be achieved. Intensive work is being carried out to produce thin SOI layers (<0.5 µm) of homogeneous thickness over the whole wafer. At present SIMOX still holds the advantage here.
- A wide variation in the thickness of the buried SiO$_2$ layer can be achieved.
- The monocrystalline SOI layer is of the same quality as the original silicon wafer, because the BESOI technique does not cause crystal defects, unlike the SIMOX process.

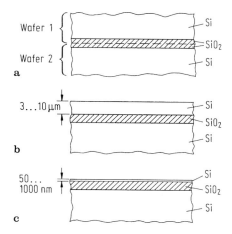

Figure 3.1.24 a–c. Diagram showing processing sequence for producing BESOI layers (BESOI= *B*onded *E*tched-Back *S*ilicon *O*n *I*nsulator). Van der Waals' forces are responsible for the "adhesion" between wafers 1 and 2. **a** Two oxidized wafers are brought into direct contact (wafer bonding), **b** The thickness of wafer 1 is reduced to 3–10 μm by lapping or polishing, **c** The thickness of wafer 1 is reduced to 50–1000 nm by controlled back-etching

Regarding their commercial use, both SOI techniques are in their infancy. They could have a very important future, however, because of the potential benefits of SOI layers (for both low-power and high-power applications, stability at high operating temperatures or under cosmic radiation, reduction in processing steps). Other techniques for creating SOI layers, such as the SOS technique (Silicon-On-Sapphire) or the recrystallization of polysilicon using a scanning laser beam (see Sect. 3.1.26), may be superseded by the BESOI or SIMOX methods.

3.1.8
Annealing techniques

In silicon technology the term annealing refers to the treatment of silicon wafers at raised temperatures in an inert atmosphere[10] (nitrogen, argon, hydrogen, forming gas[11]). There is no growth of new layers in the process, and no material is removed, but the layers that already exist, and even the silicon substrate itself, may be changed significantly. The most important changes brought about by annealing, some wanted and some not, are given below.

---

10 The term "annealing" is often used in a broader sense. For instance CVD deposition or thermal oxidation at 900 °C constitutes "annealing at 900 °C" for the silicon substrate itself and for those layers on the silicon wafer that are not involved in the CVD deposition or oxidation.
11 Forming gas refers to a mixture of 90 % nitrogen with 10 % hydrogen.

- Activation of doping materials
- Diffusion of doping materials, heavy metal atoms, alkali ions, oxygen, hydrogen etc.
- Removal (repair), or growth, or creation of crystal defects in the monocrystalline silicon
- Removal (repair) of fixed interface states and rechargeable interface states at Si-SiO$_2$ interfaces
- Intermetallic reactions (e.g. aluminium-silicon reaction in contact holes, silicide formation)
- Grain growth in polycrystalline layers (polysilicon, aluminium)
- "Densification" of deposited films (CVD films, cross-linkage of polyimide films, curing of photoresist films)
- Film flow (phosphorus glass, boron-phosphorus glass, photoresist)

These effects are looked at in more detail in the relevant sections.

Annealing is usually performed at atmospheric pressure in the same type of tube furnaces used in thermal oxidation (cf. Sect. 3.1.14). The relatively slow heating and cooling of the wafer in this type of oven process means that the cycle time often lasts for thirty minutes or more, even if the actual annealing time required is much shorter.

Due to the limited thermal budget of advanced integrated circuits during manufacture RTA techniques (RTA = *R*apid *T*hermal *A*nnealing) are becoming more important [3.20]. Figure 3.1.25 shows four different rapid thermal annealing processes. These fall into two categories, depending on whether the silicon wafer is heated as uniformly as possible over a few seconds (rapid isothermal annealing, Fig. 3.1.25 a and rapid optical annealing, Fig. 3.1.25 b) or whether the thermal energy is directed onto the wafer (via a heating filament, Fig. 3.1.25 c, laser beam, Fig. 3.1.25 d, or electron beam) using a scanning technique to cover the whole wafer surface.

The localized absorption of energy lasting just milliseconds means that the scanning techniques can also be used to melt small areas of polysilicon layers, and thus perform a type of float zone process on the silicon surface as it scans [3.21] (Fig. 3.1.26). If contact between the polysilicon film and the monocrystalline substrate is provided at several points, then the polysilicon film takes on the monocrystalline structure of the silicon substrate when it recrystallizes. Thus in principle it is possible to integrate transistors into several planes one above the other (three dimensional integration, see Sect. 3.3.1).

The localized laser heating shown in Fig. 3.1.25 d has found important applications. By localized vaporization of polysilicon patterns lying on an SiO$_2$ film with poor thermal conductivity, "fuses" can be burnt through. This can be used e.g. to disconnect defective parts of a memory cell array (see Sect. 8.4.2), as well as to connect up redundant parts, thereby improving the yield. Another use for selective material vaporization by laser beam is the repair of defective chromium masks (see Sect. 4.2.8).

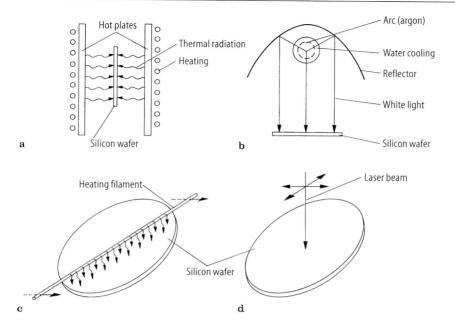

Fig. 3.1.25 a–d. Rapid thermal annealing techniques. **a** Heating the silicon wafer between two heated graphite plates (rapid isothermal annealing); **b** heating the silicon wafer by a high-power lamp (rapid optical annealing); **c** heating in strips using a heating filament (strip heater) scanned over the silicon wafer; **d** spot-type heating using a laser beam which is scanned over the silicon wafer in the x and y direction

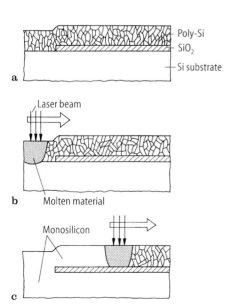

Fig. 3.1.26 a–c. Creating a monocrystalline silicon layer on a SiO$_2$ layer using a scanned laser beam. The processing sequence is shown in the diagrams. **a** Producing a polysilicon film on an SiO$_2$ layer containing gaps through to the monocrystalline substrate; **b** a laser beam causes localized melting of the polysilicon layer. The gaps in the SiO$_2$ layer act as centres for the recrystallization of the molten silicon; **c** if the laser beam is moved across the polysilicon surface, then that part of the polysilicon layer over which the laser beam has passed, takes on the monocrystalline structure of the silicon substrate

3.2
The monocrystalline silicon wafer

Manufacture of an integrated circuit starts from a monocrystalline silicon wafer. Its most important properties are summarized below.

3.2.1
Geometry and crystallography of silicon wafers

From a productivity perspective there is a continuous trend toward increasing the silicon wafer diameter. Figure 3.2.1 shows how the wafer diameter has grown over the years. Each increase in the wafer diameter brings its own problems to the manufacture of integrated circuits, not only to the production of the monocrystalline silicon wafers themselves, but also at the processing steps. These problems mainly relate to the increasing weight[12] of the silicon wafers and to the ever tighter tolerances and conditions of uniformity that are required (e.g. lower temperature gradients during high temperature processes).

In the manufacture of integrated circuits, the silicon wafers are normally cut from a single crystal ingot in such a way that the wafer surface is a crystallographic ⟨100⟩ or ⟨111⟩ plane. It can be useful, e.g. for better epitaxial growth conditions, to deviate by just a few degrees from this angle. In order to identify the main crystallographic orientations in the wafer plane, the wafer manufacturer provides a so-called flat at the wafer periphery. The edges of the rectangular circuit structures usually run parallel or at right angles to the flat. A second (shorter) flat can be used to identify whether it is a ⟨100⟩ or ⟨111⟩ slice, or whether it is p or n type, according to the position of the short flat relative to the long flat.

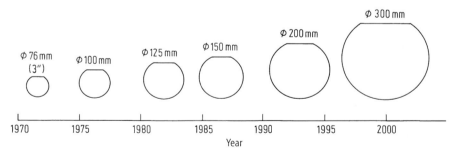

Fig. 3.2.1. The increasing size of silicon wafers used in the manufacture of integrated circuits

---

12   The wafer thickness has to be increased in line with wafer diameter for reasons of mechanical strength. A 76 mm wafer is typically 375 μm thick, a 100 mm wafer 450 μm, a 150 mm wafer 675 μm and a 200 mm wafer 725 μm thick. A thickness of 770 μm is the target for 300 mm wafers.

Particular demands are placed on the flatness of the wafer surface, because light-optical projection lithography can only produce a sharp image over a very small focal range. For instance, if 1 μm features are to be produced, then the wafer thickness must not vary by more than 1 μm over an area of 2 cm × 2 cm (LTV = Local Thickness Variation, see Sect. 4.2.6). This assumes that the back of the silicon wafer is held in close contact with a flat table by vacuum suction.

3.2.2
Doping of silicon wafers

When the silicon crystal is grown, the correct amount of doping material (boron, phosphorus, arsenic, antimony) is added to the molten material to achieve the required concentration in the single crystal. For integrated circuits that do not require an epitaxial layer, the base doping (also called substrate doping) of the silicon wafers is the lowest doping concentration that occurs in the circuit, lying in the range 10^{13}–10^{16} cm^{-3}. If, on the other hand, an epitaxial layer will be added to the wafer later, higher substrate doping levels may be desirable of over 10^{20} cm^{-3} (low resistivity substrate, see Sect. 3.3). Figure 3.2.2 shows the relationship between substrate doping level and resistivity for p- and n-type doping.

3.2.3
Monocrystalline silicon growing techniques

The vast majority of monocrystalline silicon is today manufactured using the so-called Czochralski technique (CZ process) [3.22]. The float-zone technique (FZ process) is comparatively less important.

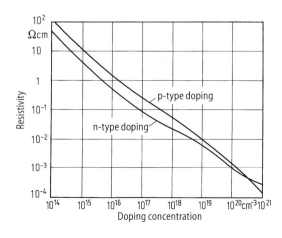

Fig. 3.2.2. Relationship between substrate doping and resistivity

The main difference between single crystals manufactured by these two techniques is the oxygen content[13]. Whilst FZ silicon has insignificant oxygen concentrations (approx. 10^{15} cm^{-3}), CZ silicon typically contains oxygen concentrations between $5 \cdot 10^{17}$ and $2 \cdot 10^{18}$ cm^{-3}. The oxygen originating from the crucible material has a critical effect on the formation of crystal defects in the subsequent processing steps.[14]

At the silicon solidification temperature, which is the point at which oxygen becomes embedded in the silicon, the solubility of oxygen in silicon is significantly higher ($2 \cdot 10^{18}$ cm^{-3}) than at subsequent processing temperatures (e.g. 10^{17} cm^{-3} at 1000 °C). This means that when the temperature falls, the oxygen tends to precipitate out on suitable condensation nuclei. These nuclei are not initially present in the monocrystalline silicon, however. Thus the majority of oxygen in silicon wafers at delivery is dissolved oxygen, i.e. it sits in interstitial sites. Depending on the sequence and duration of high-temperature processes during integrated circuit manufacture, however, oxygen clusters can form in the silicon. They act as sinks for diffused heavy metal atoms (getter centres). During thermal oxidation these oxygen clusters, which may contain heavy metal atoms, cause weak spots in the SiO_2 layer, impairing the breakdown performance and long-term stability (see Sect. 3.4.3).

The unwanted effect of the oxygen in the silicon can be turned to advantage by suitable pre-processing of the silicon wafers. Annealing over a long period (e.g. 1 day) at about 700 °C, creates nuclei in the silicon lattice onto which the oxygen will subsequently condense (Fig. 3.2.3). A second annealing stage at about 1100 °C has two effects: whilst oxygen diffuses out of the near-surface layer, oxygen is precipitated in the bulk of the silicon wafer.[15] As mentioned above, these oxygen precipitates act as getter centres for heavy metal atoms and point defects. This time, however, the effect is very welcome, because the getter centres are located in the electrically inactive part of the silicon wafer (intrinsic gettering), whilst the region close to the surface (a few µm deep), which is crucial to the electrical operation of the circuits, is depleted of oxygen (denuded zone).

The denuded zone, which is essential for high-grade gate oxides on CZ silicon wafers, is either pre-produced by the wafer manufacturer, or it is formed at the start of integrated circuit manufacture. If wells are to be produced in the silicon, then the required high temperature stage can be combined with this process (see Sect. 8.2.1).

---

13 Apart from oxygen, the silicon wafers also contain carbon. The carbon content of FZ and CZ silicon is about 10^{16} cm^{-3}. What effect the carbon has is still not fully understood. It may act as a nucleus for oxygen precipitation.
14 The oxygen also has the effect of creating so-called thermal donors, which are only significant at high oxygen concentrations and low substrate doping levels.
15 The sequence of annealing stages is often performed with oxygen out-diffusion first at about 1100 °C, followed by nuclei formation at 700–800 °C and finally oxygen precipitation at about 1000 °C.

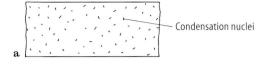
Condensation nuclei

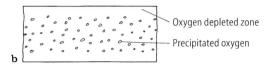
Oxygen depleted zone
Precipitated oxygen

Fig. 3.2.3 a, b. Schematic diagram of the formation of a near-surface region depleted of oxygen (denuded zone), and the precipitation of oxygen inside the silicon wafer. a First step: annealing at 700 °C; b Second step: annealing at 1100 °C.

To increase the getter effect away from the near-surface zone, the back of the wafer is often given a gettering layer. For FZ silicon where there is no opportunity to produce intrinsic getters, this back gettering is particularly important. Some of the measures for creating a dense network of crystal defects (getter centres) on the back of the wafer include mechanical damage, high dose argon implantation, scanning with a high-energy laser, high dose phosphorus diffusion and the deposition of a phosphorus-doped polysilicon layer in direct contact with the back of the wafer. The last measure has proved particularly effective. It is also of interest because it can be produced at no extra cost if the manufacturing process already involves phosphorus-doped polysilicon.

3.3
Epitaxial layers

Epitaxy refers to the monocrystalline growth of a single layer on top of a monocrystalline substrate[16].

3.3.1
Uses for epitaxial layers

Figure 3.3.1 shows a schematic view of the two most important uses of epitaxial silicon layers on monocrystalline silicon wafers. In the first application (Fig. 3.3.1 a) the wafers are produced by the manufacturer with a light to moderately doped epitaxial layer (10^{14}–10^{16} cm^{-3}) already grown on the heavily doped substrate (10^{18}–10^{20} cm^{-3}). This type of silicon wafer is very useful in eg CMOS circuits, where the minority charge carriers diffusing from the substrate

---

16 This book shall only deal with silicon epitaxial layers on monocrystalline silicon wafers. The SOS technique (SOS = Silicon On Sapphire), where a monocrystalline silicon layer is grown on an insulating monocrystalline sapphire substrate, is not covered here.

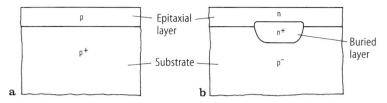

Fig. 3.3.1 a, b. The most important applications of epitaxial silicon layers on monocrystalline silicon wafers. **a** Moderately doped epitaxial layer (e.g. 10^{16} cm^{-3}) on a heavily doped silicon substrate (e.g. 10^{19} cm^{-3}); **b** moderately doped epitaxial layer (e.g. 10^{16} cm^{-3}) on a lightly doped substrate (e.g. 10^{15} cm^{-3}) with a heavily doped buried layer (e.g. 10^{20} cm^{-3})

to the surface need to be kept to a minimum (e.g. in dynamic memory cells), and where the latch-up effect must be suppressed as far as possible (see Sect. 8.3.2). In the other important case (Fig. 3.3.1 b), which is typical of most bipolar circuits, a light to moderately doped epitaxial layer (10^{15}–10^{17} cm^{-3}) is produced on substrates containing heavily doped (10^{19} to 10^{20} cm^{-3}) buried layer islands which act as low resistance collector leads for the bipolar transistors (see Table 8.5). The thickness of the epitaxial layers varies between 0.5 and 20 μm.

At present, epitaxial layers are predominantly produced by CVD deposition of silicon on monocrystalline silicon wafers at temperatures above 800 °C (see Sect. 3.1.1). At these high temperatures, the deposited Si layer assumes the crystal orientation of the Si substrate (solid epitaxy). The CVD reactors used for silicon epitaxy are the same as those types shown in Fig. 3.1.4 a, b and c.

SiCl$_4$ (see Sect. 3.1.1), SiH$_2$Cl$_2$ or SiH$_4$ are used as the starting gas in the reaction. Temperatures of more than 1100 °C are needed until SiCl$_4$ will break down sufficiently into sub-molecular components (including SiCl$_2$). Thus typical processing temperatures for SiCl$_4$ epitaxy lie in the 1150–1250 °C range. With SiH$_2$Cl$_2$ one can work from 1050–1150 °C, whilst with SiH$_4$ the processing temperature can be reduced even further to below 900 °C. SiH$_2$Cl$_2$ and SiH$_4$ epitaxy are best performed at a pressure of 0.1–1 bar [3.23] (SACVD = Sub Atmospheric Chemical Vapour Deposition, see Sect. 3.1.1). Under these conditions, selective epitaxy (see Sect. 3.1.3) and SiGe hetero-epitaxy (see Sect. 8.3.9 for use) are possible.

The epitaxial layers usually need to be doped with particular foreign atoms during deposition. This is achieved by adding a gas containing the foreign atoms (e.g. AsH$_3$), so that these atoms become embedded in the growing layer as the silicon atoms are deposited (Fig. 3.1.1).

Two other techniques for creating monocrystalline silicon layers need to be mentioned in addition to CVD deposition, even though they are still rarely used at present. These are molecular beam epitaxy and recrystallization of polysilicon. Both techniques are described in Sects. 3.1.3 and 3.1.8 respectively.

Molecular beam epitaxy can be used to create more abrupt doping gradients than with gas phase epitaxy, because of the lower deposition temperatures. Another possible application for molecular beam epitaxy is the formation of

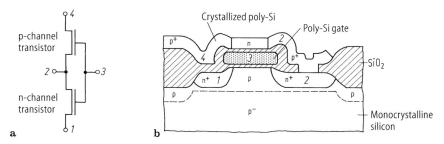

Fig.3.3.2 a, b. CMOS inverter in three-dimensional integration technology. **a** Circuit diagram; **b** diagram showing cross-section through the inverter structure. *1* ground connection; *2* inverter output; *3* inverter input; *4* supply voltage connection

monocrystalline silicide layers (NiSi$_2$, CoSi$_2$) on monocrystalline silicon substrates. If one then grows a silicon epitaxial layer on top of the silicide film, the silicide becomes a low resistivity buried layer.

The recrystallization of a polysilicon layer to form a monocrystalline film on an SiO$_2$ layer (see Fig. 3.1.26) can be used for so-called three-dimensional integration. This refers to the arrangement of transistors in two or more isolated planes lying one above the other. Figure 3.3.2 shows an example of a CMOS inverter integrated in three dimensions. This kind of arrangement not only saves surface space, but also avoids the latch-up effect. Three-dimensional integration also allows new types of components to be created. For instance a transistor in recrystallized polysilicon can be controlled from both above and below using gate electrodes. If one can tolerate a poorer cut-off performance for the p-channel transistor in Fig. 3.3.2, then recrystallization of the polysilicon is not necessary. It is then referred to as a Thin Film Transistor (TFT). TFTs can be used to approximately halve the surface area occupied by e.g. an SRAM memory cell (see Fig. 8.4.1) [3.24].

3.3.2
Diffusion of doping atoms from the substrate into the epitaxial layer

At the high temperatures of CVD epitaxy, there is a certain amount of diffusion of doping atoms from the substrate into the epitaxial layer, and vice versa. The diffusion is particularly significant if the substrate is heavily doped, whether generally or in specific areas (Fig. 3.3.1).

Section 6.3.2 deals with the diffusion of doping atoms. In order to calculate the doping atom profile in the epitaxial layer resulting from diffusion out of the substrate, one can assume that the situation approximates that shown in Fig. 6.3.2, where the doping atom concentration remains constant at the interface between substrate and epitaxial layer.[17] For instance, if one wishes to esti-

---

17 This assumption is not just a valid approximation for a uniformly doped substrate, but also applies to a buried layer, since buried layer doping is normally forced as far down as possible into the substrate (e.g. 2 µm) before epitaxy.

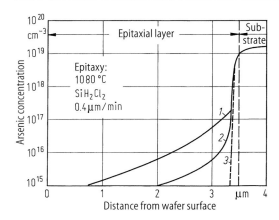

Fig. 3.3.3. Arsenic concentration profiles in an epitaxial layer, on a substrate doped with $1.6 \cdot 10^{19}$ cm^{-3} of arsenic for epitaxial deposition occurring at atmospheric pressure (curve 1) and at 0.13 bar (curve 2). Curve 3 shows the predicted arsenic concentration profile for purely solid-state diffusion of the arsenic atoms.

mate the distance z from the epitaxy/substrate interface at which the doping atom concentration has fallen by four orders of magnitude, then one can read off directly the value $z/\sigma = 2.75$ for $c(z)/c_0 = 10^{-4}$ from Fig. 6.3.2. If an epitaxial layer is deposited at e.g. 1150 °C, then one can read off a value of $5 \cdot 10^{-14}$ cm^{2}s^{-1} for the diffusion constant from Fig. 6.3.1, assuming that the diffusing doping atoms are arsenic. If t, the time that temperature processing lasts, is known, then the diffusion length $= 2\sqrt{Dt}$ can be calculated. For example, if epitaxial deposition lasts 10 min, then $\sigma = 110$ nm, and $z = 2.75\sigma = 304$ nm. If boron or phosphorus are the diffusing dopants, then one would obtain $z = 850$ nm.

In reality the substrate dopant can be found to extend much further into the epitaxial layer. The measurements shown in Fig. 3.3.3 illustrate this particularly clearly. The effect, which is known as autodoping, is caused by constant evaporation and re-incorporation of arsenic during the period of epitaxial growth. If the substrate contains buried layer islands (see Fig. 3.3.1), then a raised doping level is also found above the space between the islands (lateral autodoping). The following factors affect autodoping.
- For atmospheric pressure CVD epitaxial deposition, autodoping is less pronounced if antimony is used instead of arsenic as the buried layer or substrate dopant, and if deposition occurs at high temperatures.
- The autodoping effect can be significantly reduced by performing CVD epitaxial deposition at a reduced pressure (e.g. SACVD as in Fig. 3.3.3 curve 2).

The last finding is the main reason why lower temperature epitaxial deposition (below 1100 °C) is frequently performed at a reduced pressure. It has also been observed that another phenomenon called pattern-shifting can be practically eliminated at reduced pressure. Pattern-shifting refers to the way a step moves as it is reproduced in the layer above during epitaxial deposition – instead of progressing perpendicular to the surface it experiences a lateral shift. Steps in the surface are needed when the epitaxial layer is being deposited on a substrate

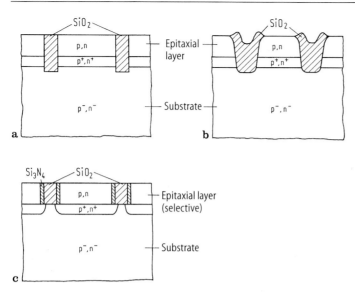

Fig. 3.3.4 a–c. Alternative techniques for creating p^+ and n^+ doped buried layer islands. **a** Etching of deep trenches and refilling with insulating material; **b** etching trenches of about half the depth then localized thermal oxidation in trench area, preferably by high-pressure oxidation; **c** creating SiO_2 ridges, doping the buried layer regions and selective growth of an epitaxial layer above the buried layer regions.

with buried layer islands (Fig. 3.3.1). This is because the subsequent pattern planes must be aligned with the buried layer plane, and visible features are needed to achieve this, i.e. steps on the silicon surface.

There is, however, another way of completely eliminating both the buried layer alignment problem and lateral autodoping. Figure 3.3.4 shows three alternatives. In the procedure of Fig. 3.3.4 a, the epitaxial layer is grown on a blanket buried layer. Trenches are then etched through the epitaxial layer down to the buried layer, at those places where the buried layer needs to be broken. These trenches are then refilled with insulating material (cf. Sect. 3.5.4). Another option uses a combination of etching and thermal oxidation (Fig. 3.3.4 b). In the last technique (Fig. 3.3.4 c) a thick SiO_2 layer is first grown on the substrate. The SiO_2 is then etched off everywhere except at those points where the insulating ridges are required. Buried layer doping (e.g. using ion implantation) can then be carried out using the SiO_2 ridges as a mask. Finally, the epitaxial layer is grown selectively on the monocrystalline silicon areas (see Fig. 3.1.3 a). In this so-called selective epitaxy the gas phase reaction is held near equilibrium (e.g. by adding HCl during $SiCl_4$ epitaxy). Thus, because of the lower energy required to form a nucleus on the silicon, nuclei for the adsorption of reactions gases only occur on the silicon, and so silicon only grows on silicon areas [3.25].

3.4
Thermal SiO₂ layers

As described in detail in Sect. 3.1.2, thermal SiO_2 layers are created by oxidation of the surface of a silicon wafer or silicon layer at temperatures of between 700 and 1200 °C in an oxidizing atmosphere. The thermal SiO_2 layers are used as isolating and passivating layers in semiconductor circuits. During the circuit manufacturing process they also often perform a masking function, e.g. against ion implantation.

3.4.1
Uses of thermal SiO₂ layers

The thickness of silicon converted into SiO_2 in thermal oxidation equals 45 % of the grown SiO_2 thickness (cf. Fig. 3.1.8). This means that the Si/SiO_2 interface moves into the monocrystalline silicon during oxidation. Thus the interface properties are less affected by the cleanliness of the Si surface before oxidation than when the SiO_2 layers are deposited, where the position of the Si surface is unchanged.

For this reason the thermal technique is preferred for producing those SiO_2 layers in direct contact with the monocrystalline silicon, in particular the gate oxide in MOS circuits (Sect. 3.4.3) and the LOCOS oxide (Sect. 3.4.2).

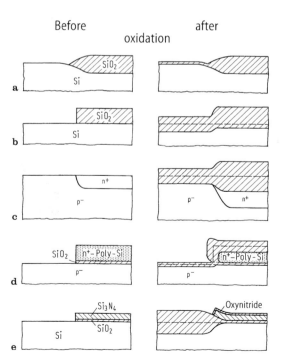

Fig. 3.4.1 a–e. The most important applications of substrate-dependent SiO_2 growth by thermal oxidation. The dashed lines in the right-hand sub-diagrams indicate the original Si surface. The different growth is caused in cases **a** and **b** by an SiO_2 layer already being present, and in cases **c** and **d** by different doping levels of the silicon regions, whilst in case **e** the silicon nitride layer acts as an oxidation barrier.

In addition, thermal oxidation is used where a substrate-dependent oxide growth rate is required, which is typical of thermal oxidation.

The most important examples of substrate-dependent layer growth are shown in Fig. 3.4.1. Figure 3.4.1 a illustrates gate oxidation where a LOCOS field oxide layer is already present. There is also an SiO_2 structure already present in Fig. 3.4.1 b. In the subsequent thermal oxidation process, the existing SiO_2 layer only grows slightly, whilst the growth rate above the oxide-free areas is much larger. This reduces the step between the two areas, whilst a step arises in the Si-SiO_2 interface. The latter factor is significant, for instance in optical edge detection for mask alignment. Figure 3.4.1 c shows an application where the SiO_2 layer grows faster on heavily doped areas of silicon than on lightly doped areas (cf. Fig. 3.1.11). One use of this effect is in self-aligned masking for boron source/drain implantation in CMOS circuits. Figure 3.4.1 d shows a growth characteristic that is frequently exploited in MOS and bipolar technology, with a relatively thick SiO_2 layer forming on the heavily doped polysilicon whilst a thinner SiO_2 layer grows on the lightly doped silicon substrate. Finally, Fig. 3.4.1 e shows local oxidation of silicon (LOCOS) as an example of substrate-dependent oxide growth, where a silicon nitride mask acts as an oxidation barrier. The LOCOS technique is dealt with in detail in the next section.

3.4.2
The LOCOS technique

As Fig. 3.4.2 illustrates, integrated MOS circuits require a thick field oxide (> about 0.4 µm) with a higher doping level beneath the field oxide. The LOCOS technique (Local Oxidation of Silicon, Fig. 3.4.1 e) is currently used almost exclusively for producing such field regions. It is therefore one of the most important key technologies for the manufacture of integrated circuits [3.26].

Figure 3.4.3 shows the detailed LOCOS manufacturing sequence and the physical processes going on. The white-ribbon[18] effect is particularly relevant here. To prevent the gate oxide thinning, an additional thermal oxide layer is grown and removed again (sacrificial oxide). Any doping atom implantation that may be required in the channel region of the MOS transistors (e.g. for adjusting the threshold voltages or preventing drain-source punch-through) should occur through this extra layer. Depositing the gate poly-Si immediately after gate oxidation has proved vital to maintaining the quality of the gate oxide.

The transition region between the field oxide and gate oxide is described as a bird's beak, because of its characteristic profile. The gradual edge transition is useful both in the monosilicon (it acts as an alignment edge, see Sect. 4.2.7, and

---

18 The term white ribbon effect originates from the inventor of the LOCOS technique, E. Kooi [3.26], who in the early stages of LOCOS technology identified the bright stripes along the LOCOS edges seen under the light microscope as gate oxide thinning.

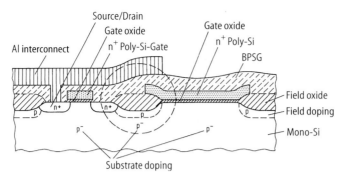

Fig. 3.4.2. Illustration of the need for a thick field oxide region and increased field doping using two adjacent n-channel MOS transistors as an example (in the right-hand MOS transistor the conduction channel runs perpendicular to the image plane). The main reason for requiring a field oxide of sufficient thickness (> 0.4 µm approx.) is to mask the field regions during source/drain doping. The field regions need to be more heavily doped than the substrate and channel areas, firstly to inhibit punch-through between the two transistors, and secondly to keep the parasitic field oxide transistor (outlined by the dashed circle) turned off in all possible operating conditions. The edge of the field doping must coincide with the edge of the field oxide (self-alignment), otherwise the channel width of the active MOS transistors would not be defined, as can be seen from the right-hand transistor.

prevents any excessive mechanical stresses) and at the SiO_2 surface (e.g. avoiding poly-Si residues after anisotropic poly-Si etching, see Fig. 5.3.1).

The disadvantage of the bird's beak is that it grows into the active transistor region (see Fig. 3.4.3 e) thus converting valuable active areas into inactive field oxide regions.[19]

Figure 3.4.4 illustrates the two most important processing steps for producing both a shorter bird's beak and a lateral shift toward the field oxide. This is done by increasing the Si_3N_4/SiO_2 layer thickness ratio and by thinning the gate oxide by over-etching the whole SiO_2 surface once the nitride mask has been removed. The trend towards lower processing temperatures makes it harder to shorten the bird's beak, because the length increases as the temperature of field oxidation decreases.

The thin SiO_2 layer below the nitride is used to isolate the silicon from the strong mechanical stresses exerted by the nitride. For the layer thickness combination shown in Fig. 3.4.4 b, however, the critical shear stress can still be exceeded in localized areas of the silicon, which can then cause dislocations. Several solutions have been put forward to avoid this, whilst still producing a short bird's beak.
– One inserts between the SiO_2 and nitride layers a polysilicon layer (e.g. 200 nm) which is completely oxidized away during LOCOS oxidation (PBL =

---

19 At 90° corners, the bird's beak is longer than along straight edges. This effect can lead to rounding off of corners, which is particularly pronounced at e.g. the end of narrow, elongated geometries.

3.4 Thermal SiO$_2$ Layers

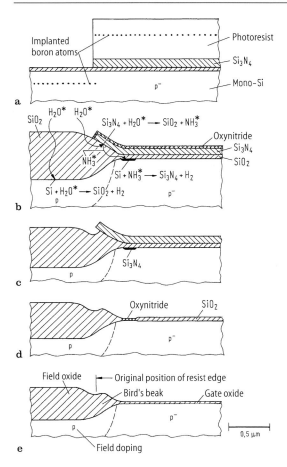

Fig. 3.4.3 a–e. Processing sequence to produce LOCOS field oxide regions with self-aligned field doping. Sub-diagram **b** shows the processes involved in creating the white ribbon effect. Three oxide over-etching procedures produce a marked reduction in the field oxide thickness. **a** After anisotropic etching of the Si$_3$N$_4$ layer and boron implantation for self-aligned field doping; **b** towards the end of field oxidation. N$_2$O* and NH$_3$* indicate the thermally excited molecules that are chemically very reactive; **c** after isotropic SiO$_2$ over-etching to remove the oxynitride; **d** after removing the nitride and the thin oxide, then (wet) reoxidation; **e** after reoxidized layer removal and gate oxidation

Poly Buffered LOCOS). Since the nitride is lying directly on top of silicon, the bird's beak is short. The polysilicon also acts as a buffer against the mechanical stresses exerted by the nitride, so that the SiO$_2$ layer, which simply serves as an etch stop layer here, can be made very thin (e.g. 20 nm) (Fig. 3.4.5).
- CVD oxynitride can be used instead of the SiO$_2$ layer, with a slower oxygen diffusion rate during LOCOS oxidation.
- The SiO$_2$ layer can be omitted entirely if the nitride layer is made so thin (< 30 nm) that it no longer exerts excessive stresses in the monosilicon. The problem here, however, is to remove the nitride without damaging the monosilicon surface.

The simplest way of compensating for the bird's beak is to make the photoresist mask in Fig. 3.4.3 correspondingly wider. This method only works, however, if one is not already working at the minimum pattern width determined by the lithography technique. For smaller structures the so-called spacer technique and

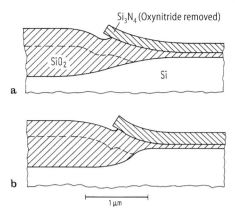

Fig. 3.4.4 a, b. Shortening the bird's beak created by the LOCOS process, by increasing the Si_3N_4/SiO_2 layer thickness ratio. The relatively long bird's beak in **a** results for a layer combination of 50 nm SiO_2 and 140 nm Si_3N_4 for the oxidation mask. In **b** the layer thicknesses are 30 nm SiO_2 and 200 mm Si_3N_4. If, after removing the Si_3N_4 mask, 300 nm of SiO_2 are etched off the 900 nm thick field oxide (dashed line), then the lateral offset of the field oxide edge from the (original) nitride edge equals 0.5 µm in case a and only 0.3 µm in case **b**. In both cases a temperature of 970 °C is assumed for thick oxidation (profiles to scale)

the thermal oxidation of a polysilicon layer provide a means of displacing the structure edges laterally by a defined amount. Figure 3.4.6 shows three examples. Section 3.5.3 describes how spacers are produced.

Apart from the options illustrated in Figs. 3.4.5 and 3.4.6, several other modifications of the LOCOS process have been proposed to prevent the field regions growing at the expense of the active regions, for instance the PELOX technique [3.28] (PELOX = Polysilicon Encapsulated Local Oxidation). Even when all possible measures are employed, field oxide land widths of about 0.5 µm have until now been considered the limit of the LOCOS technique. This limit is realistic when based on a minimum field oxide thickness of 0.4 µm (cf. Fig. 3.4.2). If, however, one can tolerate a LOCOS thickness of e.g. 0.1 µm[20], then field oxide lands with widths of even 0.1 µm are conceivable. Finally, the PBL technique (Fig. 3.4.5) provides an option for producing active areas without losing space. This is achieved by the thin oxide layer and the poly-Si above it (Fig. 3.4.5 b) serving the dual purpose of gate oxide and the lower section of the gate poly-Si respectively.

The alternative to the LOCOS technique is trench isolation (see Sect. 3.5.4). The trench isolation technique involves etching narrow trenches in monocrystalline silicon and then refilling them with insulating material. This also enables

---

20 Thin LOCOS layers are acceptable for e.g. thin SOI films (see Sect. 3.1.6) or in such cases where a high ion implantation dose for the source/drain regions is not required (e.g. in the memory cell field of DRAM memories, see Fig. 8.4.2).

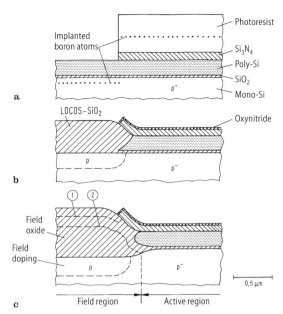

Fig. 3.4.5 a–c. Processing sequence for the PBL technique (PBL = *Poly Buffered LOCOS*), showing the change in the field oxide profile. **a** After photoresist masking of the active regions, Si$_3$N$_4$ etching and boron implantation, **b** during LOCOS oxidation, up to the point where the poly-Si has just been completely oxidized, **c** after LOCOS oxidation. The dashed lines 1 and 2 indicate the oxide surface after 1 removal of the oxynitride and 2 removal of the nitride and poly-Si layer in the active region.

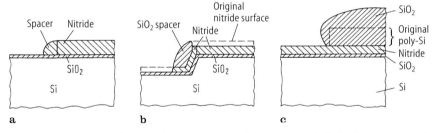

Fig. 3.4.6 a–c. Options for lateral displacement of the LOCOS bird's beak. The spacer in diagram **a** can be made of nitride or polysilicon. Boron implantation in the thick oxide regions is carried out after the spacer has been formed. In diagram **b**, the SiO$_2$ spacer is used to extend the nitride masking over the edge of the silicon. The SiO$_2$ spacer is removed again before thick oxidation [3.27]. In **c** a poly-Si structure is converted entirely into SiO$_2$ by thermal oxidation. This creates an SiO$_2$ edge that is displaced laterally with respect to the original poly-Si edge, and this acts as the masking edge for boron implantation and nitride etching.

the formation of narrow isolating ridges extending deep into the silicon substrate (several μm deep), which are ideal for isolating closely-packed bipolar transistors (Fig. 8.3.10) or p- and n-channel MOS transistors in CMOS circuits.

The process referred to as the recessed-LOCOS technique can be considered as a combination of LOCOS and trench isolation (Fig. 3.4.7), and is used for oxide isolation of bipolar transistors. As Fig. 3.4.7 shows, a planar surface can be obtained by etching into the monosilicon in the thick oxide areas. However "bird's heads" are formed in this approach.

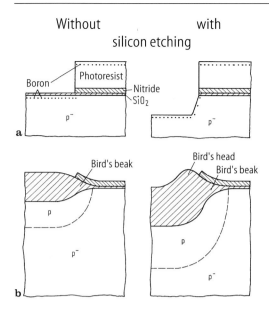

Fig. 3.4.7 a, b. LOCOS profiles with and without etching into the silicon in the thick oxide regions before LOCOS oxidation. a After boron implantation, b after thick oxidation. In the etched silicon case (right diagram), the oxide is raised up at the LOCOS edge (bird's head) because the SiO_2 takes up about twice as much space as the oxidized silicon.

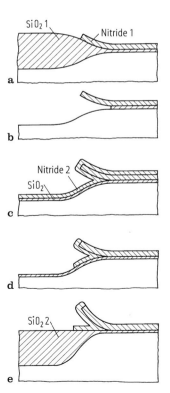

Fig. 3.4.8 a–e. Processing sequence for a LOCOS technique producing a planar thick oxide without a bird's head. a First LOCOS oxidation (SiO_2 1); b isotropic etching of SiO_2 (e.g. wet chemical); c thermal oxidation and conformal CVD nitride deposition (nitride 2); d anisotropic etching of the second nitride layer; e second LOCOS oxidation (SiO_2 2).

In bipolar technology, where oxide isolation layers are typically more than 1 μm thick, this oxide protrusion at the LOCOS edge can cause topographical problems for layer deposition, lithography and etching. The bird's head can be etched away later by the etch-back process using a resist coating (see Fig. 3.5.4). Another approach is to prevent the bird's head appearing at all by using a modified LOCOS processing sequence, such as that shown in Fig. 3.4.8. The trick here is not to etch away the silicon substrate (Fig. 3.4.7) but to oxidize the silicon to a depth equal to the etching depth, and then to etch away the oxide. This also removes the silicon in-fill beneath the nitride edge that would otherwise produce the bird's head when oxidized. Apart from avoiding the bird's head, this modified LOCOS technique has the advantage of producing a smooth step transition in the monosilicon.

3.4.3
Properties of thin thermal SiO₂ films

Of all integrated circuit applications using SiO_2 films, it is the gate and tunnel oxides which face the toughest electrical demands. This section therefore concentrates primarily on these oxide films.

Deviations from the purely dielectric behaviour of an SiO_2 layer (relative dielectric constant $\varepsilon_r = 3.9$) are attributable to fixed or mobile charges in the oxide.

Figure 3.4.9 shows schematically the different types of charges that, for instance, can cause the threshold voltage and saturation current of MOS transistors to drift [3.23].

The mobile charges in the oxide are mostly sodium ions (Na^+). These can be introduced during thermal oxidation, for instance, and may be almost eliminated by adding HCl to the oxygen during the process (see Sect. 3.1.2).

Sodium impurities which reach the wafer surface during a later stage of the manufacturing process (see Chap. 7) diffuse into the PSG or BPSG layer that is

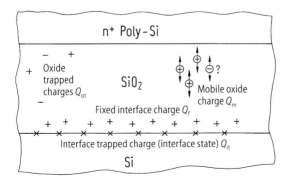

Fig. 3.4.9. Electrical charges in a SiO_2 layer

usually present, where they become bound to fixed locations ("gettered") and thus made harmless.

Of all the integrated circuit manufacturing processes, it is the plasma processes that are mainly responsible for creating fixed interface charges[21] (positive), i.e. charges bound to traps, and for producing interface states. As Fig. 3.1.5 shows, a plasma is able to produce excited atoms and molecules (radicals) that are chemically highly reactive. The generation of interface charges and states is attributed mainly to the hydrogen radical (H*), which can easily diffuse to any point of the integrated circuit, even at lower temperatures (see Figs. 6.4.1 and 6.4.2). Annealing in hydrogen or forming gas at 400–450 °C at the end of the manufacturing process can neutralize the traps and interface states, although complete removal may not always be possible. Rapid annealing at a higher temperature helps in this case (e.g. 30 s at 500 °C in forming gas).

Thus today's manufacturing techniques used in thermal oxidation allow the creation of "perfect" gate oxide layers. Unfortunately, however, the gate oxides degrade under the influence of electrons that are injected into the oxide or tunnel through the oxide. The two most significant degradation mechanisms, HE degradation (HE = *H*ot *E*lectron) and TDDB breakdown (TDDB = *T*ime *D*ependent *D*ielectric *B*reakdown) are looked at in more detail here.

When a MOS transistor is active, a peak in the field strength occurs at the drain end of the channel which can accelerate the channel electrons close to their limit velocity (10^7 cm s^{-1}). These "hot electrons" can overcome the 3.2 eV high potential barrier at the Si/SiO$_2$ interface. Some of them are caught at positive traps near the interface, and are thus neutralized. A second effect of hot electrons is that they break Si-H bonds and can thus create interface states [3.29].

The main effect of the hot electron phenomenon in the n-channel MOS transistor is a degradation of the drain current, because both the number and mobility of the channel electrons are reduced by the effect described above. LDD doping (LDD = *L*ightly *D*oped *D*rain, see Fig. 3.5.2 a) has proved an effective countermeasure, and is generally used nowadays. The gradual change in drain doping reduces the field strength peak that triggers the hot electrons. Figure 3.4.10 shows an example of drain current degradation with and without LDD. For LDD doping to be effective, it is essential that the gate electrode overlaps the LDD region properly.

In the p-channel transistor, the negative fixed interface charges at the drain end of the channel induce a hole inversion layer at the Si surface. This inversion layer means that the channel length is reduced (Fig. 3.4.11). The associated change of the threshold voltage toward more positive values leads to increased drain currents, even in the sub-threshold region. The p-channel transistor may then not be sufficiently turned off at 0 V [3.30]. A lightly doped drain (LDD) can again provide a solution here. The sub-threshold performance can also be im-

---

21 The fixed interface charges that arise during thermal oxidation can be eliminated by introducing nitrogen or argon into the oxidation furnace after oxidation (see Fig. 3.1.15).

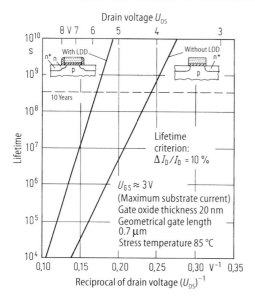

Figure 3.4.10. Lifetime of an n-channel MOS transistor under worst-case hot electron conditions (maximum substrate current) with and without Lightly Doped Drain (LDD). If these conditions were to last continuously throughout use, the permitted drain voltage for a specified life of 10 years would be 4.1 V without LDD and 5.8 V with LDD. If the worst-case conditions only apply e.g. for 10 % of the service life (duty factor 10), then from the diagram, a lifetime of 10 years is obtained for a drain voltage of 4.5 V without LDD and 6.3 V with LDD [3.31]

proved by moving from a buried channel device (with n⁺ poly-Si gate) to a surface-channel device (with p⁺ poly-Si gate) (see Table 8.8).

The second important degradation mechanism in thin oxide layers apart from hot electrons, is time dependent dielectric breakdown (TDDB). At electric field strengths near the level for dielectric breakdown, a tunnel current is created in the oxide (Fig. 3.4.12). The tunnelling charge carriers produce extra traps in the oxide, in addition to the traps already there, for both positive and negative charges. This increases the oxide conductivity and reduces the dielectric breakdown voltage. After a certain quantity of charge per cm² has flowed through, the oxide is degraded to an extent that the breakdown voltage is practically 0 V. This characteristic charge, which is only slightly dependent on the manufacturing conditions for the thermal oxide, is referred to as the charge to breakdown Q_{bd}. As the current density increases Q_{bd} drops sharply (Fig. 3.4.13). The temperature dependence follows an Arrhenius law with an activation energy of about 1 eV.

It is not immediately obvious how time dependent dielectric breakdown has a harmful role to play in real integrated circuits, since it requires field strengths to exist in the oxide close to the dielectric breakdown level. If one looks at

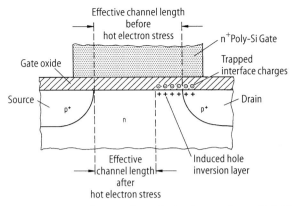

Fig. 3.4.11. Hot-electron degradation in a p-channel MOS transistor. The hot electrons injected into the gate oxide are caught in traps near the Si/SiO$_2$ interface, effectively shortening the channel length of the transistor. This is associated with an increased subthresold current, in particular for the buried-channel transistor shown here (with n$^+$ poly-Si gate).

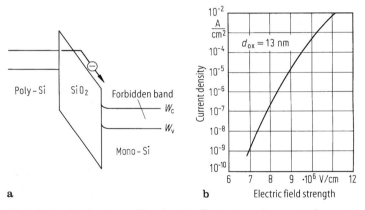

Fig. 3.4.12 a. Mechanism of Fowler-Nordheim tunnel current, demonstrated using the energy band diagram; **b** current density as a function of the field strength in a 13 nm thick SiO$_2$ layer.

Fig. 8.3.3, the field strength does not even reach half the dielectric breakdown level in the gate oxides of CMOS circuits. This would guarantee a lifetime of many thousands of years, according to Figs. 3.4.12 and 3.4.13. Even in floating-gate non-volatile memories (see Sect. 8.4.3), which are charged and discharged by a tunnel current in the tunnel oxide, one can perform approximately 10^6 write and erase operations until $1/10\ Q_{bd}$ is reached in the tunnel oxide, for instance.

The real problem is weak spots in the gate and tunnel oxides. As Fig. 3.4.14 illustrates, Q_{bd} can be reached at a localized point where the gate oxide is just

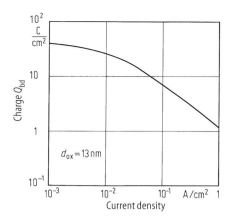

Fig. 3.4.13. Typical degradation of the dielectric breakdown characteristic of thin SiO$_2$ layers as current flows through the oxide. Q$_{bd}$ is that amount of charge, which after flowing through the oxide layer produces a dielectric breakdown voltage of 0 V for this layer.

slightly thinner. This may even occur during the integrated circuit manufacturing process[22]. The gate of the transistor affected is then shorted to the channel, which can cause the integrated circuit to fail.

Even more troublesome are those weak spots that have only been partly degraded by the end of the manufacturing process. Assume a breakdown voltage of 6 V. Functional testing of the integrated circuit with a 5 V supply voltage does not detect any functional error. Yet under operating conditions degradation progresses, and the integrated circuit can fail during operation. In order to minimise such failures, burn-in is usually carried out before the circuits are functionally tested. In burn-in the integrated circuits are stressed for several hours at an elevated temperature (e.g. 85 °C) and increased supply voltage (e.g. 7 V). This causes any affected transistors to fail completely. This kind of burn-in is usually required in complex circuits in order to achieve a reliability of 10 Fit – that means one product out of 1000 products may fail during 10 years of operation.

Time dependent dielectric breakdown is by far the most important failure mechanism in MOS circuits. Whilst other failure mechanisms such as hot electron degradation (see above) or electromigration (see Sect. 3.11.3) can be minimized by appropriate dimensioning of the relevant structures or choice of material, time dependent oxide breakdown is triggered by localized defects. This presents a continuous challenge for every MOS process line, but also to the manufacturers of the silicon substrate material, to ensure maximum cleanliness in all processing steps preceding gate oxidation.

---

22 [3.32] shows, for instance, that a heavy metal contaminant at the silicon surface can lead to localized oxide thinning in thermal oxidation.

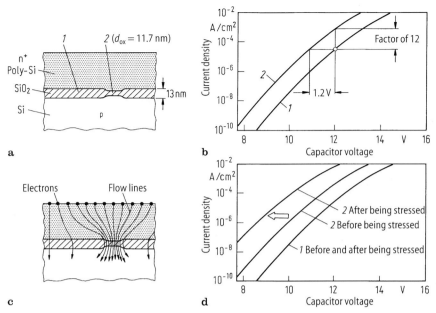

Fig. 3.4.14 a–d. Illustration of how a weak spot in an SiO$_2$ film can degrade during a processing stage that introduces charge to the poly-Si electrode (e.g. plasma etching). **a** Oxide thinning at the weak spot by 10 % is assumed, **b** the tunnel current characteristic (cf. Fig. 3.4.12) for the weak spot 2 is shifted 1.2 V to the left. At 12 V the current density would be 12 times greater, **c** if electrons are introduced onto the poly-Si electrode (which is not connected to the silicon substrate at any point) e.g. by a plasma process, then these tend to flow through the weak spot to the Si substrate because the resistance is 12 times lower here, **d** the higher current density in the weak spot means that it degrades much faster: the tunnel current characteristic shifts to the left

A higher Q_{bd} value can be obtained using a process called nitridation in a rapid annealing apparatus (see Sect. 3.1.8). Subjecting the thermal oxide layer to NH$_3$-RTN (Rapid Thermal Nitridation) treatment at 950 °C followed by O$_2$-RTO (Rapid Thermal Oxidation) treatment at 1150 °C can almost double Q_{bd}, without increasing the density of electron traps [3.33]. This is particularly useful for floating-gate memories (see Sect. 8.4.3). Nitrided oxides have a higher nitrogen concentration at the surface and close to the Si/SiO$_2$ interface. These nitride-like layers can inhibit the diffusion of e.g. boron atoms into the substrate from a poly-Si electrode doped with boron.

3.5
Deposited SiO$_2$ films

Unlike thermally grown SiO$_2$ layers, the growth of deposited SiO$_2$ films does not depend on the substrate material (except for selective SiO$_2$ deposition, Fig. 3.1.3 b), and does not modify the underlying layer.

3.5.1
Creating deposited SiO$_2$ films

The CVD technique, described in detail in Sect. 3.1.1, is by far the most commonly used deposition process. Apart from CVD SiO$_2$ films, sputtered SiO$_2$ films (see Sect. 3.1.4) and spin-on glass layers (see Sect. 3.1.5) are important.

There are several processes available for CVD SiO$_2$ deposition, the most important being listed below.

$$SiH_4 + O_2 \xrightarrow[\text{1bar}]{430°C} SiO_2 + 2H_2 \text{ (silane oxide technique)}$$

$$SiH_4 + O_2 \xrightarrow[\text{40 Pa}]{430°C} SiO_2 + 2H_2 \text{ (LTO technique)}^{23)}$$

$$Si(OC_2H_5)_4 \xrightarrow[\text{40 Pa}]{700°C} SiO_2 + \text{ gases (TEOS technique)}^{24)}$$

$$Si(OC_2H_5)_4 + O_3 \xrightarrow[\text{0,5 bar}]{400°C} SiO_2 + \text{ gases (SACVD technique)}^{25)}$$

$$SiH_2Cl_2 + 2N_2O \xrightarrow[\text{40 Pa}]{900°C} SiO_2 + \text{ gases (HTO technique)}^{26)}$$

$$SiH_4 + 4N_2O \xrightarrow[\text{plasma, 40 Pa}]{350°C} SiO_2 + \text{ gases (PECVD technique, plasma oxide)}^{27)}$$

$$SiH_4 + 4N_2O \xrightarrow[\text{plasma, ion bombardment, 40 Pa}]{350°C} SiO_2 + \text{ gases (Dep./Etch technique)}^{28)}$$

Atmospheric pressure CVD reactors such as those illustrated in Fig. 3.1.4 a, b and c, are used for the silane oxide technique. The LTO technique, however, is usually performed in a reactor such as type e in Fig. 3.1.4, whilst the TEOS and

---

23 LTO = Low Temperature Oxide
24 TEOS = Tetra-ethyl-ortho-silicate
25 SACVD = Sub-Atmospheric CVD (see Fig. 3.1.3 b)
26 HTO = High Temperature Oxide
27 PECVD = Plasma Enhanced CVD
28 See Fig. 3.1.7 b.

HTO processes mostly use the reactor type shown in Fig. 3.1.4 c or f. PECVD and SACVD oxide films are deposited in the type of CVD reactors shown in Fig. 3.1.6.

Silane oxides, LTO films and PECVD oxide films have a significantly higher etch rate in dilute hydrofluoric acid than do thermal oxides. This suggests that these films have a relatively "loose" internal structure, although annealing can lead to a certain "densification" of the layer structure. TEOS and HTO films have almost the same etch rate characteristic as thermal SiO_2 layers.

The techniques above do not provide the same edge coverage. The silane oxide process has the worst edge coverage (such as in Fig. 3.1.2 a and c), LTO and PECVD films have average edge coverage (see Fig. 3.1.7 a), whilst near-conformal coating of steps is achieved with the TEOS and HTO processes [3.36] (see Fig. 3.1.2 b and d). With the SACVD and Dep./Etch techniques, one even obtains a degree of planarization (see Figs. 3.1.3 b and 3.1.7 b).

The internal mechanical stresses within the layers can be either tensile or compressive. Most applications require compressive stress, because tensile stresses can cause cracks, particularly at steps. Silane oxide and LTO films exhibit internal tensile stresses, whereas TEOS, HTO and plasma oxide films are under internal mechanical compression.

As already shown in Figs. 3.1.1 and 3.1.5, the CVD reaction does not take place in the gas space, but on the hot surfaces of solids. If the wafers are not lying on a base, but are standing upright as in some reactor types (see Fig. 3.1.5), then the back of the wafer will always be coated with the same film thickness as the front side. In hot wall reactors the CVD reaction also occurs on the quartz boats and the internal reactor walls, since the boats and walls are at the same temperature as the wafers. As the number of coating runs rises, there is an increased risk of particles peeling off the boat and tube walls. It is for this reason that plate-type reactors such as those in Figs. 3.1.4 c and 3.1.6 a and c are becoming more popular.

3.5.2
Applications of deposited SiO_2 films

Deposited SiO_2 films are used in silicon technology where conformal surface coverage is required (e.g. in the spacer technique, Sect. 3.5.3, and for trench isolation, Sect. 3.5.4). It is also used where oxidation of the silicon substrate must be avoided (e.g. for the SiO_2 mask in trench etching), or where thermal oxidation is basically impossible (e.g. for the isolation layer in multilevel metallization, Sect. 3.5.5). SiO_2 films doped with a few percent by weight of phosphorus or boron are particularly important, and are dealt with in Sect. 3.6.

3.5.3
Spacer technology

A spacer refers to a structure that is only formed along a step. Figure 3.5.1 shows the two essential processing steps for producing SiO_2 spacers: at first SiO_2 dep-

3.5 Deposited SiO₂ films

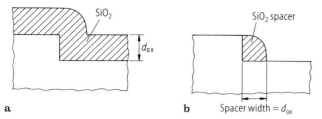

Fig. 3.5.1 a,b. Processing steps for producing an SiO$_2$ spacer at a vertical step. **a** Conformal SiO$_2$ deposition (e.g. by TEOS process); **b** anisotropic etching of the SiO$_2$ film to a depth d_{ox}.

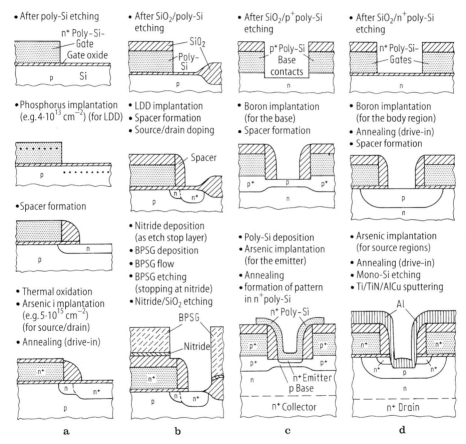

Fig. 3.5.2 a–d. Four important applications of SiO$_2$ spacers. **a** Lightly Doped Drain (LDD); **b** diffused region contact, e.g. source/drain of a MOS transistor, which overlaps the field oxide and poly-Si edge (self-aligned); **c** self-aligned poly-Si base and emitter connections for a bipolar transistor; **d** self-aligned source/body contact of a DMOS transistor.

osition by a technique that provides maximum conformal coating (e.g. TEOS) and then anisotropic etch back of the SiO_2 layer, by an amount equal to the thickness of the SiO_2 on the planar areas. A SiO_2 spacer is then left behind on abrupt steps.

In silicon technology, SiO_2 spacers are used where pattern edges need to be shifted laterally, or where edges need to be isolated. Figure 3.5.2 shows four examples of spacer applications. Diagram a shows the creation of an LDD doping profile (LDD= *L*ightly *D*oped *D*rain) for reducing the peak field strength at the drain edge of a MOS transistor (see Sect. 3.4.3), b illustrates an overlapping (self-aligned) contact [3.34], c the formation of self-aligned polysilicon base and emitter contacts for a bipolar transistor [3.35], and d the creation of self-aligned source/body contacts of a DMOS transistor (see Sect. 8.3.2).

Figure 3.4.6 shows other applications of the spacer technique. Finally, it should be mentioned that one can produce sub-µm dimensions using the spacer technique without the need of lithographic pattern generation (see Sect. 4.6). In this method, the spacers are the only structures left after etching away the pattern that they flank. This indirect method for producing a pattern does have some major limitations, however, since only short-looped structures with uniform pattern widths can be produced.

3.5.4
Trench isolation

Trench isolation refers to the lateral isolation between adjacent transistors, or other active regions, using trenches etched into monocrystalline silicon and filled with insulating material (STI = Shallow Trench Isolation). The TEOS process is ideal for filling the trenches (see Sect. 3.5.1), because near conformal deposition is possible, even in narrow trenches.

Figure 3.5.3 shows the processing sequence for a trench isolation technique (BOX technique, BOX = *B*uried *ox*ide). This process not only enables very narrow SiO_2 ribs to be produced for lateral isolation of MOS transistors (no bird's beak problem as in the LOCOS technique), but it can also be used to create isolating walls extending far into the monocrystalline silicon (e.g. 3 µm deep). These are particularly useful for advanced CMOS and bipolar circuits containing densely-packed transistors, where ideally the isolating walls should extend to the same depth as the wells (in CMOS circuits) or the buried layers (in bipolar circuits, see Fig. 3.3.4 a). Polysilicon can be used to fill the trenches instead of SiO_2 (Fig. 3.5.3 b).

3.5.5
SiO_2 isolation films for multi-level metallization

Aluminium is used almost exclusively today for the metal tracks in integrated circuits. Since the silicon-aluminium eutectic point is at 570 °C, processing tem-

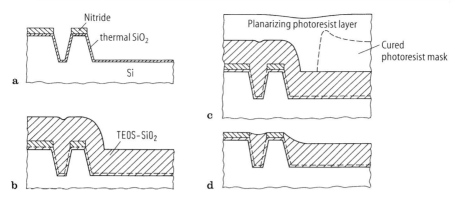

Fig. 3.5.3 a–d. Processing sequence in a trench isolation technique. **a** Etching trenches in monosilicon (e.g. with nitride mask). Thermal oxidation; **b** conformal deposition of an SiO_2 film or poly-Si film (film thickness = depth of trench); **c** applying a photoresist mask in the wide trenches (resist thickness = depth of trench). Curing of resist so that it does not dissolve into the subsequent photoresist. Deposition of planarizing photoresist layer by spin-on method; **d** etch-back of resist and SiO_2 films (resist etch rate = SiO_2 etch rate) until just before reaching the nitride mask. Chemical-mechanical polishing (CMP, see Sect. 5.1.2) of the SiO_2 down to nitride polish stop.

peratures must be kept below about 500 °C once aluminium has been deposited on the silicon wafer. Thus, of the SiO_2 CVD deposition techniques (see Sect. 3.5.1), only the silane oxide process, the LTO, SACVD and PECVD techniques and the Dep./Etch process can be considered for producing the insulating layers between the first and second aluminium planes.

The silane oxide, LTO and PECVD techniques are not suitable for IMD layers (IMD = *Intermetal Dielectric*), however, because of their non-conformal step coverage, which makes it difficult for the aluminium tracks of the second metallization plane to cross over steps (cf. Fig. 8.5.1). If one combines these techniques with planarizing processing steps (examples in Figs. 3.5.4 and 3.5.5), however, then the required planarized surface for the second metallization can be obtained (see also Fig. 8.5.3).

Only the SACVD and Dep./Etch techniques provide surfaces that are already fairly planar after deposition (see Figs. 3.1.3 b and 3.1.7 b). This is also true for bias sputtering (Fig. 3.1.19), however this method has become obsolete for productivity reasons (slow growth rate of about 1 µm/h).

3.6
Phosphorus glass films

Phosphorus glass or PSG films (PSG = *Phosphorus Silicate Glass*) are SiO_2 layers containing 2–10 % phosphorus by weight. Their importance in silicon technol-

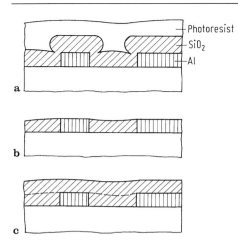

Fig. 3.5.4 a–c. Processing sequence for planarization of a PECVD oxide film to produce a step-free surface for the second interconnect plane of a multi-layer metallization. **a** Deposition of plasma oxide (PECVD) and coating with photoresist; **b** etch-back with same etching rate for both photoresist and plasma oxide; **c** deposition of plasma oxide (see also Sect. 5.3.7)

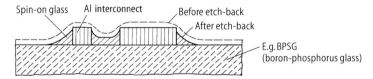

Fig. 3.5.5. Planarization of a surface containing steps, by applying spin-on glass (see Sect. 3.12.1) and etching back the spin-on glass. The etch-back can be isotropic (e.g. in diluted hydrofluoric acid) or anisotropic (with reactive ion etching). In this method the spin-on glass is simply used to round off the edges and fill in trenches, whilst the actual insulating layer above the Al interconnects is a sputtered or PECVD SiO_2 film (see Fig. 8.5.3). Polyimide can also be used instead of spin-on glass for planarization (see Sect. 3.12.2)

ogy relates to two notable properties, namely their gettering effect for alkali and heavy metals, and their ability to flow at temperatures around 1000 °C.

3.6.1
Producing phosphorus glass films

Similar to SiO_2 films, phosphorus glass films can be grown either thermally or using CVD deposition. Implanting high phosphorus concentrations into SiO_2 is also possible, but seldom done in practice.

Thermal techniques (Fig. 3.6.1) operate typically at 900 °C in an oxidizing atmosphere containing phosphorus (PH_3 or $POCL_3$) to convert the surface of existing silicon or SiO_2 layers into phosphorus glass. For practical reasons, only relatively thin phosphorus glass films are produced (0.2 µm max.). If there is an SiO_2 layer present (left column in Fig. 3.6.1), the SiO_2 layer is converted into phosphorus glass up to a certain depth (typically 0.1 µm) with negligible change

3.6 Phosphorus glass films

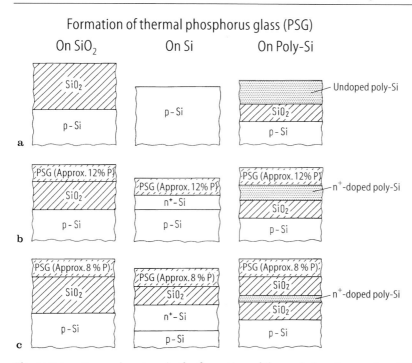

Fig. 3.6.1 a–c. Processing steps in the formation of thermal phosphorus glass (PSG) on an SiO$_2$ layer and on silicon. **a** Initial condition; **b** after thermal treatment (900 °C) in POCl$_3$ (or PH$_3$) and oxygen; **c** after thermal oxidation (1000 °C)

in the overall layer thickness. The phosphorus content is then very high (about 12 %). During subsequent thermal oxidation (or annealing in an inert atmosphere) at a raised temperature (e.g. 1000 °C), the phosphorus content in the PSG drops to almost 8 %, whilst the thickness of the phosphorus glass remains practically the same.[29]

The PSG layer that grows on a monocrystalline or polycrystalline silicon base (centre and right-hand column in Fig. 3.6.1) is practically the same as the PSG layer grown under the same conditions on an SiO$_2$ base. The Si thickness converted into PSG equals about half the PSG layer thickness. Part of the phosphorus diffuses into the silicon and forms an n$^+$-doped region there. After subsequent thermal oxidation, an SiO$_2$ layer is created between the silicon and the PSG layer, whilst the phosphorus in the silicon continues to diffuse.

---

29 At 1000 °C, phosphorus from the PSG diffuses less than 10 nm in 1 hour into the undoped SiO$_2$ below (see Sect. 6.3.6).

Deposited phosphorus glass films are just as important as those produced thermally. Any of the CVD techniques mentioned in Sect. 3.5.1 can be used for deposition. In those methods using a gaseous source, the phosphorus is introduced as phosphine (PH_3). Where $Si(OC_2H_5)_4$ is used as the liquid TEOS source (TEOS and SACVD processes), $P(OC_2H_5)_3$ or $PO(OC_2H_5)_3$ is mixed with it.[30]

Unlike silane oxide or LTO films which do not contain phosphorus, CVD-PSG films are under internal mechanical compression. This is an advantage, for instance, in preventing cracks forming at steps.

3.6.2
Flow-glass

At temperatures of about 1000 °C, phosphorus glasses begin to flow. This property is particularly exploited in MOS processes in order to planarize the surface, e.g. the polysilicon structures [3.37] (Fig. 3.6.2). An ideal topography for aluminium tracks is thus produced on the flow-glass surface.

The minimum temperature needed for the phosphorus glass to flow depends on the phosphorus content and on the gas atmosphere. The phosphorus concentration in the phosphor glass should not exceed 8 %, otherwise there is a risk that the aluminium tracks above will corrode. At this concentration, a temperature of about 1000 °C is required in an atmosphere of nitrogen or oxygen to achieve a good flow. Thermal treatment lasting e.g. 30 min at 1000 °C, however, causes significant diffusion of the doping atoms in the silicon. This means that the shallow doping profiles needed for advanced circuits can no longer be maintained. Thus there is a need for a flow-glass process that subjects the silicon wafers to lower thermal stress.

If the flow process for the PSG layer is performed in an atmosphere of water vapour, a temperature of 930 °C is sufficient, however the silicon regions beneath the phosphorus glass are thermally oxidized at these conditions, which is normally unwanted. Other options of low-temperature phosphorus glass flow, include processing at high pressure (e.g. in a high-pressure oxidation facility at 20 bar and 850 °C) or very rapid processing (e.g. 5 s at 1100 °C) in a rapid annealing system (cf. Fig. 3.1.25). Finally, the flow temperature can be reduced by adding boron (or germanium) to the phosphorus glass. The process is more complicated because another parameter, the boron concentration in the boron phosphorus glass (BPSG) has to be controlled. For BPSG containing 4 % B and 4 % P by weight, the same flow behaviour is achieved at just 900 °C as is obtained at 1000 °C for PSG containing 8 % P by weight with no added boron, with conditions otherwise the same. If the BPSG layer lies directly on top of the silicon,

---

[30] Arsenic glass (AsSG) can be produced by the TEOS process by adding $As(OC_2H_5)_3$ or $AsO(OC_2H_5)_3$ to the $Si(OC_2H_5)_4$. This technique is used for e.g. arsenic doping of trenches (see Sect. 6.1).

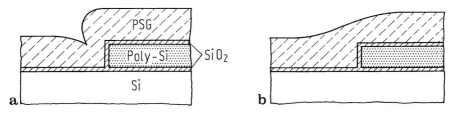

Fig. 3.6.2 a, b Smoothing abrupt steps by flow glass. **a** Surface profile after CVD deposition of the PSG film; **b** surface profile after PSG film flow (flow glass)

then at e.g. 900 °C, there is a certain amount of diffusion of both boron and phosphorus out of the BPSG into the silicon. Where this is undesirable, a thin undoped SiO_2 film can be provided between the BPSG layer and the silicon surface using thermal oxidation or CVD deposition, as shown in Fig. 3.6.2.

3.6.3
Thermal phosphorus glass

Thermal phosphorus glass, produced by converting the surface of an SiO_2 layer into PSG (Fig. 3.6.1, left-hand column), is used for instance in bipolar processes to provide a gettering, and thus stabilizing layer before applying the metallization. In such applications the PSG film thickness is approximately 0.1 µm. Using the flow process on such thin films is not effective, however, because it has almost no planarizing or edge-smoothing effect.

In early MOS processes using an aluminium gate it was also useful to convert the surface of the gate oxide into a PSG film for stabilization purposes. Where polysilicon gate electrodes with high phosphorus content are used, such gate oxide stabilization is not necessary.

Thermal phosphorus glass is also obtained unintentionally when the processing sequence in Fig. 3.6.1 (right-hand column) for thermal doping with phosphorus is used on monocrystalline silicon (e.g. for backside gettering, see Sect. 3.2.3) or on polysilicon layers. In these cases the phosphorus glass is normally removed again by diluted hydrofluoric acid after the doping process (Fig. 3.6.1 b).

3.7
Silicon nitride films

The main reason for using silicon nitride in silicon technology is that it provides an excellent barrier to any type of diffusion. Its dielectric constant ($\varepsilon_r = 7.8$) which is twice as high as that of SiO_2, also makes Si_3N_4 a useful material for the dielectric in capacitors.

3.7.1
Producing silicon nitride films

Almost all the well-known processes for producing SiO_2 films can be suitably adapted to create Si_3N_4 films as well. Currently the most important of these include the high-temperature nitride technique equivalent to the HTO process, and the plasma nitride technique corresponding to the plasma oxide process (cf. Sect. 3.5.1).

In the high-temperature nitride process, which is usually performed in CVD reactors of the type shown in Fig. 3.1.4 f, the following reaction takes place:

$$3SiH_2Cl_2 + 4NH_3 \xrightarrow[30Pa]{750°C} Si_3N_4 + \text{ gases (LPCVD process)}$$

Plasma nitride layers are produced by the following reaction:

$$3SiH_4 + 4NH_3 \xrightarrow[\text{plasma, 30Pa}]{300°C} Si_3N_4 + 12H_2 \text{ (PECVD process, plasma nitride).}$$

Diagrams of 3 types of plasma CVD reactor are shown in Fig. 3.1.6. Although the high-temperature nitride layers contain almost no hydrogen, and are also impermeable to hydrogen at temperatures below 500 °C, plasma nitride films contain a relatively high amount of hydrogen depending on the conditions of manufacture.

In a similar process to thermal oxidation, the silicon surface can be converted into silicon nitride in an atmosphere of NH_3, although a temperature of about 1200 °C is required for a 7 nm thick nitride layer [3.38]. The following reaction takes place in thermal nitridation:

$$3Si + 4NH_3 \xrightarrow[\text{1 bar}]{1200°C} Si_3N_4 + 6H_2$$

The nitridation temperature can be reduced using plasma excitation. Thermal nitridation is still under development and is thus hardly used as yet.

Nitrided oxide layers are of more relevance than thermal nitride layers, particularly for the tunnel oxide in floating-gate memories (see Sect. 3.4.3).

3.7.2
Nitride films as an oxidation barrier

Under thermal oxidation conditions, the surface of silicon nitride films produced using the high-temperature nitride process is slightly oxidized, i.e. converted into oxynitride (usually less than 10 nm). Thus even relatively thin nitride films can protect underlying silicon layers from thermal oxidation. This oxidation barrier function is exploited particularly in the LOCOS process, which was described in detail in Sect. 3.4.2. Other applications are illustrated in Fig. 3.7.1.

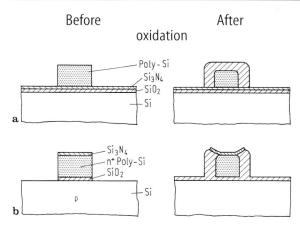

Fig. 3.7.1 a, b. Two examples of Si$_3$N$_4$ used as an oxidation barrier. **a** Enclosing polysilicon structures in a relatively thick SiO$_2$ coat, without oxidizing the Si substrate; **b** relatively thick oxidation of the side walls of polysilicon structures (e.g. for selective formation of silicide on poly-Si areas, without silicide forming on mono-Si areas; cf. salicide technique in Sect. 3.9.1).

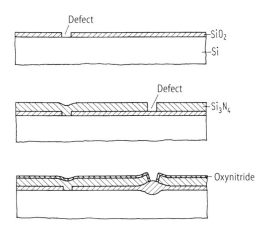

Fig. 3.7.2 a–c. Processing steps for eliminating local defects in thin double Si$_3$N$_4$-SiO$_2$ films.
a Thermal oxidation; **b** nitride deposition; **c** thermal oxidation. This results in a triple film of oxynitride/Si$_3$N$_4$/SiO$_2$ (ONO)

3.7.3
Nitride films as a capacitor dielectric

A double Si$_3$N$_4$-SiO$_2$ film (NO) or a triple SiO$_2$-Si$_3$N$_4$-SiO$_2$ film (ONO) is useful as a dielectric in the capacitor of e.g. dynamic memory cells (see Table 8.3), not just because of the high dielectric constant of the Si$_3$N$_4$, but also because it is possible to avoid almost any localized defects in the double or triple layer [3.39] (Fig. 3.7.2). Using an Si$_3$N$_4$-SiO$_2$ or SiO$_2$-Si$_3$N$_4$-SiO$_2$ film as a gate dielectric is however critical, because the high density of traps at the Si$_3$N/SiO$_2$ interface can cause electrical instabilities. In the so-called MNOS transistor (*M*etal *N*itride

Oxide Semiconductor), in which the SiO_2 thickness is only about 2 nm, it is in fact these rechargeable traps that enable non-volatile charge storage to be achieved (see Sect. 8.4.3).

3.7.4
Using nitride films for passivation

The top layer of integrated circuits is usually a plasma nitride film about 1 µm thick, containing holes for the contacts for wire bonding (see Sect. 8.5.4). The plasma nitride film is used as a passivation layer because it acts as a stable barrier against water vapour, sodium ions and other contaminating or corrosive substances.

Plasma nitride films do have a high hydrogen content however, which can lead to instabilities after long-term use of the integrated circuits. These are particularly evident if negative gate voltages on MOS transistors and raised operating temperatures are used. Oxynitride films or double layers of plasma oxide and plasma nitride appear to be a better option.

The effect of plasma nitride on polysilicon resistors is described in Sect. 3.8.3.

3.8
Polysilicon films

The introduction of polycrystalline silicon (polysilicon) into MOS technology, and more recently into bipolar technology, has provided numerous new opportunities that have still not been completely explored today. Those properties that have proved particularly useful are excellent compatibility with other materials used in silicon technology, temperature stability to over 1000 °C, ease of doping and oxidation, and the ability to produce conformal edge coverage. These properties are exploited to produce self-aligned structures, three-dimensional integrated configurations plus stable interconnections, gate electrodes, resistors and contacts.

3.8.1
Producing polysilicon films

Polysilicon films are produced almost exclusively today using the low-pressure CVD method. In this process the silane breaks down into silicon and hydrogen on the hot surfaces in the CVD reactor:

$$SiH_4 \xrightarrow[60Pa]{630°C} Si + 2H_2 \cdot$$

A hot wall reactor (cf. Fig. 3.1.4 f) is most commonly used, but also the reactor type shown in Fig. 3.1.4 c. In tube reactors such as that in Fig. 3.1.4 f, it is standard practice to compensate for the depleted silane along the tube axis by providing an axial temperature gradient (higher deposition rate at higher temperatures). This means that a uniform deposition rate can be achieved for more than 100 wafers.

Rinsing the reactor with HCl before deposition and adding a small quantity of HCl to the silane can increase the quality of the polysilicon layers. The deposition rate is around 20 nm/min, and film thicknesses of between 0.2 and 0.5 μm are typical in integrated circuits. Polysilicon layers deposited using the low-pressure CVD technique exhibit almost perfect conformal surface coverage because of the reaction-controlled CVD process (cf. Fig. 3.1.2 b, d).

Undoped polysilicon films have very high resistivity (about 10^4 Ωcm). Thus all applications in which the polysilicon acts as an electrical conductor involves doping with boron, phosphorus or arsenic. In order to save an extra doping step, it is convenient to perform doping at the same time as polysilicon deposition by adding B_2H_6, PH_3 or AsH_3 to the SiH_4. This can be done using a cool wall reactor such as in Fig. 3.1.4 c. This doping method has not really found favour in horizontal tube reactors, however, because of technical problems (difficult to achieve uniformity over many wafers, slow growth rates). The most commonly used technique for producing layers of low resistance n^+-doped polysilicon is to produce a thermal phosphorus glass film using a $POCl_3$ or PH_3 source (cf. Fig. 3.6.1, right-hand column). In all other cases (boron or arsenic doping, low doping levels) ion implantation is the preferred method (see Sect. 6.2).

3.8.2
Grain structure of polysilicon films

Polysilicon films are made up of individual grains, the sizes of which depend both on the deposition conditions and on the subsequent processing steps. The grains decrease in size as the deposition temperature in the CVD reactor falls. Whilst grain sizes of 10–50 nm are obtained at 630 °C (Fig. 3.8.1 a), films deposited below 590 °C are practically amorphous. Processing the films in inert or oxidizing atmospheres at temperatures of up to 1000 °C and higher, produces negligible change in the fine-crystalline or amorphous structure. The same applies to lightly doped polysilicon films. Films with a high concentration of dopant, particularly of phosphorus ($>10^{20}$ cm^{-3}), do experience significant increases in grain size for temperatures above 800 °C. The grains can then reach 0.5 μm in size and above, and may thus extend through the full depth of the polysilicon film (Fig. 3.8.1 b and c).

Unlike amorphous silicon films which have a completely smooth surface, the grain structure of polysilicon layers produces a rough surface.

As described in Sects. 3.1.8 and 3.3.1, a polycrystalline film can be transformed into a monocrystalline silicon film by scanning a high-energy beam across the surface, even when it lies on top of an SiO_2 layer. This makes three-dimensional integration possible (example in Fig. 3.3.2).

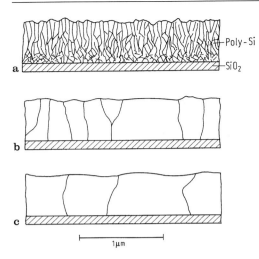

Fig. 3.8.1 a–c. Grain structure of polysilicon films. **a** After low-pressure CVD deposition at 630 °C; **b** after subsequent thermal phosphorus doping ($3 \cdot 10^{20}$ cm^{-3}) at 900 °C using a POCl$_3$ or PH$_3$ source; **c** after subsequent thermal treatment at 1000 °C

3.8.3
Conductivity of polysilicon films

As in monocrystalline silicon, it is the doping atoms (boron, phosphorus, arsenic) at active sites in the polysilicon that are responsible for the electrical conductivity. There are two main differences in the doping characteristic of polysilicon compared to monocrystalline silicon. Firstly, the diffusion constant for diffusion along grain boundaries is about an order of magnitude larger, and secondly dopant segregation occurs at the grain boundaries. The grain boundaries also constitute energy thresholds for charge transport (Fig. 3.8.2).

The first effect leads to a relatively rapid distribution of the dopant throughout the film thickness, whilst dopant segregation at the grain boundaries has the effect that part of the dopant is electrically inactive. Segregation increases sharply as the temperature falls, with the foreign atom concentration within the grains and at the grain boundaries reaching equilibrium by diffusion of the foreign atoms. The actual foreign atom concentration that results inside the grains thus depends on the sequence of temperature changes during a manufacturing process. The energy thresholds at the grain boundaries increase the resistance, as does the foreign atom segregation.

Figure 3.8.3 shows the resistance of polysilicon layers doped with relatively high concentrations of boron, phosphorus and arsenic, and annealed at 1000 °C. It shows quite clearly that the lowest resistivity is obtained with phosphorus, being practically equal to that for monocrystalline silicon (cf. Fig. 3.2.2).

Figure 3.8.4 shows the drastically different resistivity characteristic of lightly doped polysilicon films compared with monocrystalline silicon. To illustrate the sensitivity of the high resistance values to the manufacturing process used for the integrated circuit, Fig. 3.8.4 shows curves with and without plasma nitride passivation. It appears that the hydrogen contained in the plasma nitride is able

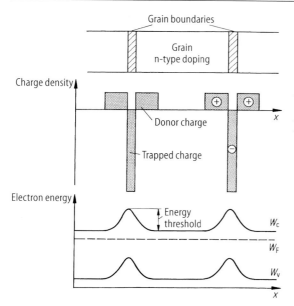

Fig. 3.8.2. Schematic diagram of the creation of an energy threshold for electrons flowing across a grain boundary in polysilicon. The energy threshold is produced by electrons trapped at the grain boundaries which result in electron depletion in those regions near the grain boundaries

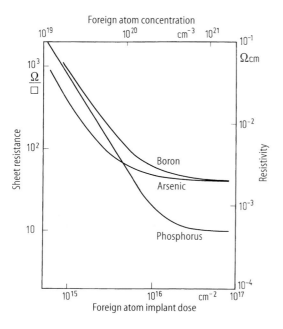

Fig. 3.8.3. Sheet resistance and electrical resistivity of 0.5 μm thick polysilicon layers with boron, arsenic and phosphorus doping after furnace-annealing at 1000 °C for 30 min

to diffuse to the grain boundaries in the polysilicon, there reducing the traps responsible for the potential barrier.

As in all physical processes in which an energy threshold must be overcome by thermal energy, the barrier model predicts an exponential temperature

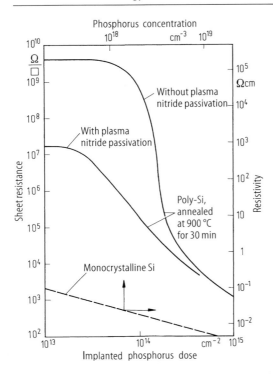

Fig. 3.8.4. Sheet resistance and resistivity of a 0.5 µm thick polysilicon layer in which a low dose of phosphorus ions have been implanted. The centre curve shows the effect of plasmanitride passivation. The curve for monocrystalline silicon is included for comparison

dependence for the polysilicon resistivity according to the Arrhenius law. This is confirmed experimentally, although the resistivity of low-resistance polysilicon films is only weakly dependent on the temperature because of tunnelling through the potential barrier.

3.8.4
Uses of polysilicon films

Polysilicon did not exist at all in the early bipolar and MOS circuits. The first application of a polysilicon gate in the MOS process (around 1970) was a crucial breakthrough for MOS technology, because it allowed to exploit the major advantages of polysilicon. These advantages are that it can tolerate high temperatures (which enables self-alignment of source and drain with the gate electrode), that it can be doped and thermally oxidized, that it forms an ohmic contact with aluminium and monosilicon, and also forms an electrically stable gate electrode over a thin gate oxide. Furthermore, it can provide an extra connection plane (although its resistivity is three orders of magnitude larger than that of aluminium), and finally, conformal step coverage is possible with polysilicon using low-pressure CVD deposition. The benefits of polysilicon are exploited in numerous further applications.

The fact that polysilicon can be thermally oxidized is widely exploited, with examples given in Figs. 3.4.5 and 3.4.6. Polysilicon is also being used increasingly in modern bipolar technology. Emitter efficiency can be increased using an n^+-doped polysilicon emitter (constant current gain over several decades of collector current). If the base connections are also made of polysilicon (p^+-doped), then self-aligned emitter/base configurations can be achieved. The emitter/base structure then consumes a minimum of space, thereby producing very low capacitance, fast bipolar transistors (see Figs. 3.5.2 c and 8.3.8).

Lightly doped Poly-Si films can be used as high-value resistors in bipolar and MOS circuits, particularly as load elements in inverter circuits (poly loads), which take up less surface area.

MOS transistors in poly-Si (TFT= *Thin Film Transistor*) do have worse electrical properties, but are useful in specific applications, e.g. as load transistors in inverters or SRAM memory cells (see Sect. 8.4.1).

Polysilicon is also the preferred material for programming circuits. In circuits with built-in redundancy, polysilicon links are selectively broken (by localized vaporization using a current pulse or a finely focused laser beam) to isolate defective parts of the circuit, or made locally conductive (e.g. by localized dopant activation using a finely focused laser beam after ion implantation) in order to connect up redundant working circuit sections. The same technique can be used for the programming of ROMs (ROM = *Read Only Memory*). In EPROMs and E^2PROMs (EPROM = *Erasable Programmable* ROM, E^2PROM = *Electrically Erasable Programmable* ROM) the programmable elements are MOS transistors with a floating (i.e. electrically isolated) polysilicon gate, which is charged at high field strengths by the tunnel current flowing through the thin oxide (see Sects. 3.4.3 and 8.4.3).

The contact from polysilicon to monocrystalline silicon areas in the Si-gate MOS process is referred to as a buried contact. Figure 3.8.5 shows the main processing steps for producing a buried contact. Only partial coverage of the contact hole by the poly-Si is typical for this type of contact. This is necessary because otherwise the overlap of phosphorus- and arsenic-doped regions required for a low-resistance connection could not be guaranteed. The disadvantage with this buried contact process is that the sensitive gate oxide is exposed to several processing steps before being covered with polysilicon, which can cause local defects in the gate oxide. Also, as there is no etch stop over the monosilicon during etching of the polysilicon (Fig. 3.8.5 c), etching inevitably progresses into the monosilicon. A buried contact process such as that shown in Fig. 3.8.6 overcomes both problems.[31]

In both buried contact processes shown in Figs. 3.8.5 and 3.8.6, it is assumed that the polysilicon layer is produced before source-drain doping, as is necessary for self-aligned gate electrodes. If, on the other hand, an extra polysilicon film is

---

31 Both problems can also be avoided using the processing sequence of Fig. 3.8.5 by depositing a portion (e.g. 0.1 μm) of the subsequent gate poly-Si layer after gate oxidation.

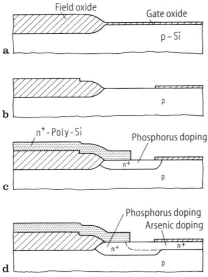

Fig. 3.8.5 a–d. Processing sequence for producing a buried contact. **a** Initial condition after gate oxidation; **b** etching of buried contact hole; **c** poly-Si deposition, phosphorus doping, etching of poly-Si; **d** arsenic implantation for source/drain, drive-in of doping atoms

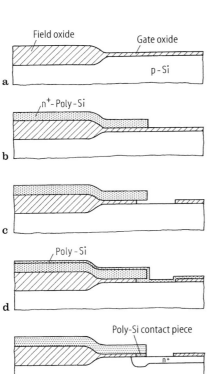

Fig. 3.8.6 a–e. Processing sequence for a buried contact, without the disadvantages of the buried contact process shown in Fig. 3.8.5. The key processing step is step d, in which the hollow space under the polysilicon edge is filled with polysilicon thanks to conformal deposition. **a** Initial condition after gate oxidation; **b** n^+ poly-Si deposition and etching; **c** isotropic SiO_2 etching; **d** conformal poly-Si deposition; **e** Overall removal of thin poly-Si film by etching, followed by source/drain doping

3.8 Polysilicon films

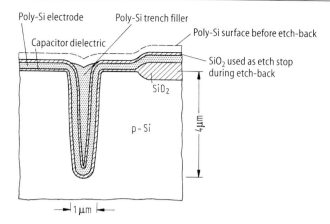

Fig. 3.8.7. Cross-section through the capacitor in a dynamic trench memory cell

provided in the MOS process solely as an interconnect layer and not as a gate layer (e.g. for line transmission in a dynamic memory circuit), then standard contact technology can be used where the contact hole is completely covered by the polysilicon. Figure 3.5.2 b shows such a contact. In this case the contact can also be designed to overlap a first polysilicon pattern (self-alignment). All the polysilicon/monosilicon contacts described have the common feature that for n^+-doped polysilicon, ohmic contacts can only be formed with n^+-doped monosilicon regions.

The uniform surface coverage possible with polysilicon, even in narrow, deep trenches, can be used to fill trenches for dynamic memories for example. Figure 3.8.7 shows the cross-section of the capacitor in a dynamic memory cell, where the surface area of the capacitor is increased by the formation of a deep trench. If, as is assumed in Fig. 3.8.7, the poly-Si electrode is not thick enough to completely fill the trench with polysilicon, the trench can be completely filled with a second poly-Si deposition. In this case a thin SiO_2 film can be used as an etch stop during etch-back of this poly-Si layer (also cf. the trench isolation technique in Sect. 3.5.4).

Section 8.6.4 discusses the processing sequence for producing the trench capacitor of a memory cell of only 0.6 μm^2 in size. Unlike the trench capacitor in Fig. 3.8.7, the trench wall of this capacitor is structured to produce the buried plate, collar and buried-strap contact (cf. Fig. 8.4.3 b). Standard lithographic technology (Chap. 4) is unable to produce these features, however. Instead such techniques as oblique implantation and spacer formation (see Sect. 3.5.3) are used in its place, as well as resist thinning to specific trench depths and partial etch-back (recess etching) of the trench previously filled with e.g. poly-Si.

Vertical MOS transistors are another practical example of this trench processing [3.40, 3.41].

The completely uniform deposition of poly-Si, even in narrow gaps with large aspect ratios (= ratio of gap depth to gap width), makes its use also ideal for stacked capacitors (see Fig. 8.4.3 c and d).

Finally, the optical properties of polysilicon should be mentioned, since these play an important role in lithography (see Chap. 4). The large difference between the refractive indices of silicon and SiO_2 (4.75 and 1.45 respectively) means that light is strongly reflected at every Si/SiO_2 interface. This reflection is exploited when detecting alignment marks (see Fig. 4.2.24 a). On the other hand, light reflection can be a disadvantage because it causes variations in line widths (see Sect. 4.2.3). A thin amorphous Si film sputtered onto aluminium acts as an anti-reflective layer (see Sect. 4.2.3).

3.9
Silicide films

Silicides are metal/silicon compounds used in silicon technology for thermally-stable low-resistance tracks and contacts. The silicide films are typically 0.1–0.2 μm thick. The most commonly used silicides are $MoSi_2$, WSi_2, $TaSi_2$ and $TiSi_2$ (for tracks) and $PtSi$ and Pd_2Si (for contact films). Other silicides include $CoSi_2$, $NbSi_2$, $NiSi_2$ and the rare earth silicides, but these remain almost unused as yet. Table 3.1 summarizes the properties of the most important silicides.

Table 3.1 Properties of the most important silicides

	$MoSi_2$	WSi_2	$TaSi_2$	$TiSi_2$	$PtSi$	Pd_2Si
Resistivity (μΩ cm)[a]	40–110	30–100	35–70	15–25	30–35	30–35
Sheet resistance (Ω/□) for a 0.2 μm thick silicide film	2–5.5	1.5–5	1.75–3.5	0.75–1.25	1.5–1.75	1.5–1.75
Schottky barrier height on n-Si (mV)	570	650	600	600	880	730
Chemical resistance	very good	very good	good	poor (soluble in dilute hydrofluoric acid)	very good	medium (soluble in HNO_3 and HF+ HNO_3)
Temperature stability in °C	>1000	>1000	>1000	800	800	700
Temperature stability of Al-silicide contacts in °C	500	500	500	450	300	300

[a] The resistivity depends on the temperature conditions when the silicide was formed (silicidation). For example, the resistivity of WSi_2 with annealing at 900 °C is 100 μΩ cm, and 40 μΩ cm at 1000 °C.

3.9.1
Producing silicide films

The three major techniques for producing silicide films are simultaneous sputtering of metal and silicon, sputtering of metal alone followed by silicide formation (silicidation), and finally CVD deposition [3.42].[32]

The simultaneous sputtering of metal and silicon was described in Sect. 3.1.4. This technique is preferred for the disilicides $MoSi_2$, WSi_2, $TaSi_2$ and $TiSi_2$. The metal and the silicon must be applied to the silicon wafers in the correct proportion. Usually the aim is to achieve the stoichiometric ratio, i.e. two silicon atoms to one metal atom. This can be achieved either by co-sputtering, or sputtering of a mosaic target or sinter target (mixed target), as shown in Fig. 3.1.16.

In co-sputtering the silicon wafers are repeatedly passed under the two targets, each target sputtering rate being separately controlled. Thus a multilayer film is achieved containing alternating layers of metal and silicon. It is possible to vary the basic metal/silicon ratio with co-sputtering, and to choose which shall be the first and last layer of the multi-layer film. The disadvantage, however, is that precise simultaneous control of two targets is required. In contrast to co-sputtering, the film composition in mosaic or sinter target sputtering is determined by the target. Ideally, the two components are thoroughly mixed in the sputtered layer.

In all three versions described, amorphous films are produced after sputtering. The actual silicide formation (silication) occurs at temperatures of 600–1000 °C.[33] This process causes shrinkage of the material (about 25 %), producing relatively high tensile stresses within the silicide film (about 10^5 N cm^{-2}). The silicide films have a fine-crystalline structure.

Figure 3.9.1 illustrates another process where only the silicide-forming metal is sputtered onto the surface. The actual silicide reaction then involves the underlying silicon. In those areas where the metal film lies on an SiO_2 surface, the metal remains practically unchanged by silicidation annealing,[34] and can be etched off (e.g. with H_2O_2 where titanium is used). The result is a selective (self-aligned) silicide film on the areas of exposed silicon (salicide = *S*elf-*al*igned sil*icide*). The amount of silicon that is consumed varies between 50 and 100 % of the silicide film thickness.

A critical element of this process is the thin layer of natural oxide (0.8–1.8 nm) on the silicon, which prevents uniform silicide formation. The natural

---

32 Section 3.3.1 also referred to the possibility of producing monocrystalline silicide films using molecular beam epitaxy.
33 Silicidation can also be performed using the rapid annealing process (see Sect. 3.1.8). This is advisable because it minimizes the unwanted diffusion of doping atoms out of the silicon into the silicide.
34 This does not apply to titanium. This is converted to TiN in a nitrogen atmosphere (e.g. at 700 °C). A thin TiN film is formed on the TiSi2.

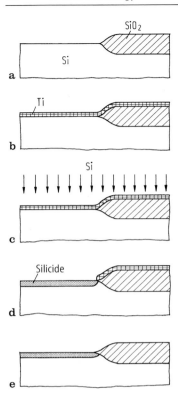

Fig. 3.9.1 a–e. Processing sequence for producing silicide films that only develop on exposed areas of silicon (salicide technique). Process step c can be omitted (see text). a Initial condition; b sputtering of metal film (e.g. Ti); c metal/silicon mixing by ion implantation (e.g. Si); d silicidation by annealing (e.g. 900 °C); e etching off metal film on SiO_2 areas

oxide therefore has to be removed by back-sputtering before metal sputtering occurs (see Sect. 3.1.4). Ion implantation e.g. of silicon or arsenic is another option for improving the uniformity of the silicide film, and thus reducing the contact resistance (ITM = *I*mplantation *T*hrough *M*etal, see Fig. 3.9.1 c). Despite the thin layer of natural oxide, the momentum of the implanted ions (recoil) brings about some degree of mixing of the metal and silicon atoms (IBIM = *I*on *B*eam *I*nduced *M*ixing). Thus a uniform silicide film may be produced during silicidation annealing.

The CVD process is becoming increasingly important for producing silicide films as an alternative to the sputtering technique. The following gas phase reactions are given as examples:

$$4SiH_4 + 2WF_6 \xrightarrow[30Pa]{400°C} 2WSi_2 + 12HF + 2H_2,$$

$$4SiH_2Cl_2 + 2TaCl_5 + 5H_2 \xrightarrow[60Pa]{600°C} 2TaSi_2 + 18HCl,$$

$$2SiH_4 + TiCl_4 \xrightarrow[plasma, 30Pa]{450°C} TiS_2 + 4HCl + 2H_2$$

Cool wall reactors (such as in Fig. 3.1.4 c) and plasma-CVD reactors (as in Fig. 3.1.6 c) are used here.

Unlike sputtering, CVD deposition produces silicide films which already have a fine-crystalline structure. Step coverage is also better than for sputtering, but the main advantage of CVD deposition is that it is more capable of producing pure layers. In sputtering, the actual target (particularly sinter targets) constitutes an impurity source.

By varying the CVD deposition conditions, CVD silicide films can be grown either over the whole surface, or selectively on exposed areas of silicon. Selective CVD silicide films have an advantage over selective silicide films obtained by metal sputtering and silicidation (see Fig. 3.9.1), in that no silicon is used up in the silicon areas.

3.9.2
Polycide films

An important application of silicide is in polycide films, which consist of two layers, the first being a polysilicon film (approx. 0.3 μm thick) usually heavily doped with phosphorus or boron, and the layer above being silicide (approx. 0.2 μm) which is normally produced directly after deposition and doping of the polysilicon. Polycide films can reduce the resistance to about one tenth of the value for heavily doped polysilicon. Whilst 0.4 μm thick polysilicon films heavily doped with phosphorus have a sheet resistance of 15–25 Ω/□, polycide films with a 0.2 μm thick silicide layer have sheet resistances of 1–5 Ω/□[35]) (cf. Table 3.1). The contact resistance with Al (1 % Si) metallization is also significantly lower, in particular for small contact holes (Fig. 3.9.2).

At present, polycide films are used almost exclusively as low-resistance alternatives to heavily doped polysilicon. In either case, temperatures of 900 °C or above may still arise after applying the polycide layers (e.g. during source/drain dopant activation, or in the glass flow process). The silicides must therefore remain stable at these temperatures. This rules out the silicides PtSi and Pd$_2$Si, which means that polycide films are almost exclusively made of MoSi$_2$, WSi$_2$, CoSi$_2$, TaSi$_2$ and TiSi$_2$ (cf. Table 3.1).

Polycide structures can also be thermally oxidized. The silicon consumed in the oxidation process comes from the polysilicon layer, diffusing through the silicide film to reach the surface. Thus the thickness of the polysilicon layer reduces, whilst that of the silicide remains almost unchanged.

Electromigration has been detected in polycide structures carrying large currents, but under real operating conditions it is unlikely to occur.

The function of the polysilicon in polycide layers is to maintain the proven benefits of polysilicon gate electrodes, such as excellent stability on thin SiO$_2$

---

35 As a comparison: a 0.8 μm thick aluminium film with 1 % silicon added has a sheet resistance of 40 mΩ/□.

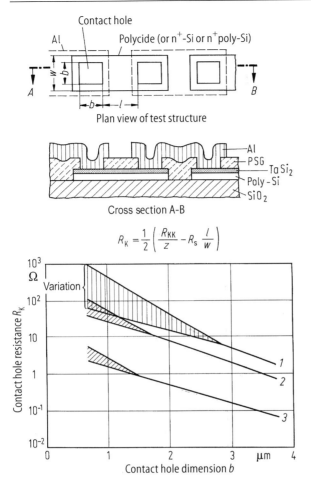

Fig. 3.9.2. Contact hole resistance for aluminium metallization (with 1 % added silicon) on 1 n$^+$-doped monocrystalline Si, 2 n$^+$-doped polysilicon and 3 polycide (poly-Si+TaSi$_2$), for different sizes of square contact holes. The contact hole resistance R_K was found by measuring the resistance R_{KK} of a contact hole chain containing z contact holes. R_s is the sheet resistance (in $\Omega/\square$) of the polycide or the n$^+$-doped silicon or the n$^+$-doped polysilicon. The sheet resistance of the aluminium metallization (approx. 30 m$\Omega/\square$) can be ignored here.

films, good step coverage, the buried contact and ease of thermal oxidation. Attempts are also being made, however, to deposit silicide films directly onto gate oxides. The main reason for this is that TaSi$_2$, for instance, has a work function difference φ_{MS} that is about 0.4 eV more positive than n$^+$-doped polysilicon. This means that channel regions of n-channel MOS transistors would not require such heavy boron doping.

3.9.3
Silication of source/drain regions

The minimum sheet resistance that can be achieved for n^+- and p^+-doped regions – the source/drain regions in MOS circuits – is constantly increasing as the penetration depth of doped regions falls (see Fig. 6.3.7). In miniaturized MOS circuits, the relatively large resistance of the source-drain regions may reduce the transistor transconductance, and the longer RC time constants can degrade the operating speed.

By selective silicide formation over the source/drain regions, a low sheet resistance of a few $\Omega/\square$ can be achieved even for small penetration depths (e.g. 0.2 μm). TiS_2, produced using the process described in Fig. 3.9.1, is the most common choice here, although the selective CVD technique mentioned in Sect. 3.9.1 is also possible.

Selective silicidation, where the source/drain regions need to be exposed whilst keeping the gate electrodes completely isolated at the same time, can be performed using the technique described in Fig. 3.5.2 b or 3.7.1 a. If the polysilicon gate structures are also to undergo silicidation, the poly silicon side-wall isolation (needed to prevent gate-drain short circuits as the silicide grows together), can be produced by a technique such as that in Fig. 3.5.2 a or 3.1.7 b.

A few extra processing steps enable silicide formation over selected areas of SiO_2 regions as well. This means that silicide connections between adjacent silicided regions can be created, thus providing e.g. enlarged contact surfaces for contact holes. To do this, a lithographic step is used to produce patterns of amorphous silicon or polysilicon, about 0.1 μm thick, at selected portions before applying the metal film (step a in Fig. 3.9.1). During silicidation (step d in Fig. 3.9.1), these silicon patterns are then also converted into silicide patterns ("straps"). This strap technique can also be used to create low-resistance connections between n- and p-doped regions.

3.10
Refractory metal films

The refractory metals of molybdenum, tungsten, tantalum and titanium are not only used in silicon technology in their silicide forms, but are also finding increasing application in their pure form. Titanium and tungsten are particularly important.

Double Ti/TiN layers (approx. 20 nm Ti and approx. 100 nm TiN) have now become the standard intermediate layer between silicon and aluminium or tungsten for metallizations in the sub-micron region (cf. Table 8.10). The thin Ti film forms a low-resistance $TiSi_2$ contact with the Si, whilst the TiN acts as a metallurgical barrier layer between the Si and Al or W. Either the PVD (sputtering) or CVD process is used to produce the TiN film.

A nitrogen saturated titanium target is used for the sputtering process. Using a TiN collimator (see Sect. 3.1.4) also helps to improve coverage of the bottom of contact holes (see Sect. 3.11.4). The electrical resistivity of sputtered TiN films varies between 100 and 400 µΩ cm.

The MOCVD technique (MOCVD = Metal Organic CVD) can be used to produce TiN with excellent conformal step coverage [3.43]. In a CVD reactor such as that in Fig. 3.1.6 c, a remote microwave plasma can be used to generate hydrogen radicals (H*) which trigger the following reaction on the hot wafer surface:

$$Ti[N(CH_3)_2]_4 + 9H^* \xrightarrow[40Pa]{400°C} TiN + gases.$$

The high resistivity of CVD TiN films (approx. 10^4 µΩ cm) is rather beneficial, because it prevents current crowding in contact holes and vias (cf. Fig. 3.11.4).

TiN films are not just important as barrier layers. Used over Al films, they also act as anti-reflection coatings for optical lithography and are used to suppress hillock growth (see Sect. 3.11.2).

In sub-micron silicon technology, tungsten has found its widespread use as a contact hole filler, but also as a material for tracks (see Fig. 8.5.3).

In order to prevent unwanted tungsten-silicon reactions (encroachment, wormholes), a Ti/TiN layer acts as a contact and barrier layer between the silicon and the tungsten (see above). CVD deposition is performed in two stages [3.44]. First, a seed layer about 50 nm thick is deposited, followed by the actual tungsten layer (e.g. 0.5 µm thick). The following reactions take place inside a reactor such as the type in Fig. 3.1.4 c:

$$WF_6 + SiH_4 \xrightarrow[500Pa]{470°C} W + gases \text{ (seed layer),}$$

$$WF_6 + H_2 \xrightarrow[10^4 Pa]{470°C} W + gases \text{ (bulk layer).}$$

Due to the excellent conformal deposition, contact holes and vias can be filled without voids forming (see Fig. 3.1.2 d) even if the contact hole walls are vertical.

With 9 µΩ cm, the resistivity of CVD tungsten is only three times larger than that of aluminium. For this reason, tungsten is also used for tracks e.g. for local interconnects (see Fig. 8.5.5). An etching process (see Sect. 5.3.4) or the Damascene technique (see Sect. 8.5.1) is suitable for producing the tungsten interconnection patterns. The advantage of using tungsten instead of aluminium for tracks is its greater stability at high temperatures (processing temperatures up to 600 °C are possible, and operating temperatures over 300 °C), and its stability at high current densities (no electromigration).

If the tungsten is simply being used as a contact hole filler, selective deposition can also be used, where the seed layer is selectively formed on the exposed silicon regions.

3.11
Aluminium films

The preferred metal for the low-ohmic interconnections of integrated circuits is still aluminium, because no other metal meets the essential characteristics required of such track layers, these being:
- low resistivity (3 µΩ cm)
- ohmic contact to p- and n^+-doped regions of silicon
- good adhesion on SiO_2, phosphorus glass and boron phosphorus glass
- bonding possible using standard wire bonding techniques (gold wires by thermo-compression, aluminium wires by ultrasonic compression)
- suitable for multilayer metallization

Aluminium does have several limitations, however, which have either been tolerated in the past or improved by specific measures. These negative properties include limiting the processing temperature to below 500 °C, unsatisfactory mechanical and chemical stability, significant electromigration, poor edge coverage, hillock growth on the film surface, reactivity with silicon in the contact holes and too high a contact resistance for small contact holes. These properties and the associated counter measures are discussed in more detail below.

3.11.1
Producing aluminium films

Although vapour deposition by thermal, inductive or electron beam vapour deposition (see Sect. 3.1.3) was the main production method until about 1980, sputtering now predominates (see Sect. 3.1.4). The disadvantages of the vapour deposition process, discussed in Sect. 3.1.3, are the reason for this change. These drawbacks are worse edge coverage and poorer reproducibility of defined film compositions, in particular for the typical 1 % silicon in aluminium combination (see Sect. 3.11.4).

Sputtering of the Al (1 % Si) film is usually performed from an alloy target (99 % Al, 1 % Si). This is also the case when other alloys are required in the Al film (e.g. Cu or Ti to increase the resistance to electromigration). The silicon wafers can be held at a raised temperature (200–400 °C) during sputter coating. At these temperatures the atoms striking the surface have a certain degree of mobility, so that narrow cracks at e.g. structure edges can be avoided.

The contact resistance can be lowered by brief back-sputtering (sputter cleaning) before applying the aluminium coat. Edge coverage is improved by bias-sputtering and collimated sputtering (see Sect. 3.1.4). The problems associated with inadequate edge coverage in small contact holes are illustrated in Fig. 8.5.1.

"Planarizing Al" is a new Al deposition technique, where the Al is made to flow directly after it is sputtered on. This is achieved by heating the silicon wafers to 550 °C ("hot Al") or placing them under a high pressure ("force fill"). The Al flows into the contact holes and fills them up (see Fig. 8.5.3). The barrier film in the contact holes (generally TiN) has to withstand severe stresses, particularly in the hot Al technique. A CVD TiN film is beneficial because of its excellent edge coverage (see Sect. 3.10).

Once the aluminium film has been laid down, it has to be patterned to produce tracks. After patterning annealing is performed in hydrogen or forming gas, typically at 450 °C. This creates low-resistance aluminium-silicon contacts, and anneals the radiation damage caused by the plasma during sputtering (see Sect. 3.1.8).

Although CVD deposition of aluminium has proven successful, it has not yet established itself. Problems arise in dealing with the unstable and highly explosive reactant (e.g. tri-isobutylaluminium), and in obtaining a reproducible film composition when silicon or other substances are added to the aluminium.

3.11.2
Crystal structure of aluminium films

Aluminium films sputtered at a temperature below 100 °C have a fine-grained structure with grain sizes of 50–100 nm. Some grains grow during annealing at 450 °C, with grain sizes reaching about 0.5 µm. If the aluminium film is sputtered at substrate temperatures of 200–400 °C, the grains reach several µm in size after annealing. They are then larger than the track widths in VLSI circuits. This is referred to as a bamboo structure, and is ideal for resisting electromigration (see Sect. 3.11.3).

Aluminium films sputtered at substrate temperatures below 100 °C have a smooth surface, while hot-sputtered aluminium films (200–400 °C) have a surface roughness of a few tenths of a µm. During annealing (450 °C) protrusions ("hillocks") are growing the film surface, sometimes up to a height of more than 1 µm. In multilayer metallization the hillocks may create short-circuits between the two track planes if the intermetal insulating layer does not completely cover the hillocks.

Hillocks can be almost totally suppressed by adding small amounts of copper to the aluminium (about 0.5 % Cu). A TiN layer on top of the Al film also inhibits hillock formation.

Aluminium is relatively soft and thus prone to mechanical damage, e.g. when handling the silicon wafers or the individual chips. A protective cover film usually made of a double layer of plasma oxide and plasma nitride (see Sect. 8.5.4) prevents mechanical damage.

Aluminium tracks not only have limited mechanical stability, but also limited chemical stability. They corrode in a damp atmosphere in the presence of e.g. phosphate ions (from phosphorus glass) or fluoride and chlorides (from impurities during processing). A thick passivation layer can again provide protection in this case.

3.11.3
Electromigration in aluminium interconnections

Electromigration refers to the migration of material within the interconnections under the influence of an electric current. The physical cause for material migration is the impact of accelerated electrons on the positive metal ions of the crystal lattice. Accordingly, the material always migrates in the direction of the electron flow and thus in the opposite direction of the current (defined as positive). Localized depletion of material can produce a break in a interconnection, and hence lead to total failure of the integrated circuit.

The Black relationship

$$\mathrm{MTF} \sim j^{-2} \exp(-W_A / kT)$$

can be used to give a good approximation for the lifetime MTF (MTF = Mean Time to Failure), where j is the current density and W_A an activation energy, found to have a value equal to 0.65 eV. This energy corresponds to grain boundary diffusion, which has been found to be the main mechanism of electromigration. This also explains why narrow aluminium interconnects with a bamboo grain structure (see Sect. 3.11.2) are more resistant to electromigration (Fig. 3.11.1).

Apart from the bamboo grain structure, there are several options for increasing the resistance to electromigration of the aluminium interconnects. Adding 0.5-2 % copper to the aluminium[36] increases the lifetime by about a factor of 10 for otherwise unchanged conditions. A similar effect is found for a titanium alloy (0.2 %). Sandwich layers of Al–Cu, Al-Ti or Al–TiS$_2$ also have a higher resistance to electromigration. A similar effect is provided by a TiN layer beneath or above the Al interconnects (see Fig. 3.11.1).

It is not just the aluminium tracks that are subject to electromigration. The aluminium-silicon contact holes can also be affected by it. As a result of current crowding at one contact edge (see Sect. 3.11.4), localized peaks in current densi-

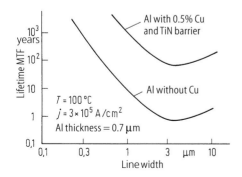

Fig. 3.11.1. Lifetime MTF (Mean Time to Failure) of sputtered aluminium interconnects for different line widths. The longer lifetime of the narrow interconnects is caused by the bamboo grain structure of the interconnects. The lifetime can be increased by some two orders of magnitude by adding copper to the aluminium and by using a TiN barrier

---

36 Aluminium interconnects with these copper concentrations are more prone to corrosion, however.

ty can be large enough for silicon atoms to migrate into the aluminium. This migration of material can be suppressed by providing a diffusion barrier between the silicon and the aluminium.

There is another migration effect in aluminium interconnects apart from electromigration. This is stress migration, which refers to the migration of aluminium atoms under the influence of internal mechanical tensions at higher temperatures (e.g. 150 °C). Breaks in interconnects can also occur from stress migration just as they do from electromigration, particularly for very narrow interconnects with widths of less than 1 μm. Adding a small amount of copper to the aluminium (about 0.5 % Cu) has proved an effective countermeasure to stress migration.

3.11.4
Aluminium-silicon contacts

If pure aluminium is used for the interconnects, then a reaction with the silicon occurs in the contact holes during annealing (typically 450 °C). A silicide is not formed, but instead the silicon diffuses into the aluminium until the solubility limit is reached (0.5 % at 450 °C). The diffusion length after 30 min at 450 °C is approx. 40 μm. The silicon does not diffuse uniformly out of the contact hole. There are two reasons for this. First, the natural oxide on the silicon (approx. 1 nm thick), which inhibits diffusion, is non-uniformly consumed by the aluminium. Second, a particularly large amount of silicon migrates out of the contact hole at that edge lying in the direction where the interconnect extends the furthest, since it is here that the supply of aluminium is largest, and hence the demand for silicon to achieve a 0.5 % silicon concentration in the aluminium (Fig. 3.11.2).

Another characteristic of silicon diffusion is that the silicon dissolves more readily at surfaces with $\langle 100 \rangle$ orientation (as in alkaline etching), whilst $\langle 111 \rangle$ oriented surfaces act practically as diffusion stops. If the silicon wafer has a $\langle 100 \rangle$ surface – which is usually the case – then pyramidal craters (spikes) are formed, whose faces are $\langle 111 \rangle$ planes forming a 55° angle with the $\langle 100 \rangle$ surface

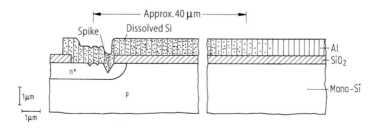

Fig. 3.11.2. Formation of spikes in a contact hole when pure aluminium is used for the interconnects. Silicon diffuses up to some 40 μm away after 30 minutes annealing at 450 °C

(Fig. 3.11.2). The spikes are not hollow spaces, however, because the lost silicon is replaced by aluminium (Kirkendahl effect). The smaller the contact hole (and thus the supply of silicon) and the greater the volume of aluminium to be saturated, the deeper the spikes that form. If a spike extends down to a pn junction, then this is shorted and the integrated circuit fails completely.

In order to overcome these problems the use of aluminium with about 1 % silicon already added was standard as early as 1970. This eliminates the driving force behind the out-diffusion of silicon from the contact holes, although the opposite effect may be observed during annealing (450 °C), with silicon precipitation being particularly pronounced in the contact holes (Fig. 3.11.3). The silicon grows epitaxially on the $\langle 100 \rangle$ oriented silicon surfaces forming pyramidal precipitates. As long as the precipitated silicon only covers a small part of the contact hole area, it has almost no detrimental effect. For small contact holes with dimensions of less than 2 µm, however, total coverage of the holes can occur. This increases the electrical contact resistance between the aluminium and n^+-doped silicon regions (cf. Fig. 3.9.2), because the precipitated silicon is doped with aluminium and is thus p-type, creating a kind of pn junction in the contact hole.

As already explained in Sect. 3.10, a $TiSi_2$ layer in the contact hole can solve the contact resistance problem. In order to totally suppress the diffusion of silicon into aluminium or of aluminium into silicon, a diffusion barrier (e.g. TiN) must be inserted between the silicide and the aluminium film.

Even with a TiN barrier layer and Si added to the Al, spiking can still arise, particularly if the TiN layer does not completely cover the base of the contact hole. This can be explained as follows: after sputtering of the AlSiCu film, the Si is distributed uniformly over the layer. This hardly changes during annealing at 450 °C because the added Si corresponds approximately to the solubility in Al. Because the solubility level reduces as the temperature falls, Si precipitates out onto the TiN film when cooled after annealing. During subsequent annealing at 450 °C (e.g. for a second Al layer), the precipitated Si apparently does not go back into solution, or only very slowly (because of $TiSi_2$ formation?). The unsaturated Al therefore "sucks" out the Si at breaks in the TiN film, so that spikes form at these locations (cf. Fig. 3.11.2).

The current density in a contact hole through which current is flowing is usually not uniform over the whole contact surface. Those contact holes with uniform current flow are most likely to be those where the current continues to flow

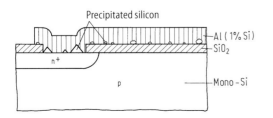

Fig. 3.11.3. Precipitated silicon when using aluminium with 1 % added silicon

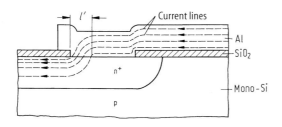

Fig. 3.11.4. Current crowding at a contact hole edge

vertically into the silicon, e.g. in the collector contacts of bipolar transistors. In other cases where the current in the silicon flows parallel to the surface, current crowding occurs at those contact hole edges facing the silicon region through which the current flows (Fig. 3.11.4). The reason for this is that the current takes the path of minimum resistance, and thus stays as long as possible in the aluminium, which has a resistivity some three orders of magnitude lower.

A contact resistance at the junction between aluminium and (heavily doped) silicon needs to be taken into account. The specific contact resistance ρ_c with p$^+$ and n$^+$ doped silicon regions equals approx. $10^{-6}\,\Omega\,\text{cm}^2$. It can be reduced to less than $10^{-7}\,\Omega\,\text{cm}^2$ by using one of the aforementioned contact films. If ρ_c and the sheet resistivity R_S of the diffused silicon layer are known, then a rough estimate can be made of the length l' of the contact hole through which the current is passing (Fig. 3.11.4) [3.45]:

$$\frac{\rho_c}{l'b} \approx R_S \frac{0.5 l'}{b}$$

$$l' \approx \sqrt{\frac{2\rho_c}{R_S}}$$

where b is the width of the contact hole perpendicular to the current flow. If $\rho_c = 10^{-7}\,\Omega\,\text{cm}^2$ and $R_S = 30\,\Omega/\square$, then l' equals approx. 0.8 µm. This means that for these resistance conditions, increasing the size of the contact hole length above 0.8 µm has no real effect electrically.

Contacts between aluminium and lightly n-doped silicon act as a diode (Schottky contacts). The properties of Schottky contacts are described in detail in [3.46].

3.11.5
Aluminium-aluminium contacts

Where there are two or more interconnect planes (multilayer metallization), contacts must also be produced between the interconnects of the different conductor planes. These contacts are called vias.

In order to produce a low resistance contact, the natural Al$_2$O$_3$ layer on the interconnects in the lower plane must be removed. This can be done by back-sputtering immediately before sputtering the upper Al layer, or by hot-sputtering (approx. 350 °C), or by inserting intermediate contact layers such as Ti/TiN, TiW or $TiSi_2$ (cf. Fig. 3.10).

As the via cross-sections become smaller, the problems of edge coverage and hole filling in vias increase, in the same way as in aluminium-silicon contacts. Possible countermeasures include tapering the via edges, Al bias-sputtering, hot-sputtering, or filling the vias with CVD tungsten (see Sect. 3.10) or Al (using the hot Al technique).

3.12
Organic films

Today's integrated circuits are almost exclusively the domain of inorganic films, although organic layers can be applied using the cheap spin-on process (Sect. 3.1.5). The reason for this dominance lies in the better thermal and electrical stability of inorganic films. Only polyimide and spin-on glass have gained major importance in integrated circuits.

Further applications for organic films refer to cases where they play a temporary role in the integrated circuit manufacturing process. The resist materials, which are fundamental to the lithographic process, are obvious examples here. These are dealt with comprehensively in Sect. 4.2.1.

Another application is the etch-back technique used on a surface containing many steps (Fig. 3.5.4 or 5.3.6).

3.12.1
Spin-on glass films

The starter material for a spin-on glass film is silicon tetra-acetate, which is dissolved in a solvent and spun onto the silicon wafers. If the film is heated to about 200 °C, then an SiO_2-type layer is produced by cross-linkage. This film has shrunk in size compared to the original film, but is far more loosely packed than thermal or CVD SiO_2 films (higher etch rate in diluted hydrofluoric acid).

Problems with spin-on glass films include poor adhesion to the previous layer (an adhesion agent may be required), a tendency to crystallization out of solution, the likelihood of cracks forming because of shrinkage, the poorer electrical quality of the SiO_2 compared with thermal or CVD SiO_2 films, and so-called via poisoning (see Fig. 8.5.4).

Spin-on glass layers are used for topography planarization (Figs. 3.5.5 and 8.5.3), as an etching mask in the tri-level resist technique (Sect. 4.2.2) and as a doping source for foreign atom doping of silicon (Sect. 6.1). In the latter case, the required doping material is added to the resist.

3.12.2
Polyimide films

A chemical precursor of polyimide, dissolved in a solvent, is used as the starter material for a polyimide film. After spinning the coat onto the silicon wafers, the solvent is evaporated at about 100 °C. The actual conversion into a polyimide film (cyclization) occurs at a temperature of 300–400 °C, and is accompanied by shrinkage of the film thickness.

Three different methods are used for structuring the polyimide film (Fig. 3.12.1). The first method (Fig. 3.12.1 a) exploits the fact that the precursor material is soluble in the alkaline developer of the positive photoresist. This technique is unsuitable for fine patterns, however, because of the isotropic etching of the polyimide. In the second method (Fig. 3.12.1 b) ionic etching is used, which works anisotropically. Since it is practically impossible to achieve a different etch rate between the photoresist and the polyimide in this case, an inorganic layer (e.g. spin-on glass or plasma CVD nitride) must be inserted between the polyimide film and the photoresist. In the third method for creating a pattern in the polyimide (Fig. 3.12.1 c), the precursor material itself is made photosensitive by adding suitable materials. Those areas that have not been exposed are then dissolved in a suitable developer, whilst the exposed areas are insoluble because of photocuring (negative resist function[37]). In the first and third method, the precursor material is not converted into polyimide until the pattern has been created.

Polyimide films have found several applications in silicon technology. Used as a protective layer, they act as a stress buffer between the chip and the molding

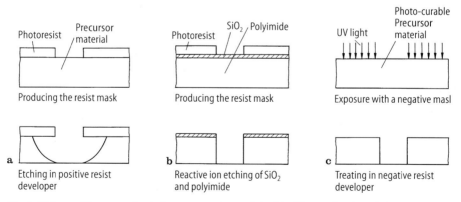

Fig. 3.12.1 a–c. Three methods for structuring polyimide films. **a** "Etching" the precursor material in an alkaline medium (e.g. positive resist developer); **b** reactive ion etching of the polyimide using an auxiliary inorganic mask (e.g. SiO₂); **c** using a photo-curable precursor material so that it itself acts as a photoresist

---

37 There are also positive acting photosensitive insulators based on polybenzoxazol [3.47].

material of the package (see Sect. 8.5.4). They are also used as mechanical protective layers because of their high resistance to scratches. They do not, however, provide a moisture-proof seal, so they are only of limited use as a single layer for chip passivation.

The same is true when polyimide is used as an insulating layer (dielectric constant $\varepsilon = 3.6$) between the conductor planes in a multilayer design, where an inorganic insulating layer (SiO_2 or Si_3N_4) is usually inserted either above or below the polyimide film. There is also an interesting option, where the polyimide helps to round off edges and fill trenches (cf. Fig. 3.5.5), while the basic intermetal insulation is provided by an SiO_2 film. This version has the advantage that during via etching only the SiO_2 needs to be etched, and that no extra steps are needed to produce complete coverage of the separate polyimide islands with moisture-proof SiO_2. Thus there is no risk of moisture penetrating vertically or laterally into critical parts of the integrated circuit. The limited long-term electrical stability of polyimide, even without the influence of moisture, presents a specific problem, which prevents the use of polyimide as a dielectric for multi-level interconnection schemes of MOS circuits.

Where polyimide is simply used as a temporary processing aid and then removed again (e.g. by plasma ashing) the electrical stability is irrelevant. Examples of polyimide being used in this way, are as the bottom layer in a multilevel resist process (Sect. 4.2.4), as a planarizing layer in the etch-back technique (Sect. 3.5.4) and as a thermally-stable masking layer (up to 400 °C) in highly stressful processing stages, such as during high-dose implantation, during reactive ion etching processes or during a liftoff process using high-temperature vaporization or hot-sputtering of the metal.

It is likely that polyimide will find other applications in the future, particularly those that exploit the temperature stability of the polyimide as well as the high etch rate selectivity between polyimide and the inorganic layers (SiO_2, Si_3N_4, Si, Al) during reactive ion etching.

4
Lithography

The films used in silicon technology (see Chap. 3) must be patterned on the silicon wafers into numerous separate regions to form tracks or other features. This pattern generation process is today performed almost exclusively by lithography (Fig. 4.1). The basic element of this process is a radiation-sensitive resist layer. This is exposed to radiation in selected areas so that when placed in a suitable developer, the exposed (or unexposed) areas are removed. The resist pattern that results then acts as a mask in the subsequent processing step, e.g. etching or ion implantation. Finally, the resist mask is removed again. It therefore only performs a temporary function, and does not form part of the final integrated circuit (exception see Fig. 3.12.1 c).

Direct pattern-generating techniques, which do not use resist masks, but instead start by applying or etching the films only in the desired portions of the substrate surface, have not established themselves as yet for productivity reasons (examples of these techniques are radiation-induced deposition or etching, and ion beam writing of doping atoms). Such processes are, however, used for local disconnection or connection of lines (e.g. for programming circuits or for

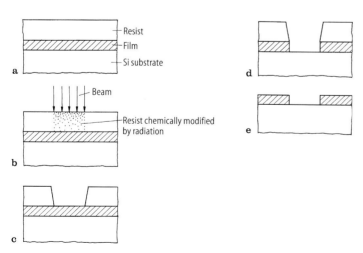

Fig. 4.1 a–e. Principle of Lithography. **a** Apply resist film and prebake; **b** expose selected areas; **c** develop the resist and postbake; **d** etch the film; **e** remove the resist

activating redundant circuit elements), and for labelling silicon wafers (e.g. using a laser beam).

Ultraviolet light, x-rays and accelerated electrons and ions are the various forms of radiation used in lithography to expose the resist. These different lithographic techniques are discussed consecutively in Setcs. 4.2-4.5.

4.1
Linewidth dimensions, placement errors and defects

The aim of lithography is to produce the numerous different resist features on a silicon wafer within dimensional tolerances, in the correct position and without defects. The effectiveness of a lithographic process is assessed according to the smallest pattern dimension that can be achieved, the line width variation, the placement error variation and the defect density.

Figure 4.1.1 explains the commonly used terms relating to resist pattern dimensions and their line width variance. The permitted line width variance typically equals ±10 to ± 30 % of the minimum pattern dimension (critical dimension CD).

For placement errors and their variation (Fig. 4.1.2), only the relative position is significant, i.e. the position of a pattern e.g. in pattern plane B with respect to the position of a pattern in pattern plane A. Thus the expressions for the overlay error and the edge placement error (Fig. 4.1.2) always refer to differences in positions and not to the positions themselves.

The edge placement error is the parameter of most interest to the circuit designer. In Fig. 4.1.2, feature a may be the polysilicon gate electrode of a MOS transistor, for example, and feature b the contact hole to the drain region of this transistor. If the designer places the left-hand edge of the contact hole at a distance $\Delta x_E(3\sigma)$ from the right-hand polysilicon pattern edge, then assuming a Gaussian distribution for the overlay error variation and for the line width variation, the designer can assume the contact hole to overlap the polysilicon in 0.3 % of the chips. The aluminium metallization in the contact hole means that the gate and drain are shorted, and thus the chip suffers complete failure. The permitted 3σ edge placement error typically equals 30–50 % of the minimum

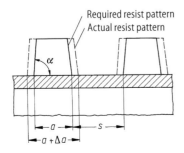

Fig. 4.1.1. Explanation of commonly used terms for resist pattern dimensions and their line width variation. a minimum resist pattern width; s minimum resist pattern separation; $a+s$ minimum pitch; Δa line width error; α sidewall angle of resist features

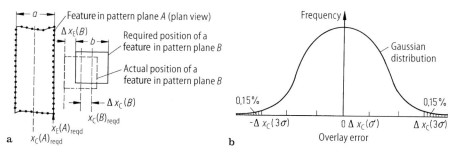

Fig. 4.1.2 a, b. Explanation of commonly used terms when specifying placement errors. $x_C(B)_{reqd} - x_C(A)_{reqd}$ = required distance between centre positions of features a and b in pattern planes A and B respectively;

$\Delta x_C = \Delta x_C(B) - \Delta x_C(A)$ = overlay error

$\pm \Delta x_C(3\sigma) = 3\sigma$ variation of overlay error

$\Delta x_E = \Delta x_E(B) - \Delta x_E(A) = \Delta x_C - \frac{1}{2}\Delta a - \frac{1}{2}\Delta b$ = edge placement error

$\pm \Delta x_E(3\sigma) = \sqrt{\left[\Delta x_C(3\sigma)\right]^2 + \left[\frac{1}{2}\Delta a(3\sigma)\right]^2 + \left[\frac{1}{2}\Delta b(3\sigma)\right]^2}$

= 3σ variation of edge placement error

feature size. The permitted overlay error varies between 20 and 40% of the minimum feature size.

Although the assumption that placement errors have a statistical (Gaussian) distribution is not always valid in practice, it has proved to be a close approximation to actual circumstances. It is a particularly useful approximation when individual amounts can be treated as statistically independent quantities and summed up by taking the root of the sum of the squares[1] (see Fig. 4.1.2).

A defective resist pattern can arise if, for instance, a particle is lying on the resist area to be exposed during exposure (see Chap. 7). If the particle is larger than the width of the area to be exposed, then adjacent resist patterns are "shorted" together (cf. Fig. 4.1), which usually means complete failure of the circuit concerned (e.g. metallization short-circuit). The same is true for complete breaks in resist patterns, which can be caused by mechanical stress (scratches, handling damage) for example. Pattern defects that are smaller than the mini-

---

[1] The vectorial statistical addition of x- and y-position errors also produces the (obvious) result that the same placement error exists in the 45° direction, with the same variation $\Delta d(3\sigma)$, as in the x- and y-direction:

$$\Delta d(3\sigma) = \frac{1}{\sqrt{2}} \sqrt{\left[\Delta x(3\sigma)\right]^2 + \left[\Delta y(3\sigma)\right]^2} = \Delta x(3\sigma)$$

mum pattern dimension can also cause the whole circuit to fail. For instance in a polysilicon gate electrode where the gate length has been reduced locally by 30 % because of a pattern defect, the sub-threshold current increases and early drain-source punchthrough may occur. Both phenomena can cause the circuit to fail.

A simple example can illustrate how to estimate a Figure for the allowed defect density. If a 70 % chip yield based solely on pattern defects[2] is specified, then for a 1 cm^2 chip with 4 critical pattern planes, only about 0.1 killing pattern defects per cm^2 are allowed, or in other words one killing pattern defect on 10 cm^2 of each pattern plane. To achieve such a low defect density, extremely high demands have to be placed on the purity of the materials used and on clean working practices for both equipment and operating staff (also see Chap. 7).

4.2
Photolithography

4.2.1
Photoresist films

Resists are basically classified as positive or negative according to whether the exposed or unexposed areas respectively are dissolved away by developing. The negative resists previously in widespread use have now been superseded by the positive resists for pattern dimensions below 3 µm. This is mainly because negative resists have a poorer contrast characteristic and have a tendency to swell. The focus of this section shall therefore be on positive resists, with a brief look at interesting recent advances in negative resists.

The most commonly used positive resists contain three essential components: a novolak resin used for layer formation, the photo-sensitive compound and a solvent. Diazonaphthoquinone is the photo-sensitive compound used most frequently in positive resists. Figure 4.2.1 shows the characteristic chemical reactions of diazonaphthoquinone which occur during exposure and development of the positive resist films, and which ultimately lead to the removal of the exposed areas.

The photo-chemical conversion of the diazonaphthoquinone in the ultraviolet light (wavelength 300–450 nm) involves the absorption of light. Thus as the exposure time progresses, light absorption decreases because the number of unconverted molecules is constantly dropping. This is referred to as bleaching of the resist film. Figure 4.2.2 shows the absorption coefficient as a function of the exposure dose (given by the product of the light intensity in the resist and the exposure time) for a typical positive resist.

---

2 There are other defects apart from pattern defects, e.g. weak spots in layers, that have nothing to do with lithography (see Chap. 7).

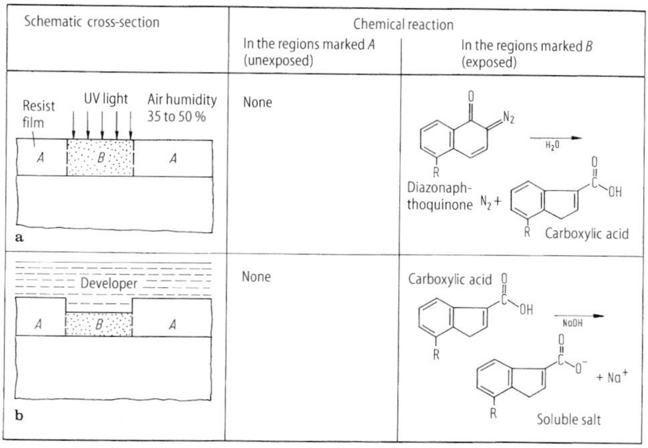

4.2 Photolithography

Schematic cross-section	Chemical reaction In the regions marked A (unexposed)	In the regions marked B (exposed)
a Resist film, UV light, Air humidity 35 to 50%, A B A	None	Diazonaph-thoquinone N_2+ → N_2 + H_2O → Carboxylic acid
b Developer, A B A	None	Carboxylic acid + NaOH → Soluble salt + Na^+

Fig. 4.2.1 a, b. Chemical reactions during exposure (a) and development (b) of a photoresist containing diazonaphthoquinone as the photo-sensitive component. During exposure, nitrogen is released if moisture is present in the resist, for which a specific air humidity is required (35 to 50% relative humidity). The lower cross-sectional diagram shows a temporary condition during the development process. At the end of development the area B is completely dissolved away. The reaction between an acid (carboxylic acid) and an alkaline developer (NaOH here) causes portion B to dissolve

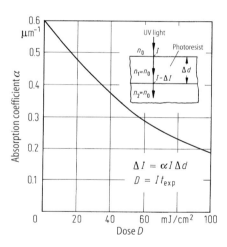

Fig. 4.2.2. Absorption coefficient α of a typical photoresist with respect to the exposure dose D. The illumination wavelength is 436 nm. The cross-sectional diagram in the inset shows the experimental conditions when measuring α (small resist thickness Δd, matched refractive indices $n_0 = n_1 = n_2$ to avoid reflections at the boundaries). $t_{\exp}$ is the exposure time

This curve can also be interpreted as a measure of the resist sensitivity, because it indicates how fast the photo-reactive compound in the resist is being converted during exposure. It is more usual, however, to define the resist sensitivity in terms of the residual resist thickness after exposure with a specified dose and subsequent developing. Figure 4.2.3 shows this residual resist thickness as a function of the exposure dose for a typical positive resist. The dose D_0

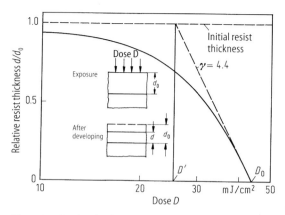

Fig. 4.2.3. Resist thickness after developing (fixed development time) as a function of the exposure dose for λ = 436 nm in a typical positive resist. The dose D_0 is a measure of the resist sensitivity. The gradient γ at the foot of the curve is called the resist contrast

at the foot of the curve is that dose which is just sufficient to completely remove the resist under the given development conditions. This dose is referred to as the "resist sensitivity".

Another important parameter characterizing the resist is the resist contrast γ. Referring to Fig. 4.2.3, γ is the gradient at the foot of the resist thickness curve[3]

$$\gamma = \left[\log(D_0 / D')\right]^{-1}.$$

Both the resist sensitivity and the resist contrast can be varied within a given range by changing the processing conditions during resist prebake before exposure, and during resist development. The resist sensitivity generally increases (i.e. smaller D_0 values) for lower prebake temperature, higher developing concentration and higher developing temperature[4], whilst the resist contrast increases with lower prebake temperature, lower developing concentration and lower developing temperature[4] [4.1].

In modern production lines, both the resist coating and the developing processes are fully automated, with each wafer taken one after the other through the subsequent processing stations on so-called tracks. Either belt conveyors or air cushions are used to transport the wafers.

The first processing station on the resist coating track is usually a hot plate, on which the wafers are heated to about 200 °C in order to evaporate off any water molecules adsorbed at the surface (dehydration bake). The next step is to ap-

---

3   Sometimes γ is calculated from the natural logarithm of the dose instead of the log to the base ten. γ values are then smaller by a factor of 2.3.
4   This is only true for developers containing metal

ply a film of adhesion agent just a few atoms thick. Both this and the dehydration step are performed to improve the adhesion of the resist layer to the underlying layer. HMDS (hexamethyldisilazane) is often used as the adhesion agent. It is either applied by gas phase deposition or dropped onto the wafer in solution and then spun on. The resist itself is then applied by spin-coating (see Sect. 3.1.5), at a speed of about 5000 rpm. The resist thickness may vary between 0.5 µm for flat or planarized surfaces, and 2 µm for surfaces containing many steps. Then the solvent is completely expelled at about 100 °C on a hot plate (resist prebake).

The resist is now a solid film, but it is still extremely susceptible to mechanical stresses. These can cause the film to peel away from the layer beneath at localized areas, and to break into tiny splinters of resist which can settle elsewhere as unwanted particles. Since the wafer edge is particularly at risk from mechanical damage (wafer chucks, carriers), the photoresist at the wafer edge is often removed during the actual resist coating stage, by directing a fine jet of solvent at the wafer edge when still on the turntable for resist coating.

After the resist prebake, alignment and exposure are performed in a wafer exposure unit (see Sect. 4.2.5). Then the wafers are placed back on a track for the photoresist to be developed[5]. This involves spraying developer fluid[6] onto the resist surface from jets (spray development), or simply dropping the developer onto the surface until it is totally covered (puddle development). Once the specified development time has elapsed[7], the developer is rinsed off with water, and the wafers then spun dry by centrifuging. A postbake step on a hot plate ensures that the photoresist is completely dry. This curing process also increases the chemical resistance. The postbake temperature varies between 100 and 180 °C, depending on what demands will be placed on it in the subsequent processing step (wet etching, plasma etching, ion implantation). It should be noted, however, that the resist starts to flow above about 120 °C, causing the resist pattern to deform. Resist flow can be reduced considerably by hardening the resist surface before the postbake.[8] Irradiating with shortwave ultraviolet light (wavelength approx. 250 nm) or plasma treatment are two ways of hardening the resist.

After the postbake, the wafers are inspected for dimensional conformance, placement accuracy and possible defects in the resist patterns. This is either done manually or automatically using high-resolution light microscopes or scanning electron microscopes.

---

5 An extra heating stage can be performed before developing if necessary (post exposure bake) (see Sect. 4.2.3).
6 To avoid sodium contamination of SiO_2 films (see Sect. 3.4.3), developers are often used which contain no alkali ions.
7 Sometimes endpoint detection techniques are used, e.g. using laser interference (see Sect. 5.2.7).
8 Hardened resist patterns which are then baked at a high temperature remain stable even when a second resist film is spun on top of the resist pattern and developed. Thus it is possible to produce two resist masks one on top of the other, which is used for certain planarization techniques for instance (see Fig. 3.5.3).

During the whole process sequence from resist coating to the processing stage following photolithography (etching, ion implantation), the silicon wafers are not allowed to be exposed with daylight, because standard resists have a residual sensitivity to a wavelength of about 500 nm. Thus processing is performed in yellow-light rooms. In addition, the short-wave part of the visible spectrum must be filtered out from the light microscopes used for inspection.

After etching or ion implantation, the resist is normally removed (stripped) in two stages. First it is treated in the oxygen plasma of a barrel reactor and then it undergoes wet-chemical processing e.g. in Caro's acid ($H_2SO_4+H_2O_2$) (see Sect. 7.3). If the resist film has previously been heavily stressed (e.g. by very high dose ion implantation) or has been baked at a high temperature (e.g. 180 °C), then complete removal of the resist can be difficult. In such instances, strong alkaline solutions[9] or a chemical such as fuming nitric acid may be needed. Another possible step is to expose the resist mask after developing the resist. The conversion of the photochemically active components (diazonaphthoquinone) can limit cross-linkage of the resist at high temperatures or during heavy ion bombardment.

4.2.2
Formation of photoresist patterns

This section shall consider more closely how the exposure and developing conditions affect the formation of the resist patterns.

When a photoresist is illuminated, the transition between an exposed and unexposed region is not abrupt, but is "blurred" because of light diffraction (see Sect. 4.2.6). Figure 4.2.4 a shows this non-abrupt light intensity or dose transition (dose = intensity×exposure time) between an unexposed and an exposed region.

First let us assume that the light intensity reaching the resist surface is constant throughout the resist thickness. Making another simplifying assumption, that the developer only removes the resist in a vertical direction[10], one can construct the shape of the resist after exposure and developing (Fig. 4.2.4 c) from the dose curve (Fig. 4.2.4 a) and the residual resist thickness curve (Fig. 4.2.4 b). The foot of the resist feature is located at the position x_1, at which the irradiated dose equals the dose D_1 at the foot of the residual resist thickness curve.

If one doubles the development time, then twice as much resist is removed in the vertical direction at every point. A different curve for the residual resist thickness is then obtained (Fig. 4.2.4 b, dashed curve), and the resist linewidth de-

---

9   Alkaline solutions cannot be used, however, if the resist is lying on aluminium or silicon, because aluminium and silicon would be attacked.
10 In reality, the developer always removes the resist normal to the instantaneous resist surface, the removal rate being a function of the exposure dose at the point concerned. These actual conditions are taken into account in simulation programmes [4.2] used to calculate resist profiles. The isotropic behaviour of the developer means that the resist is removed more heavily in a lateral direction than is shown in Fig. 4.2.4.

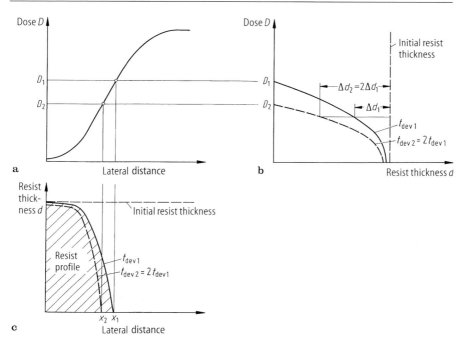

Fig. 4.2.4 a–c. Constructing the resist profile (c) from the dose curve (a) and curve (b) showing the residual resist thickness with respect to dose (cf. Fig. 4.2.3). D_1 is the dose at which the resist film is just fully removed after a development time t_{dev1}. The dashed curves in (b) and (c) are obtained if the development time is doubled. The foot of the resist profile is then shifted from x_1 to x_2.

creases (Fig. 4.2.4 c, dashed) with its foot shifted from x_1 to x_2. This means that the line width of the resist patterns has reduced in size by $2(x_1 - x_2)$. Using Fig. 4.2.4, one can deduce that the change in line width is smaller the steeper the residual resist thickness curve, i.e. the higher the resist contrast γ (cf. Fig. 4.2.3). This also produces a steeper edge to the resist pattern, so that near-vertical resist edges can be obtained even for smooth dark/bright transitions, which is particularly important for reactive ion etching. Another benefit of a high γ value, which can be inferred from Fig. 4.2.4, is that the line width of the resist patterns depends only slightly on the resist thickness for the same illumination dose.

To summarize, one can state that a high resist contrast guarantees steep resist edges and an insensitivity of the resist line width to variations in resist thickness and development time. A high γ value does not, however, compensate for variations in the exposure dose within the resist as explained in Fig. 4.2.5. These variations lead to line width fluctuations that depend on the slope of the dose curve in the resist, irrespective of the value of γ. If the exposure dose D_{max} varies by ΔD_{max} (Fig. 4.2.5 a), then the resultant reduction in resist line width is given by

$$2\Delta x = \frac{-2\left(\dfrac{\Delta D_{\max}}{D_{\max} + \Delta D_{\max}}\right)}{\left[\dfrac{1}{D}\left(\dfrac{dD}{dx}\right)\right] \text{at the resist edge}}$$

The term in the numerator is usually called the exposure contrast, and is mainly determined by the optical imaging system (see Sect. 4.2.6). The exposure contrast in the resist can be increased somewhat compared to the effective exposure contrast at the resist surface using contrast enhancing layers (see Fig. 4.2.15).

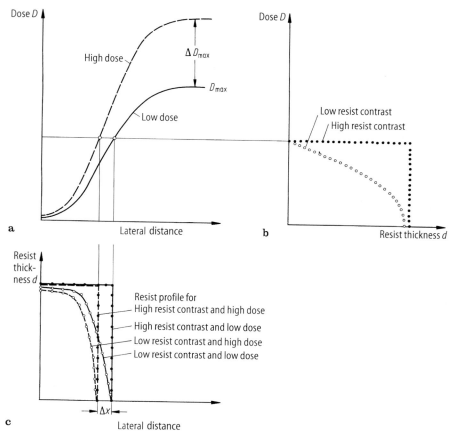

Fig. 4.2.5 a–c. Constructing the resist profile (c) from the dose curve (a) and the residual resist thickness curve (b). In (a) a high and a low maximum exposure dose are shown, and in (b) a high and a low resist contrast are assumed. The dose change $\Delta D_{\max}$ causes the foot of the resist profile to move by Δx, irrespective of the resist contrast. In this case the only effect of the higher resist contrast is to increase the steepness of the resist edges

In practice, there can actually be strong local fluctuations of the light intensity I_{max}, and thus the exposure dose D_{max} in the resist ($D_{max} = I_{max} \cdot t_{exp}$), even if the intensity of the light, and thus the dose, is uniform over the surface of the wafer. This problem is dealt with in the next section.

4.2.3
Light intensity variation in the photoresist

In the wafer steppers used today (see Sect. 4.2.5), narrow-band light from a high-pressure mercury vapour lamp or a laser is used for illumination.[11]

Pronounced interference effects therefore arise between incident and reflected light waves. Figure 4.2.6 shows the effect of wave interference on the intensity profile in the photoresist for single-wavelength exposure.

The two extreme cases are of interest, as all other possibilities lie between these two limits. At one extreme, the resist thickness d_r satisfies the condition

$$d_r = k \frac{\lambda}{2n_r} \text{, for k = 1, 2, 3,...}$$

where λ is the wavelength of the light in a vacuum and n_r is the refractive index of the photoresist, which is smaller than the refractive index n_s of the substrate. In this case (Fig. 4.2.6 a) waves *1* and *3* are in phase opposition at every point in the resist (as are waves *2* and *4*), firstly because the path difference for the waves is λ/n_r, i.e. a whole wavelength, and secondly because there is a phase shift of π at the resist/silicon boundary.

The other extreme case arises if

$$d_r = (k-1/2) \frac{\lambda}{2n_r} \text{, with k = 1, 2, 3,...}$$

In this case (Fig. 4.2.6 b) waves *1* and *3* are in phase, as are waves *2* and *4*.

In both cases, pronounced local variations in intensity occur in the resist in a direction perpendicular to the surface, because of the interference between coherent waves travelling in opposite directions. Corrugated resist sidewalls are the result of this standing wave effect. Most of the corrugation can be avoided if the resist is annealed at about 100 °C after exposure (post exposure bake). During annealing the photochemically modified molecules in the resist diffuse to a distance of about 50 nm, smoothing off the initially pronounced maxima and minima in the concentration of the photochemically converted molecules. This annealing stage has the same effect as if the resist were exposed to light of constant intensity throughout its depth at a level equal to the mean value of the localized intensities. This mean value is shown dashed in Fig. 4.2.6 a and b.

---

11 The major illumination wave lengths are 436 nm (g-line), 365 nm (i-line), 248 nm (deep UV) and 193 nm (far UV).

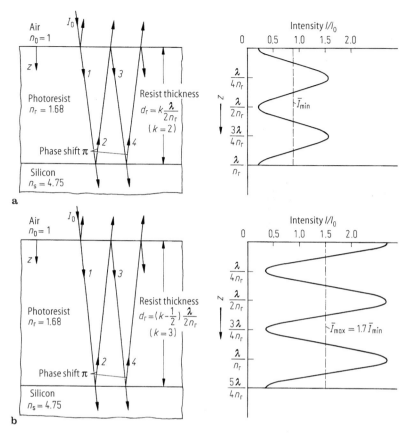

Fig. 4.2.6 a, b. Intensity profiles in a photoresist layer (refractive index $n_r = 1.68$) on silicon ($n_s = 4.75$ at $\lambda = 436$ nm) for two different resist thicknesses. **a** Assuming a resist thickness $d_r = \lambda/n_r = 260$ nm. In this case wave 3 has the opposite phase to wave 1. This results in a minimum mean intensity $\bar{I}_{min}$ through the resist thickness; **b** the resist is thicker than in **a** by $\lambda/4n_r = 65$ nm. Since waves 1 and 3 are now in phase, the mean intensity in the resist $\bar{I}_{max}$ is approximately 1.7 times larger than in **a**. It is assumed that the light intensity I_0 falling on the resist is identical in each case. For reasons of simplicity, light absorption in the resist has been ignored. It is assumed that the incident light is normal to the resist surface; it is only shown at an angle for clarity

The difference in the mean values of the intensity $\bar{I}_{min}$ and $\bar{I}_{max}$ for the two extreme cases is much more important than the standing wave effect [4.3]. The difference arises because wave 3 is opposite phase with wave 1 in one case, whilst in the other case they are in phase. This also applies to waves 2 and 4. The phenomenon is physically identical to the effect in Newton's interference fringes ("colours of thin films"). It shall therefore be referred to here as Newton's interference effect. For $\lambda = 436$ nm, the ratio $I_{max}:I_{min}$ equals 1.7 for a silicon substrate (see

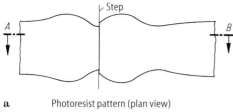

Fig. 4.2.7a. Schematic representation of the line width variations of a photoresist pattern running across a step in the surface; b the line width variations are caused by changes in the resist thickness around the step [4.3]

a   Photoresist pattern (plan view)

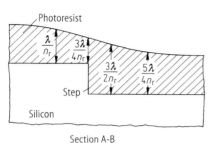

b   Section A-B

Fig. 4.2.6), and 2.8 for a substrate with 100 % reflection (see Fig. 4.2.9). The values are larger for shorter wavelengths.

As was shown in Fig. 4.2.5, a different dose or intensity in the resist, such as that produced by Newton's interference effect when generating patterns with monochromatic light, leads to different resist pattern widths. Larger or smaller resist pattern widths are obtained, depending on whether the local resist thickness produces a maximum or minimum average intensity in the resist. The difference in resist thickness between the two extreme cases equals $\lambda/4n_r$, which is about 65 nm for $\lambda = 436$ nm. If one can keep the variation in resist thickness on the substrate to less than ±15 nm around one extreme value (e.g. $5\lambda/4n_r =$ 325 nm), then line width variations are small. This can be achieved for flat substrates (chromium-coated glass plates for masks, unstructured silicon wafers) and thinner resist films (<0.5 µm) for instance.[12] Line width variations are unavoidable, however, when the resist patterns must be produced on surfaces containing steps (Fig. 4.2.7), which is generally the case in wafer processing except for the first pattern level.

There are several options for reducing Newtonian interference itself and its effect on line width variations.

As can be deduced from Fig. 4.2.5, the line width variations caused by the interference effect would be reduced for a high exposure contrast, i.e. a sharp dark/bright transition at the boundary between exposed and unexposed regions during resist exposure. Section 4.2.6 explains that in projection exposure a

---

12 Those resists with a tendency to produce surface "striations" are undesirable. The striations are slight undulations in the resist surface that can form when the resist film is spun on.

shorter wavelength, a higher numerical aperture of the imaging system and exact positioning of the resist layer in the focal plane are all particularly important factors for a steeper dark/bright transition.

The effective exposure contrast in the resist can also be increased using special resist techniques (Sect. 4.2.4).

If the resist is exposed to wide-band illumination or a source having several wavelengths, then the individual Newtonian interference effects at the different wavelengths can cancel each other out to some extent.

The Newtonian interference effect itself can only be reduced by attenuating wave 3 in Fig. 4.2.6. This can be achieved by several measures discussed below.

Suitable anti-reflective films between the substrate and the photoresist layer can attenuate the light wave reflected back from the substrate into the resist (wave 2 in Fig. 4.2.6), and thus also attenuate wave 3.[13] Sputtering a thin amorphous silicon film between a highly-reflective aluminium surface and the photoresist film, for instance, can reduce the intensity of the reflected light by more than one order of magnitude. Similar anti-reflective films can also be inserted between the resist film and other layers used in silicon technology [4.4]. Antireflective coating (ARC) can also be used to reduce exposure of supposedly unexposed areas as a result of light reflection at steps (Fig. 4.2.8), since wave 2 (Fig. 4.2.6) is also attenuated in the process.

The general drawback of non-absorbing anti-reflective films, however, is that their optical thickness must meet very tight tolerances, and the films must be matched to the substrate. Such films also tend to increase the cost and complexity of the integrated circuit manufacturing process, because an extra layer has to be applied, etched and possibly completely removed again.

The light absorption in the resist can also be increased by using a thicker resist layer (Fig. 4.2.9), by adding absorbers to the resist (dyed resist) and by exposing the resist to a wavelength that it absorbs more strongly. For instance the standard positive photoresists have stronger absorption at the mercury spectral lines of 356 nm and 405 nm than at 436 nm.

The light attenuation in the resist cannot be increased exessively, however, because otherwise that part of the resist near the surface would be exposed too strongly, whilst that part near the substrate would not be exposed enough. This would lead to long exposure times, tapered resist sidewalls and increased linewidth for thicker resists.

Newton's interference effect may be quantitatively defined by a factor a_N, which equals the ratio of the maximum to the minimum intensity coupled into the resist:

$$a_N = \left(\frac{I_{RO}(\max)}{I_{RO}(\min)} \right)_{Newton}$$

---

[13] An anti-reflective film on top of the resist surface can also attenuate wave 3. This option has not been implemented in practice, however.

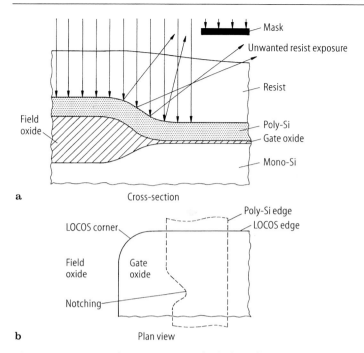

Fig. 4.2.8 a. Unwanted resist exposure by light reflection at a LOCOS edge. **b** The effect is particularly pronounced at a LOCOS corner, for instance, which acts like a focussing device, causing notching in the poly-Si structure

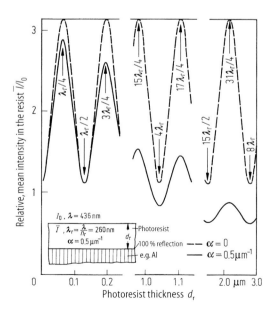

Fig. 4.2.9. Intensity in the photoresist (averaged over the resist thickness) for a 100% reflective substrate, as a function of the photoresist thickness. The dashed curve represents negligible light absorption in the resist. In this case the intensity varies by a factor of 2.8. The unbroken curve is for an absorption coefficient $\alpha = 0.5\ \mu m^{-1}$ (cf. Fig. 4.2.2). The exposure wavelength is taken to be $\lambda = 436$ nm. One can see how a thick resist reduces interference thanks to its strong attenuating effect. An intensity variation of just 40 % is obtained for a resist thickness of 2 μm

In the worst case (100 % reflecting substrate and non-absorbing resist, see Fig. 4.2.9), a_N = 2.8 for λ = 436 nm. For the case of "non-absorbing photoresist on silicon" (see Fig. 4.2.6), then a_N = 1.7. Lower a_N values are obtained with strongly absorbing resists (dyed resists) or with thicker photoresist layers (cf. Fig. 4.2.9). a_N values of 1.2–1.0 can be obtained using anti-reflection layers. The ideal case is a top-surface imaging resist system (see next section) with a 100 % absorbent bottom resist. Here a_N = 1, irrespective of the layer structure and the topography beneath the resist layer. This technique will be looked at more closely below, together with other special resist techniques.

4.2.4
Special photoresist techniques

Sections 4.2.1–4.2.3 have identified that an ideal photoresist would meet the following requirements:
- high resist contrast (γ)
- high sensitivity (short exposure time)
- low light absorption
- exhibit no Newtonian interference effect
- stability of resist patterns during reactive ion etching and high-dose ion implantation
- easily removable resist mask

Attempts to find a solution approaching such an ideal photoresist system have focused on a multilevel resist system, or on a single layer resist allowing top surface imaging.

It must be stressed, however, that all these resist techniques have one main drawback. The larger number of individual processing steps means they are more expensive, and may have a higher defect density than the single resist technique described in Sect. 4.2.1.

A top-surface imaging resist technique called the trilevel resist technique shall be considered first. Figure 4.2.10 shows one of the many versions that have been proposed. The three layers in the trilevel resist system are made up of a so-called bottom resist layer, an intermediate spin-on glass film (see Sect. 3.12.1) and a so-called top resist layer. In all trilevel resist techniques, the top resist acts as the photochemically active layer. The bottom layer can be e.g. a positive resist or its resin, which can be made highly absorbent by adding an absorber, or by high-temperature baking (approx. 200 °C). This type of bottom resist layer not only ensures that practically no light is reflected from the substrate into the top resist, but is also made thick enough (e.g. 1.5 µm) to smooth off the steps that may exist on silicon wafers. This means that the top resist layer can be spun on at a uniform thickness irrespective of the steps on the surface. Last but not least, the bottom resist layer also acts as the actual masking film when etching the pattern into the underlying layer (e.g. SiO_2), which means that the top resist does

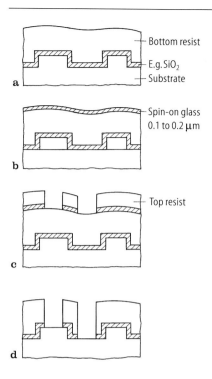

Fig. 4.2.10 a–d. Cross-sectional diagram of the processing steps in a trilevel resist technique. **a** Spin-on of bottom resist layer and annealing at approx. 200 °C; **b** application of spin-on glass layer and annealing at approx. 200 °C; **c** producing a resist mask (top resist) and etching of spin-on glass layer; d reactive ion etching of bottom resist (and top resist mask) in O$_2$ plasma, and reactive ion etching of the SiO$_2$ layer (and the spin-on glass mask)

not have to perform this role. The top resist must simply be sufficiently resistant to the etching of the thin spin-on glass film. The latter acts as an etching mask during anisotropic reactive ion etching of the bottom resist in the oxygen plasma (see Sect. 5.3.7).

If it is possible to make the top resist mask resistant to the bottom resist etching, then there is no need for the spin-on glass film. This less costly technique is called a bilevel resist technique and is illustrated in Fig. 4.2.11 [4.5]. In this case the etch resistance is achieved by silylation of the top resist pattern, i.e. by introducing Si into the top resist. When silylation is performed in a Si-containing liquid (e.g. hexamethyldisilazane HMDS), the top resist patterns swell[14]. This expansion of the top resist patterns is actually welcome, because it can compensate for reduced dimensions caused by necessary over-exposure (see Sect. 4.2.6) and etching processes.

Another example of a bilevel resist technique is shown in Fig. 4.2.12. Here the top resist is an inorganic resist system applied by vapour deposition or sputtering. Since these coating techniques cannot provide adequate smoothing of steep steps, the planarizing bottom resist layer is needed. Figure 4.2.12 demonstrates

---

14 In another proposed option [4.7], the Si introduced is initially into the top resist.

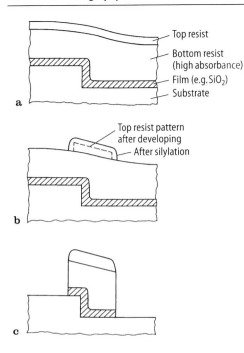

Fig. 4.2.11 a–c. Key processing steps in a bilevel resist technique, in which the top resist patterns are made etch resistant by silylation. **a** Applying the bottom resist and top resist. The bottom resist is baked at about 200 °C to prevent the top resist dissolving into it. **b** Exposure, developing and silylation (introduction of Si) of the top resist. **c** Reactive ion etching of the bottom resist (in O_2 plasma) and of the SiO_2 layer

the mechanism in the most well-known inorganic resist, which is the $Ag_2S/GeSe$ system [4.6]. This resist system has several interesting features: all processing stages can be performed dry, including developing, high values of resist contrast have been obtained, and it appears to provide somewhat enhanced exposure contrast. This resist technique has not really gained acceptance yet, however, mainly because of the risk that the silicon will become contaminated by the silver (see Sect. 6.4).

Figure 4.2.13 shows a single-level resist technique [4.8]. This gives more or less the same result as the bilevel technique in Fig. 4.2.11, with the resist portion near the surface assuming the role of the top resist. This resist also has the benefit of resist silylation being performed dry. Its disadvantage compared to the bilevel technique in Fig. 4.2.11, is that there are fewer degrees of freedom in the conditioning of the bottom and top resist.

The idea of starting with a highly absorptive resist so that the latent image is only created near the surface, can be applied generally to negative resists. The faster removal rate in the developer of unexposed regions of a negative resist compared with exposed areas, means that the whole resist thickness can be removed right at the development stage without having to use reactive ion etching. This is not the case with highly absorptive positive resists.

Figure 4.2.14 shows the processing sequence and the chemical reactions taking place in a negative resist technique that starts from a positive resist (image

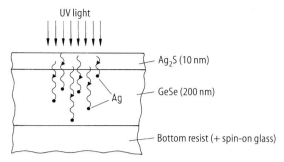

Fig. 4.2.12. Mechanism in the inorganic resist system Ag₂S/GeSe. Under UV illumination, the silver migrates out of the Ag₂S into the underlying GeSe layer. The system is developed in a CF₄/O₂ plasma, in which the GeSe which has not been doped by silver has a high removal rate, whilst the silver-doped parts of the GeSe layer are retained. The Ag₂S/GeSe resist system therefore acts as a negative resist

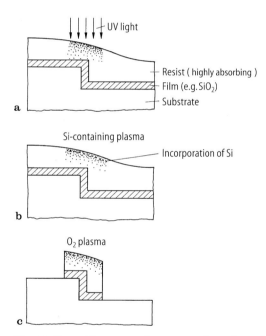

Fig. 4.2.13 a–c. Key processing stages of a resist technique in which silicon is incorporated in the exposed areas, so that these regions are etch resistant to reactive ion etching in an oxygen plasma (negative resist). The density of the dots in the exposed areas indicates the concentration of photochemically converted photoactive molecules. **a** Near-surface exposure of the highly absorbing resist; **b** silylation of the exposed near-surface regions in a plasma containing silicon; **c** reactive ion etching of the resist in an O₂ plasma (= developing of resist) and etching of SiO₂ layer

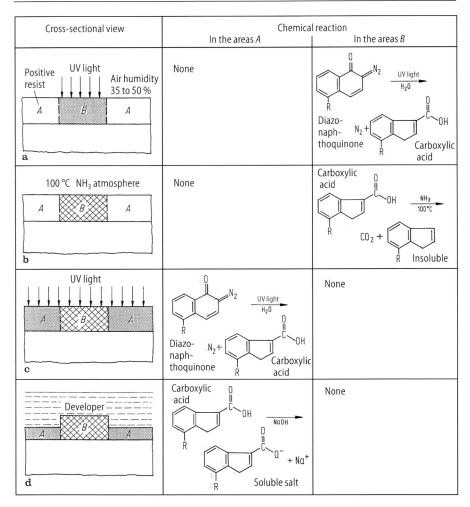

Fig. 4.2.14 a–d. Chemical reactions taking place during processing in the image reversal technique, where a reverse image can be obtained using a standard positive resist. The key processing step causing image reversal is the annealing in ammonia gas. This destroys the carboxylic acid in the exposed areas, so that these areas remain insoluble in the alkaline developer. **a** Selective exposure; **b** annealing in ammonia gas; **c** blanket exposure; **d** developing in the alkaline developer

reversal) [4.9]. It is useful to compare the chemical reactions with those in a standard photoresist (Fig. 4.2.1). Annealing in ammonia gas[15] is the essential

---

15 There are also resists which contain an alkaline component, so that annealing can be performed without an ammonia gas atmosphere.

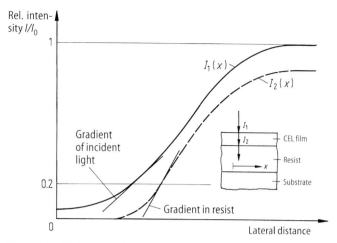

Fig. 4.2.15. Enhancing the exposure contrast (intensity gradient) by the CEL technique (Contrast Enhancing Layer). The gradients of the intensity curves at $I/I_0 = 0.2$ are important because the edge of the resist pattern is located at about 20% of the maximum exposure intensity I_0 (see Sect. 4.2.6). The gradient of the intensity at this point is critical to the dimensional accuracy of the resist patterns

image reversal step: it converts the carboxylic acid produced in exposed areas into a compound that is insoluble in the alkaline developer. In the subsequent blanket exposure, the diazonaphthoquinone in the previously unexposed areas is converted into carboxylic acid, whilst in the areas that were exposed earlier there is no diazonaphthoquinone left to convert. Thus the previously unexposed areas are removed in the alkaline developer, and the previously exposed areas are retained.

The image reversal technique has the advantage that the resist patterns increase in width with overexposure, unlike positive resist patterns which become narrower. A certain amount of overexposure is often deliberate in order to obtain the smallest possible variation in line width (see Sect. 4.2.6). Thus widening of the resist patterns by overexposure is usually an advantage, because this can compensate for the pattern shrinkage which usually occurs in the processing stages following lithography (etching, LOCOS process).

The resist techniques described so far (except for the inorganic resist process) are unable to increase the exposure contrast in the resist. The CEL Technique (CEL = Contrast Enhancing Layer) aims at increasing the contrast, however [4.10]. The principle behind the CEL technique is the introduction of a photosensitive film (CEL film) containing a very high concentration of photoactive molecules. This film is applied to the actual resist layer before exposure.[16)] If

---

16 The 0.5 μm thick CEL film is applied using the spin-on technique. It is removed after exposure.

such a double layer is exposed, a small amount of light initially penetrates the actual resist layer, because the high concentration of photoactive molecules in the CEL film makes it highly light absorptive (cf. Fig. 4.2.2).[17] As the exposure dose increases, however, the concentration drops and light absorption decreases as in a normal photoresist. The exposure contrast is enhanced, i.e. the intensity gradient increases at the transition between a dark and bright region, because the areas receiving a higher dose start to transmit more light than the areas receiving a lower dose (Fig. 4.2.15).

One disadvantage of the CEL technique is that longer exposure times are necessary, since it takes time to bleach the CEL film and to overcome the residual absorption in the CEL film.

4.2.5
Optical exposure techniques

The role of exposure in lithography is to project the desired light pattern onto the surface of a substrate coated with a film of photoresist. The resolution of the imaging system (this is equivalent to the intensity gradient at the pattern edges) is one of the four key criteria used to assess a lithographic system in practice. (cf. Sect. 4.1). The other three criteria are the achievable overlay accuracy for superimposed patterns, the achievable defect density and the imaging costs. The last item depends, amongst other things, on the equipment purchase price and the number of silicon wafers that can be exposed per hour.

Figure 4.2.16 shows the procedure used today, for converting the pattern data stored on magnetic tape (e.g. dimensions and centre points of rectangles) into a correctly placed light pattern on the resist surface.

In the older optical pattern generators, UV light is used to project the aperture of a rectangular shutter through a projection lens onto a glass plate coated with a chromium film plus a layer of photoresist. The aperture dimensions are varied in the x and y direction according to the magnetic tape data, so that differently sized rectangular images can be produced subsequently. The required position of the rectangle is then obtained by moving the table carrying the glass plate, again controlled by the magnetic tape data. Since a wafer generally contains several hundred chips, direct wafer exposure would take a very long time, because the mechanical movements of the aperture and the table are relatively time-consuming. For this reason the patterns of only a few chips are generated in 10×, 5× or 4× magnification[18]. These patterns are then repeated on the wafer using another piece of equipment called a wafer stepper (see Fig. 4.2.16).

---

17 The absorption coefficient equals 0.6 μm^{-1} for a normal resist, whilst in CEL films it varies between 1.5 and 5 μm^{-1}.
18 Once the exposure procedure in the pattern generator is complete, the resist is developed and the chromium film not covered in resist etched away to produce a chromium mask (reticle).

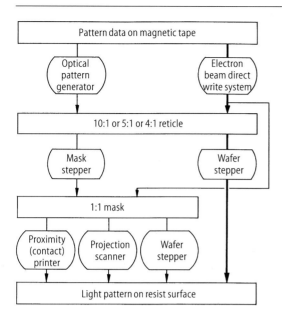

Fig. 4.2.16. The major pieces of equipment used today to convert the pattern data stored on magnetic tape into the accurately placed light features required on the resist surface. The thick arrow highlights the established procedure for sub-micron lithography. All pieces of equipment shown are optical devices, except for the electron beam direct write system (top right-hand side)

For pattern generation electron beam direct writing has become the established technique [4.11]. The system is described in more detail in Sect. 4.4.2. More recently, a rapid optical pattern generator has been developed, which operates on the principle of an electro-optically controlled laser beam.

As can be seen in Fig. 4.2.16, steppers can be used at various stages in the process. All steppers work on the common principle that the required repeating of a pattern, is achieved by moving a precisely controllable table carrying the substrate into all positions where the pattern is required. High-resolution lenses are used for the repeated pattern transfer in the optical steppers shown in Fig. 4.2.16. The full resolving power of the lens is only obtained when the substrate surface is located within its depth of focus. The stepper table is therefore moved automatically before each exposure in a direction perpendicular to the table surface, and also tilted if any unevenness in the substrate surface needs to be compensated for (levelling). Each of the lateral table positions is set up with a positioning error of < 0.1 µm using laser interferometers. Wafer steppers, unlike mask steppers, need an alignment device in order to position the reticle pattern exactly with respect to the patterns already on the wafers. Alignment is either performed before each exposure (site-by-site alignment) or only once per wafer before the first exposure. Modern wafer steppers can achieve an alignment accuracy of better than ±50 nm.

Figure 4.2.17 shows the design of a wafer stepper [4.12]. The wafer stepper technique is the most expensive one among the wafer exposure techniques, but it offers the best performance in terms of resolution, placement accuracy and defect density. They are therefore widely used for pattern dimensions below 2 µm.

The number of chips that are placed on the reticle of a wafer stepper (block reticle) corresponds to the maximum of chips within the image field of the projection lens (typically 30 mm ∅), in order to shorten the duration of wafer exposure. This enables modern wafer steppers to achieve a throughput of more than 50 wafers (200 mm ∅) per hour.

Improvements in the projection lenses to produce finer patterns over ever larger image fields involve a drastic increase in technical complexity. A new type of wafer stepper, called a scanning wafer stepper, is therefore used for pattern sizes below 0.3 µm. As Fig. 4.2.18 illustrates, the scanning wafer stepper requires a much smaller projected image field than the standard wafer steppers. This is possible because the reticle and wafer are moved (scanned) beneath an exposure slit (e.g. 30 mm×5 mm) during illumination, so that the whole reticle field is ultimately projected onto the wafer. This exposure technique relaxes the demands placed on the projection lens, but has to be paid for by a complicated scanning mechanism, which must produce precise simultaneous movement of reticle and wafer at different speeds (reticle 4 times faster for 4 : 1 projection). The step-and-scan principle also achieves better focusing of the wafer surface, because re-focusing is performed continuously during the scan ("on the fly").

Early exposure techniques used a full-wafer 1:1 pattern transfer, e.g. by proximity printing or 1:1 scanning projection (cf. Fig. 4.2.16).

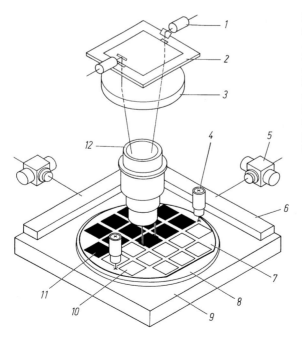

Fig. 4.2.17. Diagram of an optical wafer stepper [4.12].
1 alignment optics; *2* reticle; *3* lens for telecentric imaging; *4* pre-alignment optics; *5* laser interferometer; *6* reflecting surface; *7* wafer; *8* wafer chuck; *9* x-y table; *10* exposed image site; *11* image site that has not yet been exposed; *12* lens system

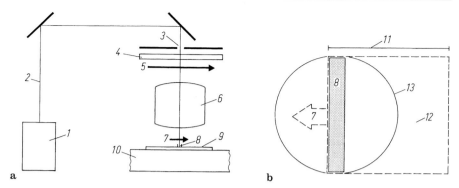

Fig. 4.2.18 a. Diagram of a scanning wafer stepper [4.13]; **b** Illumination field produced on the wafer. *1* light source; *2* optical path; *3* illumination slit at the reticle; *4* 4:1 reticle; *5* reticle scanning; *6* projection lens; *7* wafer scanning; *8* illumination slit at the wafer; *9* wafer; *10* x-y table; *11* wafer scanning length; *12* wafer area exposed after scanning; *13* image field of projection lens

In proximity printing, masks and wafers are held apart at a distance of 5–40 µm, whereas in contact printing the mask and wafer are pressed against each other by applying a certain degree of pressure. Proximity printers are not used for pattern sizes below 3 µm because of resolution problems (see Sect. 4.2.6). With contact printing, however, the risk of mechanical damage to the resist film and mask is so severe that it is only used where a large defect density can be tolerated.

1:1 scanning projection systems operate similar to the scanning steppers (cf. Fig. 4.2.18), with the only difference that the length of the exposure slit is the same size as the wafer diameter. If one then moves the slit at right angles to its length over the mask and wafer (scanning), then a sharp image of the pattern is produced over the whole wafer. The most commonly used versions of this type of equipment do not actually move the arc-shaped illumination region (ie the projection optics are not moved). Instead, the mask and wafer are rigidly linked to each other and moved jointly normal to the optical axis.

4.2.6
Resolution capability of optical exposure techniques

The resolving power of an imaging system is physically defined as the shortest distance between two points or two lines for which an intensity difference can still just be detected at the central position between the points or lines. This definition is not useful however, when applied to resist pattern formation. The intensity curve at the transition between an exposed and an unexposed region (dark/bright transition) is of more relevance (see Sect. 4.2.2).

In proximity exposure this intensity curve depends on the light diffraction at a mask pattern edge (Fresnel diffraction). Figure 4.2.19 shows the intensity curves for three different mask and wafer separations s. The intensity gradient at the point where the pattern edge should be (x = 0), is given by

$$d\left(\frac{I}{I_0}\right)/dx \approx \frac{0.7}{\sqrt{\lambda s}}.$$

This expression for the intensity gradient assumes that the curve is linear. Thus the width Δb of the dark/bright transition region equals the reciprocal of this expression, or

$$\Delta b \approx 1.5\sqrt{\lambda s}.$$

Both these equations apply in principle to an isolated dark/bright transition, i.e. dark and bright regions extending to infinity. In reality they are also a good approximation for the patterns found in practical applications. However, if the width of the dark and/or bright region is smaller than the width Δb of the transition region, the intensity gradient at x = 0 decreases. This limiting case gives an approximate figure for the practical resolution limit for pattern formation. The minimum pattern width that can be reproduced is thus

$$b_{min} \approx 1.5\sqrt{\lambda s}.$$

For proximity printing at a wavelength of λ = 400 nm and a distance s between the mask and wafer of 10 μm, b_{min} = 3 μm. This is about the limit of proximity printing in practice.

For contact printing a value of about half the resist thickness is used for s. Assuming a resist thickness of 1 μm, then for λ = 400 nm one obtains b_{min} = 0.7 μm. Despite this excellent resolution, the contact printing technique is hardly used for VLSI circuits because of its inherent high defect density.

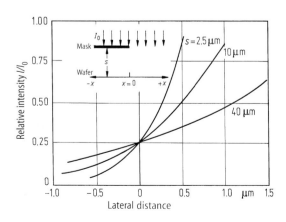

Fig. 4.2.19. Simulated intensity curve at a pattern edge in proximity exposure (wavelength λ = 400 nm) for various mask-wafer separations s

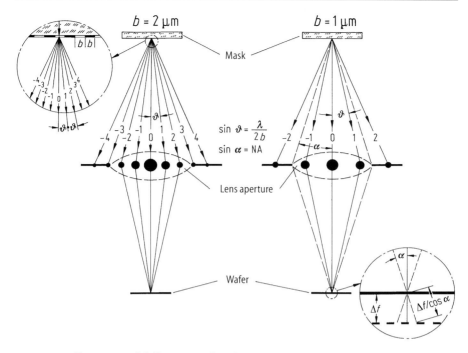

Fig. 4.2.20. Illustration of diffraction orders (0, 1, −1, 2, −2, ...) in the lens aperture and defocusing in diffraction limited projection exposure. The area of the filled circles in the lens aperture indicates the intensity of each diffraction order. Δf is the defocusing distance. The path difference between the waves incident at angle α and the central wave is $\Delta f/\cos\alpha - \Delta f$

In both proximity and contact printing, the resolving power is reduced if the wafer surface and the mask surface reflect incident light back onto the resist. Chromium masks with a low reflection coefficient ("black chromium") are a useful option here.

The resolving power of projection systems is analysed using Fig. 4.2.20, making the basic assumption that there are no lens errors or other degrading effects. The mask is assumed to contain lines and spaces at a pitch of $2b = 4$ μm (2 μm wide chromium lines each separated by 2 μm) and $2b = 2$ μm respectively. In projection systems the light source is imaged in the lens aperture by an optical illumination system. The light is diffracted at the line patterns on the mask. Thus not only does the actual image of the light source appear in the lens aperture plane (zero order diffraction), but also the higher order diffraction images. The angle ϑ between each diffraction order is given by

$$\vartheta \approx \sin\vartheta = \frac{\lambda}{2b},$$

assuming ϑ is small.

Those diffraction orders which fall within the lens aperture are those for which:

$$|k\vartheta| \leq \alpha \quad (k = 0, \pm 1, \pm 2, \ldots)$$

where k is the diffraction order and α is half the aperture angle as seen from the mask. α or sin α is called the numerical aperture (NA) of the lens.[19]

$$\alpha \approx \sin\alpha = NA$$

The lens projects an image of the mask pattern onto the surface of the silicon wafer where the resist film is located. The more diffraction orders that fall within the lens aperture, the more accurate the image produced, since each diffraction order carries part of the image information (although higher orders contain successively less information). Figure 4.2.20 shows an example for an actual set of conditions where the light wavelength λ = 436 nm and the numerical aperture NA = 0.28. In this case the diffraction orders 0, 1, −1, 2 and −2 fall within the lens aperture for a 4 µm line grid, whilst only orders 0, 1 and −1 are included for the 2 µm grid. The limit of resolution is obviously reached when the two first diffraction orders just disappear from the lens aperture. This occurs when $\vartheta = \alpha$. With $\vartheta \approx \lambda/2b$ and $\alpha \approx$ NA, the minimum pattern width b_{min} is given by

$$b_{min} \approx \frac{\lambda}{2(NA)}.$$

If one starts with a point light source, then the diffraction images in the lens aperture are also point images. This means that the first two diffraction orders would pass rather abruptly out of the lens aperture if the pattern size on the mask approaches b_{min}. Such jumps are not desirable in practice. Standard projection imaging equipment therefore works with an extended light source[20], so that the zero diffraction order fills e.g. 50 % of the lens aperture diameter. One then refers to a fill factor σ = 0.5.

If extended diffraction images (high σ values) are used, part of the first diffraction order already lies outside the lens aperture before the limit of resolution is reached, and so cannot be used to reproduce the image. This means that the intensity gradient is slightly reduced at the dark/bright transitions of the image patterns.

Figure 4.2.21 shows the curve $I(x)$ of the light intensity incident on the silicon wafer at a dark/bright transition for a fill factor σ = 0.5 [4.15]. The required position of the resist edge is where $I \approx 0.25\ I_0$. The intensity gradient at this point is given by

$$\frac{dI}{dx} \approx \frac{2(NA)}{\lambda} I_0 \quad \text{for } \sigma = 0.5$$

---

19 Where a reduced image is produced, the numerical aperture of the lens is given by half the aperture angle seen from the silicon wafer.
20 Another reason for using an extended light source is the higher exposure intensity that can be obtained.

For a high-contrast resist (see Sect. 4.2.2), a change in intensity ΔI_0 causes the resist edge to shift (see also Sect. 4.2.2) by an amount Δx:

$$\Delta x \approx \frac{\lambda}{2(\text{NA})} \cdot \frac{\Delta I_0}{4(I_0 + \Delta I_0)} \quad \text{for } \sigma = 0.5$$

The minimum pattern dimension $b_{\min}$ that can be reliably transferred is that dimension for which adjacent dark/bright transitions just begin to overlap (Fig. 4.2.22) [4.16]:

$$b_{\min} \approx \frac{\lambda}{2(\text{NA})} \quad \text{for } \sigma = 0.5$$

If the resist layer does not lie precisely in the focal plane of the lens, then the resolution will be worse and the intensity gradient at the dark/bright boundaries decreases [4.16]. The bottom insert in Fig. 4.2.20 shows the situation for an image plane located at a distance Δf from the focal plane. The path difference Δl between the central beam and the beams incident at angle α is given by

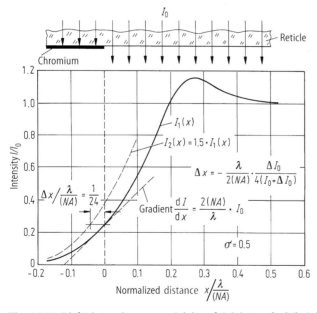

Fig. 4.2.21. Light intensity curves $I_1(x)$ and $I_2(x)$ at a dark/bright boundary (pattern edge) for the two incident intensities I_0 and $1.5 I_0$. λ is the exposure wavelength, NA the numerical aperture of the projection lens and σ the fill factor. Increasing the incident intensity by 50 % shifts the resist edge by

$$-\Delta x = \frac{1}{24} \cdot \frac{\lambda}{(\text{NA})}$$

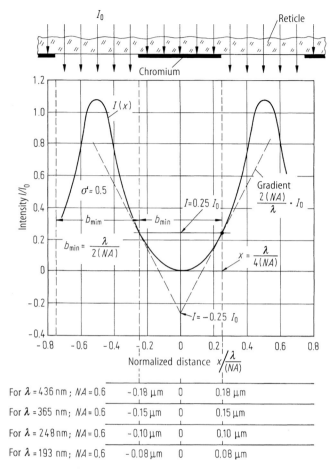

Fig. 4.2.22. Light intensity $I(x)$ when the "minimum" line grid is being projected, with line and space width both equal to b_{min}. At b_{min} the intensity gradient at the pattern edge is still just equal to that at a single dark/bright transition (see Fig. 4.2.21). The lower part of the diagram shows the values of $\pm b_{min}/2$ for 4 important exposure wavelengths

$$\Delta l = \frac{\Delta f}{\cos\alpha} - \Delta f \approx \frac{1}{2}\Delta f \alpha^2 \quad \text{for small } \alpha.$$

The image starts to degrade when the path difference Δl approaches $\lambda/4$ (the Rayleigh criterion). Inserting this value in the equation, a depth of focus $\pm\Delta f_R$ (Rayleigh depth) can be defined within which the image still appears "sharp":

$$\Delta f_R = \pm\frac{\lambda}{2\alpha^2} \approx \pm\frac{\lambda}{2(\text{NA})^2} \quad \text{for } \Delta l = \frac{\lambda}{4}$$

Table 4.1. Minimum feature size b_{min} and allowable intensity variation $\Delta I_0/I_0$ (exposure latitude) for the allowable defocus distance $2\Delta f_R$ (focus latitude) for the 4 major exposure wavelengths $\lambda = 436$ nm, 365 nm, 248 nm and 193 nm. The numerical aperture is taken to be NA = 0.6. A fill factor of $\sigma = 0.5$ has been assumed

Exposure wavelength λ	Focus latitude $2\Delta f_R$	Minimum feature size b_{min}	Exposure latitude $\Delta I_0/I_0$ for a line width variation $< 0.2\, b_{min}$
436 nm (g-line)	1.2 μm	0.51 μm	37 %
365 nm (i-line)	1.0 μm	0.43 μm	37 %
248 nm (deep UV)	0.7 μm	0.29 μm	37 %
193 nm (far UV)	0.55 μm	0.22 μm	37 %

The following relationships are obtained for a defocusing distance $\Delta f_R = \pm\lambda/2(\text{NA})^2$ [4.16]:

Intensity gradient at the pattern edges:

$$\frac{dI}{dx} \approx 0.7 \cdot \frac{2(\text{NA})}{\lambda} \cdot I_0 \text{ for } \Delta f = \Delta f_R, \sigma = 0.5$$

Minimum feature size:

$$b_{min} \approx 1.4 \cdot \frac{\lambda}{2(\text{NA})} \text{ for } \Delta f = \Delta f_R, \sigma = 0.5$$

Change in line width for an intensity change ΔI_0:

$$2\Delta x \approx \left[-0.1 - \frac{\Delta I_0}{2(I_0 + \Delta I_0)}\right] \cdot b_{min} \text{ for } \Delta f = \Delta f_R, \sigma = 0.5$$

Table 4.1 shows these parameters for 4 important exposure wavelengths at numerical aperture NA = 0.6.

Table 4.1 does not specify the line width variation. Instead it specifies the exposure latitude, which is the variation in intensity that is allowed if the line width variation is not to exceed 20 % of the minimum line width for a permitted defocusing range of $2\Delta f_R$ (focus latitude). Using the equation above, this intensity variation $\Delta I_0/I_0$ equals 37 %.

The main contributor to the local intensity variation in the resist ΔI_{R0} is Newtonian interference. In Sect. 4.2.3, the intensity variation $\Delta I_{R0}/I_{R0}$ is given as 70 % for the example "non-absorptive photoresist on poly-Si" (see Fig. 4.2.6). Suitable layers inserted between the poly-Si and resist can act as anti-reflection layers (SiO_2, Si_3N_4, amorphous Si, TiN), and thus reduce the intensity variations to within the range 10–40 %. Values in the 10% region can also be obtained using a TSI resist technique (TSI = *Top Surface Imaging*, see Sect. 4.2.4), with a highly absorptive bottom layer.

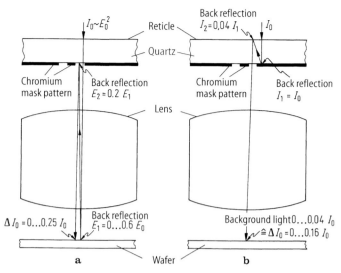

Fig. 4.2.23 a, b. In projection exposure, reflections from the reticle interfaces have an additive effect on the light intensity reaching the resist. **a** The light pattern projected onto the wafer is reflected back to the reticle and then re-reflected back from here onto the wafer. With the reticle and wafer being in focus, the back-projected light is coherent with the direct incident light. The field strengths E, not the intensities ($\sim E^2$), are therefore superimposed; **b** in this case the pattern from the chromium mask is projected out-of-focus onto the wafer because of the longer optical path length. The background light of 0.04 I_0 max. at a pattern edge is equivalent to a value of ΔI_0 of 0.16 I_0 max. (cf. Fig. 4.2.21). The back reflections from the chromium mask are ignored here because the chromium is assumed to be non-reflecting ("black chromium")

Another source of local variations in the exposure intensity in the resist are reflections at steps, which can lead to localized notching of the resist pattern at the worst affected points (see Fig. 4.2.8).

Yet another cause for intensity variations is the light reflected from the reticle. Figure 4.2.23 explains how these reflections are superimposed to produce an increased intensity at the resist ΔI_0 of 0-0.25 I_0 depending on the wafer reflectance (Fig. 4.2.23 a) and of 0–0.16 I_0 depending on the area percentage of chromium on the reticle (Fig. 4.2.23 b). The maximum background light intensity in Fig. 4.2.23 b of 0.04 I_0, equivalent to a ΔI_0 of 0.16 I_0 (cf. Fig. 4.2.21) applies to a bright line in a dark environment (nested line). Thus in total one can assume values for $\Delta I_0/I_0$ of between 0 and 40 % arising from reticle reflections.

The continuus trend to shrink the pattern dimensions drives the introduction of shorter exposure wavelength. One can see from Table 4.1, for instance, that 0.35 µm technology is feasible with DUV lithography, but that i-line lithography would no longer be practical.

For economic reasons, however, one would prefer to continue using existing lithography equipment as long as possible before moving to a shorter exposure

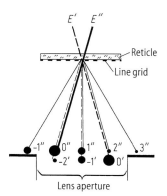

Fig. 4.2.24. Demonstration of the higher resolution achieved with off-axis exposure than with Köhler illumination (cf. Fig. 4.2.20). Two zero-order diffraction patterns are produced at the periphery of the lens aperture when a periodic reticle pattern is illuminated by the two coherent waves E' and E''. The key to higher resolution is that the second-order diffraction components also fall within the lens aperture ($-2'$ and $2''$). The areas of the circles in the diagram indicate the intensity of each diffraction order

wavelength. Furthermore, it would be an advantage to have larger values for the focus latitude than those specified in Table 4.1.

Two techniques are being used increasingly in order to achieve a smaller minimum pattern size and a larger focus latitude for the same wavelength. These are off-axis illumination (OAI) and the use of phase shifting masks (PSM).

Figure 4.2.24 considers the effect of off-axis illumination in projection printing [4.16, 4.18]. Unlike Köhler illumination, where the light source is imaged in the centre of the lens aperture (see Fig. 4.2.20), in off-axis illumination the light source is projected as a ring (annular illumination) using a fly's eye lens, or at 4 locations at the periphery of the lens aperture (quadrupole illumination). This technique enables higher diffraction orders to be included in the lens aperture, which increases the resolution.

The increased focal depth for off-axis illumination can be explained in terms of the path difference, which is now shorter for all the waves arriving at the resist surface than it is in Fig. 4.2.20. For example, first order light waves are located in the centre of the lens aperture for off-axis illumination. For Köhler illumination, however, they lie at the edge of the lens aperture, and thus have a large path difference compared with waves in the centre of the lens aperture.

The effect of off-axis illumination can be further enhanced by differently attenuating the intensities of the multiple light source images within the aperture. For example, off-axis illumination with 3 concentric annular light sources of different intensities produces a higher resolution than a single annular illumination [4.18].

A phase-shifting mask (PSM) differs from a standard chromium mask (COG = Chrome On Glass) by having two types of transparent regions. The optical path length of the light waves in the two regions differs by $\lambda/2$, which is equivalent to a phase shift of 180°. If n_{PS} is the refractive index of the phase shifting layer, then the thickness d_{PS} of the phase shifting layer is given by:

$$d_{PS} = \frac{\lambda}{2(n_{PS}-1)} .$$

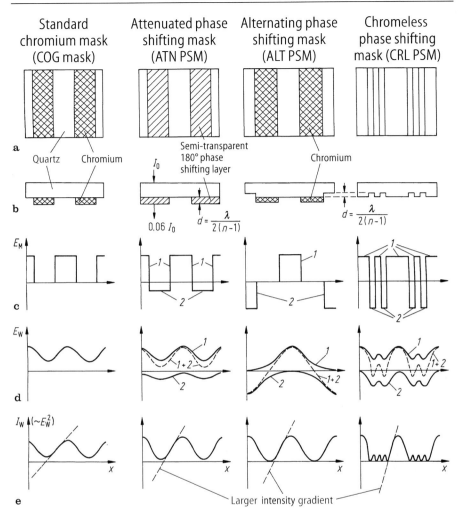

Fig. 4.2.25 a–e. Illustrating the contrast-enhancing effect of 3 major types of phase shifting masks in comparison with a standard chromium mask. **a** Plan view of masks; **b** cross-section through mask; **c** amplitude E_M of the electric field strength of the light wave directly behind the mask; **d** field strength amplitude E_W at the wafer surface; **e** light intensity $I_W \sim E^2_W$ at the wafer surface. The increased intensity gradient at the pattern edges is produced by the partial cancellation of the light diffracted into the dark areas

Figure 4.2.25 shows three different versions of a PSM and how they work. In all cases the result is a steeper intensity gradient at the pattern edges compared with the standard chromium mask. As was explained in an earlier section, an intensity gradient increased by a factor k means that the minimum achievable feature size is then reduced by the same factor k.

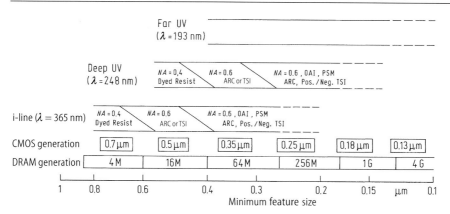

Fig. 4.2.26. Lithography scenario for the sub-micron region. The pattern size range for the CMOS generation (cf. Fig. 8.3.3) and the DRAM generation of chips (cf. Table 8.9) are given for reference. For a given exposure wavelength λ, finer patterns can be obtained by using a larger numerical aperture NA, by off-axis illumination (OAI), phase shifting masks (PSM) and by reflection-inhibiting measures (ARC, TSI)

An attenuated PSM (Fig. 4.2.25, 2nd column), also called a halftone PSM, contains semi-transparent (approx. 6 %) 180° phase shifting regions instead of the opaque chromium regions in a standard chromium mask (COG Mask). In an alternating PSM (Fig. 4.2.25, 3rd column), also known as a Levenson-type PSM, adjacent transparent regions have phases that differ by 180° [4.22]. With this type of phase shifting mask one is however faced with phase conflicts at special pattern configurations. The phase conflicts may be overcome by introducing 90° phase shifting regions. The chromium-free phase shifting masks consist of fully-transparent regions that have a 180° phase difference compared with the surrounding area (Fig. 4.2.25, 4th column). This type of phase mask exploits the fact that a 180° phase shift on the reticle appears as a narrow dark line having an intensity gradient twice as large as at a chromium edge [4.20]. Larger dark regions may be realized using a phase grid with a pitch smaller than 0.2λ/NA. In rim-type phase shifting masks, which are not shown in Fig. 4.2.25, a rim producing a 180° phase shift runs around the edge of the chromium features [4.21]. The disadvantage with this type of PSM is the difficulty of mask inspection and repair.

Based on current experience of off-axis illumination (OAI) and phase shifting masks (PSM), one can conclude that up to twice the pattern resolution and twice the focus latitude can be achieved if these techniques are applied appropriately (cf. Table 4.1). Figure 4.2.26 shows the lithography scenario for the sub-micron region. Further extension of optical lithography below 0.1μm may be feasible using the 157 nm wavelength. Lenses for 157 nm will be made of calcium fluoride CaF_2. Recent developments of lens makers indicate the realization of NA values up to 0.8. This would in fact extend optical lithography down to near 50nm.

4.2.7
Alignment accuracy of optical exposure equipment

In order to fully exploit the packing density possible for a minimum feature size on an integrated circuit, the relative overlay error (see Fig. 4.1.2) of two superimposed patterns (e.g. track over contact hole) must be only about one third of this minimum dimension.

There are three steps to be performed in order to align a mask pattern precisely over an existing pattern on the wafer. First, suitable alignment marks on the wafer have to be detected (by eye using a microscope, or automatically). Then the amount and direction of the relative displacement of these marks with respect to the location of relevant marks on the mask must be registered. Finally, the mask and wafer must be moved relative to each other in order to cancel out the relative placement errors.

Figure 4.2.27 shows the essential principles of alignment mark detection on a resist-coated silicon wafer. None of the methods are straightforward, because the contrast of the alignment marks depends on the surface condition of the wafers (film thickness, surface roughness, edge profile).

In the bright field version of the edge contrast method[21], problems may arise from Newton's interference fringes in the vicinity of steps because of variations in resist thickness (cf. Sect. 4.2.3). Although the dark field version of edge detection is free of interference problems, the steps often don't reflect enough light. In the phase contrast method, the resist thicknesses in regions 1 and 2 may differ by $\lambda/2n$, in which case $I_2 = I_1$ (cf. Sect. 4.2.3). The same applies to the diffraction contrast and Fresnel zone methods. In the latter case, the sum of the waves from the raised areas at the focal point can have a phase value that is exactly 180° different from the sum of the waves from the non-raised areas. Thus when the amplitudes from both components are equal, there is zero intensity at the focus.

Alignment mark detection is performed either with the same wavelength as is used for resist exposure, or with a wavelength in the visible part of the spectrum where the resist is not photoactive.

If one uses the exposure wavelength, or a wavelength that is close to it, then one has the advantage of being able to project the alignment marks onto the reticle plane through the lens (through-the-lens alignment, on-axis alignment). This means that an alignment lens placed above the reticle can be used to project the wafer alignment marks and the associated reticle alignment marks onto an optical sensor, for instance (see Fig. 4.2.17).

---

21 The larger the change in refractive index at a step, the easier it is to detect optically. For example, a step in silicon (e.g. a LOCOS step) is much easier to detect than an equivalent resist-coated SiO_2 step, because the refractive indices of SiO_2 and photoresist are only slightly different ($n_{photoresist} = 1.7$, $n_{SiO2} = 1.45$, $n_{Si} = 4.75$, for $\lambda = 436$ nm).

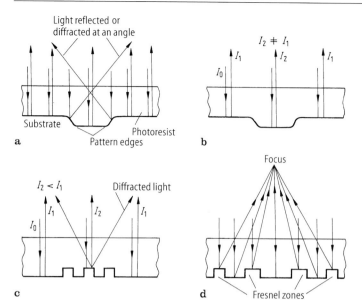

Fig. 4.2.27 a–d. Principle behind the four main optical techniques for detecting alignment marks. a Edge contrast method: the relatively small aperture angle of the alignment lens means that beams reflected at a large angles from the edges of the alignment marks do not enter the lens. The edges therefore appear as dark lines in a bright background (bright field method). If the light is incident at an angle, or if the light reflected normally is cancelled, then one sees bright lines in a dark background (dark field method); b phase contrast method: the reflected intensities I_1 and I_2 are different because of the different optical thickness in the regions 1 and 2; c diffraction contrast method: in the centre region of the grid-shaped pattern, some of the reflected light is diffracted at specific angles. The alignment marks are detected using the intensity of the diffracted light and the intensity I_2 of the normally reflected light ($I_2 < I_1$); d Fresnel zone method: the focal point is positioned at that distance from the substrate where the path difference for reflected waves from adjacent raised zones equals λ. This is also the case for the non-raised areas. The focal point lies exactly over the centre of the annular Fresnel zone pattern.

One problem with alignment mark detection at exposure wave lengths can be the low light intensity from the alignment marks. The intensity is low because the resist layer may be highly absorptive at the exposure wavelength (e.g. in top surface imaging), and because the resist should be exposed as little as possible during the alignment process.

When visible light is used to detect the alignment marks, these problems do not arise. There is another difficulty to overcome, however, relating to the lens system. In this case the image of the back-projected alignment marks is out of focus, and does not lie in the reticle or mask plane, because the lens is usually only error-corrected for the exposure wavelength. The problem is solved in various ways. One option is to align the silicon wafer using an additional alignment

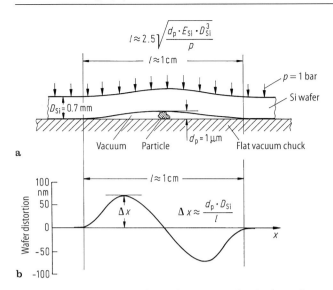

Fig. 4.2.28 a. A 1 μm particle on the vacuum chuck of a wafer stepper produces wafer warpage of about 1 cm in diameter; **b** this deflection produces a lateral wafer distortion of up to 70 nm. E_{Si} = modulus of elasticity of silicon.

microscope. The alignment microscope contains reference alignment marks located at a fixed position from the optical axis of the exposure lens (off-axis alignment). After this alignment procedure, the silicon wafer is moved by a fixed distance, so that it then lies under the exposure lens, aligned with the reticle.

This type of alignment can only be performed once per wafer (global alignment), and so cannot be used for site-by-site alignment. This may be unacceptable in some alignment and exposure equipment using visible light for alignment mark detection. If this is the case, alignment is then performed through the exposure lens, with optical correction elements used in the alignment path to compensate for the out-of-focus image of the wafer alignment marks in the reticle plane.

In the Fresnel zone alignment method, the light emerging from the focal point is projected back through the imaging lens, but not onto the mask plane. The image is projected onto a quadrant sensor. The table carrying the wafer is moved (mechanically or piezo-electrically), until the focal point lies in the centre of the quadrant sensor, i.e. all four quadrants receive the same light intensity.

The best wafer steppers currently available can achieve an alignment accuracy of ±50 nm. When mix and match equipment is used, an accuracy of ±80 nm is possible. Problems can still arise from geometric distortion of the silicon wafer however. Most wafer steppers can compensate for a linear distortion by a slight adjustment to the magnification of the projection system. This is achieved, for example, by a defined change in air pressure in the lens chamber (change in the air refractive index).

It is impossible for the alignment system to compensate for non-linear distortion within a site, however. Figure 4.2.28 shows how a non-linear distortion of several tenths of a micron can arise because of wafer warpage. The only remedy here is to remove the cause of the warpage, e.g. avoiding temperature gradients in the wafer during high-temperature processes (see Sect. 3.2.3), or by preventing particles landing on the vacuum chuck of the wafer stepper.

4.2.8
Defects occurring in optical lithography

The defect density has already been cited in Sect. 4.1 as a figure of merit for assessing the performance of a lithographic process.

Not only must the wafers and resist film be kept in perfect condition to avoid local defects, but so must the mask or reticle as well. The causes of defects on the wafer surface and how to prevent them are dealt with in Chap. 7. Where there is a risk of mechanical damage to the resist films, this has already been highlighted (e.g. in contact printing and with incorrect wafer handling).

In order to avoid defects on the masks or reticles, similar precautions to those recommended for wafers need to be taken. The allowable defect density on reticles must meet tighter demands, however, because a single mask defect will be duplicated on the wafer each time the mask is exposed. This is particularly serious in a reticle that contains e.g. two identical chips. A single fatal defect on the reticle would then cause 50 % of all the chips on the wafer to fail. Reticles must therefore be kept completely free of defects, and must stay perfect even with repeated use.

In order to produce defects free reticles, it is usual nowadays for repair work to be carried out on reticles. Repair techniques can be used to remove chromium residues, either by vaporizing them away with a fine-focused laser beam or by sputtering them off with an ion beam (e.g. gallium). Holes in the chromium film can also be repaired, e.g. by laser-induced local chromium deposition (break down of chromium carbonyl in the high-energy laser beam). Automatic techniques are used to localize the defects. The local light/dark distribution on the reticle is recorded and either compared with the object data on the magnetic tape, or compared with the corresponding distribution on an identical neighbouring chip. Any differences are then recorded as defects.

A defect-free reticle can be spoilt simply by a particle landing on its surface. The effect of these particles can be made almost negligible by placing a nitro-cellulose membrane (a pellicle) a few mm away from the reticle surface [4.23] (Fig. 4.2.29). Since the membrane and any particles that may be lying on it are positioned far outside the range of focus of the imaging lens[22], the images of the

---

22 As was explained in Sect. 4.2.6, if v is the image reduction factor, the depth of focus on the reticle side of a lens is larger by the factor v^2 compared to the depth of focus on the wafer side.

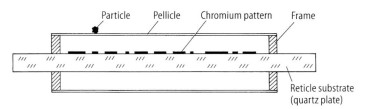

Fig. 4.2.29. Pellicle technology to cancel the detrimental effect of particles landing on a reticle. A pellicle can be placed on the back of the reticle as well

particles are so far out of focus that they simply produce a slight local drop in intensity.

The pellicle must have a uniform thickness equal to a multiple of $\lambda/2n$ over the whole reticle area (λ = exposure wavelength, n = refractive index of the pellicle). This prevents light waves reflected from the pellicle surfaces from impairing the image of the reticle patterns. An alternative is to prevent reflection by coating the two pellicle surfaces with an anti-reflective film.

There are difficulties associated with the pellicle technique, however, one being the risk of mechanical damage to the membrane. It is also harder to inspect the reticles, because one can only work with special high-resolution microscope lenses that have a large working distance.

4.3
X-ray lithography

Lithography using soft X-rays with a wavelength of around 1 nm appears, in principle, to be the obvious next step from optical lithography if the wavelength-dependent resolution is to be pushed below the 0.1 µm limit. X-rays, like light waves, are a form of electromagnetic radiation, and therefore cannot be affected by external electric and magnetic fields. There are crucial differences between optical and X-ray waves, however, with regard to their interaction with matter. For this reason not only are the radiation sources different, but the imaging principles and mask technology are fundamentally different as well [4.24]. X-ray lithography systems thus require their own specific development, drawing very little from optical experience.

Since the refractive index for X-rays is almost the same for all materials ($n \approx 1$), the reflectors and lenses common in optical imaging systems are not suitable for X-rays. Thus X-ray lithography cannot be used for pattern generators, direct write systems or projection equipment (cf. Fig. 4.2.16). Only proximity and contact printing are possible from a suitable X-ray mask.

Different materials do have different X-ray absorption coefficients, but at a wavelength of 1 nm the difference is barely two orders of magnitude (Fig. 4.3.1), compared with more than ten orders of magnitude for UV light. This means that

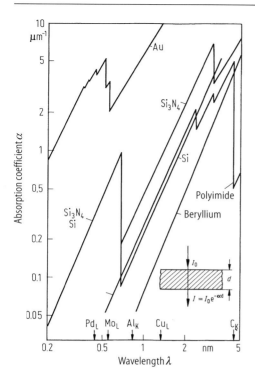

Fig. 4.3.1. Absorption of soft X-rays in several relevant materials. The wavelength of the most important characteristic X-ray radiation is also shown on the abscissa

mechanically stable mask substrates such as the glass plate in optical chromium masks cannot be used for X-ray masks, making them relatively fragile objects.

4.3.1
Wavelength region for X-ray lithography

The resolving power of proximity exposure using parallel X-ray beams depends on two factors: beam diffraction at the mask features and the scattering range of the photoelectrons released by the X-rays in the resist. The photoelectrons are responsible for the exposure of the resist.

Section 4.2.6 showed that the minimum diffraction-dependent feature size b_{min} which can be reproduced by proximity exposure is:

$$b_{min}(\text{diffraction}) \approx 1.5\sqrt{\lambda s}$$

where s is the separation between the mask and wafer. b_{min} is shown in Fig. 4.3.2 for varying X-ray wavelength λ at three different separations s [4.25]. It is assumed that the mask patterns completely absorb the X-ray radiation, and that the pattern edges on the mask are vertical.

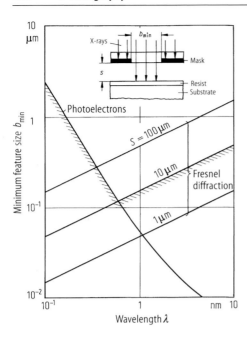

Fig. 4.3.2. The minimum feature size that can be reproduced in proximity exposure using parallel soft X-rays [4.25]. The resolution is limited on the one hand by Fresnel diffraction (separation s between mask and wafer) and on the other by the scattering range of the photoelectrons released by the X-rays in the resist. As an example, the minimum reproducible feature width b_{min} as a function of the wavelength is shown hatched for a mask separation $s = 10$ μm

If the minimum feature size is assumed to be twice the range of the photoelectrons, then one obtains the curve shown in Fig. 4.3.2, which increases for decreasing wavelength.

For a mask separation of 10 μm, a resolution of about 0.1 μm is predicted for the ideal case, i.e. for a wavelength of 0.6 nm. At smaller wavelengths the photoelectrons degrade the resolution, whilst at longer wavelengths Fresnel diffraction is predominant. Thus the region between 0.5 and 1 nm is the optimum wavelength range for X-ray lithography.

4.3.2
X-ray resists

X-rays excite photoelectrons in matter, which makes any electron-sensitive resist also suitable as an X-ray resist in principle. The positive photoresists described in detail in Sect. 4.2.1 can also be used as X-ray resists, because the chemical conversion mechanism of diazonaphthoquinone into carboxyl acid can be triggered not just by UV light, but by electrons as well. Since X-rays are only weakly absorbed by organic films, however (cf. Fig. 4.3.1), the exposure times are relatively long. Only by using a synchrotron (see Sect. 4.3.3) supplying about 200 mW/cm^3 to the resist surface, exposure times of a few seconds may be achieved even for the more insensitive resists, e.g. the positive photoresists mentioned or PMMA, which has an X-ray sensitivity of 500-1000 mJ/cm^2 [23)] [4.26].

The more sensitive X-ray resists[24] such as PBS, COP, FBM or MFA, which have an X-ray sensitivity of less than 100 mJ/cm² [4.27], have the disadvantage that they are less resistant to such processes as reactive ion etching or high dose ion implantation. A tri-level technique (see Fig. 4.2.10) offers a solution here (although a more costly option), where the bottom resist takes on the role of the etch and implantation mask.

Since in X-ray exposure there are practically no back-reflections from the substrate surface (as in optical lithography), and also no back-scattering[25] from the substrate (as in electron lithography), the geometry of the resist patterns are solely determined by the X-rays incident on the resist and by the resist contrast (for definition see Fig. 4.2.3). The theoretical feature resolutions specified in Fig. 4.3.2 have actually been achieved in practice.

4.3.3
X-ray sources

Three types of X-ray source are considered for X-ray lithography: standard X-ray tubes, plasma sources and synchrotron radiation.

Figure 4.3.3 illustrates the principle of an X-ray tube and the geometrical situation in proximity exposure. The finite focal spot size d causes a blurred intensity change at the transition from an exposed to an unexposed region. The width Δb of the transition region is given by

$$\Delta b = s \frac{d}{S}$$

where s is the mask-wafer separation and S is the distance between the focal spot and mask. If one assumes that the minimum reproducible line width is $b_{min} = 2\Delta b$, then one obtains

$$b_{min} = 2s \frac{d}{S}$$

which means there is another limit to add to the limits shown in Fig. 4.3.2. For example, when $s = 10$ µm, $d = 6$ mm[26] and $S = 30$ cm, the minimum line width is 0.4 µm.

---

23 For definition of sensitivity see Fig. 4.2.3
24 One option for increasing the X-ray sensitivity of PMMA, for instance, is to include chlorine or fluorine groups in the PMMA and thus increase the absorption of X-ray radiation.
25 If the substrate surface is made of a heavy metal (e.g. gold or tantalum), there is more back-scattering of photoelectrons into the resist layer, so that over-exposure arises at the foot of the resist feature.
26 6 mm is a typical focal spot diameter for high-power X-ray tubes, where up to 30 kW of heat loss can be removed by intensive water cooling and/or rotation of the anode.

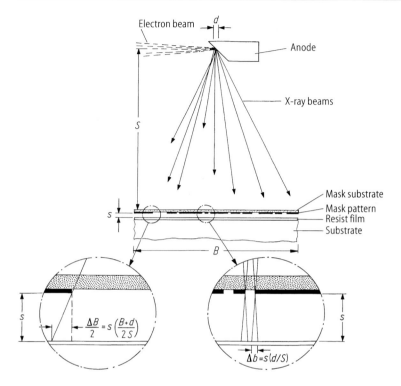

Fig. 4.3.3. Geometrical situation in proximity exposure using X-rays from an X-ray tube. The energy of the electron beam typically equals 25 keV, and the radiated power 5-30 kW. The anode is water-cooled and is made of metals such as palladium, copper, aluminium or silicon. The X-ray radiation leaves the evacuated chamber of the X-ray source through a beryllium window, entering the mask and wafer chamber, which contains helium at atmospheric pressure

Divergence of the X-rays means that the image field on the wafer is larger by ΔB than the dimension B of the site on the mask, where (Fig. 4.3.3)

$$\Delta B = s\left(\frac{B+d}{S}\right)$$

Allowance must be made during alignment for this image field magnification. A change Δs in the proximity separation from one pattern level in an integrated circuit to the next pattern level, distorts the image as given by the formula, creating a corresponding placement error. For example, a difference ΔB of about 0.7 μm is obtained when $\Delta s = 2$ μm, $B = 10$ cm, $d = 6$ mm and $S = 30$ cm. This means a placement error of 0.35 μm at the periphery of the image field, which is already more than that permitted in 1-μm technology.

The most common X-ray tubes use palladium, copper, aluminium or silicon as the anode material, and produce characteristic X-ray radiation in the range 0.4–1.4 nm (cf. Fig. 4.3.1). Even with high-power X-ray tubes (30 kW) one still only obtains an intensity of about 1 mW/cm² at a distance of 30 cm. Long exposure times (well over 1 minute) therefore have to be taken into account unless extremely sensitive resists are used (see Sect. 4.3.2).

For the limitations mentioned, X-ray lithography with an X-ray tube as radiation source, would not be able to be considered for a technology requiring features below 1 μm in size.

Smaller source diameters (about 1 mm) and higher beam intensities (approximately 10 mW/cm² time averaged intensity at a distance of 50 cm) can be obtained with plasma sources. Plasma sources, which all operate in pulsed mode (pulse length approx. 20 ns, pulse frequency 0.1–1 Hz), are not yet sufficiently advanced technologically to justify their use in X-ray lithography, however.

Synchrotron radiation represents an almost ideal X-ray source for X-ray lithography [4.28]. Synchrotron radiation is based on the principle that accelerated relativistic electrons (i.e. electrons with a velocity near light velocity) emit radiation. If the electrons in a synchrotron have reached the maximum velocity in their orbit, then the electrons are accelerated perpendicular to their direction of motion, and continuous X-ray radiation is emitted in the direction of motion (Fig. 4.3.4). According to theory, the wavelength λ_p (given in nm) at which the intensity is a maximum, is related to R, the radius of the electron orbit (in m) and E, the electron energy (in GeV) by the equation

$$\left(\frac{\lambda_p}{\text{nm}}\right) = 0.23 \left(\frac{R}{\text{m}}\right)\left(\frac{E}{\text{GeV}}\right)^{-3}.$$

The maximum electron energy E_{max} depends on the maximum achievable magnetic field B_{max} (given in T = Tesla) according to the equation

$$\left(\frac{E_{max}}{\text{GeV}}\right) = 0.3 \left(\frac{R}{\text{m}}\right)\left(\frac{B_{max}}{\text{T}}\right).$$

The Berlin synchrotron BESSY contains an acceleration ring to which are coupled three storage rings, in which the electron energy remains constant. The radius of one storage ring equals 1.8 m[27], and the magnetic field that can be obtained with conventional magnets is 1.5 T. Thus for E_{max} = 0.8 GeV one obtains λ_p = 0.8 nm, which lies in the required region (cf. Fig. 4.3.2). In fact a continuous spectrum is radiated between 0.3 and 2 nm, with the maximum lying at 0.8 nm.

---

27 The actual radius is larger (approx. 10 m), because straight sections are included between the deflecting magnets, so that the ring orbit is made up of circular orbit sections separated by straight sections. The circle that would be formed by putting together the circular pieces has a radius of 1.8 m.

The X-ray spectrum reaching the resist-coated silicon wafer, however, is different from the radiated spectrum because the X-ray beam passes through one or more membranes of beryllium, silicon or polyimide (several microns thick) and through the mask substrate (e.g. silicon). The membranes separate the different pressure stages that are needed for the transition from the ultra-high vacuum of the storage ring to the atmospheric pressure around the silicon wafer. As can be seen from the absorption curves for beryllium, silicon and polyimide (Fig. 4.3.1), attenuation increases for wavelengths above the intensity maximum at 0.8 nm (and also for those below for silicon), so that the bandwidth of the continuous spectrum is reduced. The integral intensity reaching the resist surface equals about 200 mW/cm^2 for BESSY. This intensity is sufficient to achieve acceptable exposure times, even with more insensitive resists which benefit from increased stability (see Sect. 4.3.2).

As illustrated in Fig. 4.3.4, the X-ray radiation is emitted with a certain divergence. The divergence in the x direction can be adjusted by the angular window ϑ in the storage ring. It is set so that for a tube length of 10 m (the distance between storage ring window and mask), the beam divergence in the x direction is about 5 cm, which corresponds to a maximum image field dimension of 5 cm. Thus sin ϑ has a value of about 0.005, compared with about 0.15 in proximity exposure using X-ray tubes. The associated placement errors (see above) are thus practically negligible for this low divergence angle.

The theoretical value of the divergence angle Ψ in the y-direction is given by the ratio of the electron energy E to the energy $E_0 = m_0c^2$ of the electrons:

$$\Psi = \frac{E_0}{E}$$

E_0 equals approx. 0.5 MeV, so that for $E = 0.8$ GeV one obtains $\Psi = 6.2 \cdot 10^{-4}$. For a tube length of 10 m, the divergence of the beam in the y direction equals

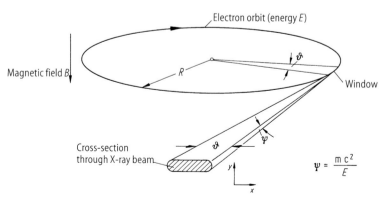

Fig. 4.3.4. Principle of radiation from a circular electron orbit

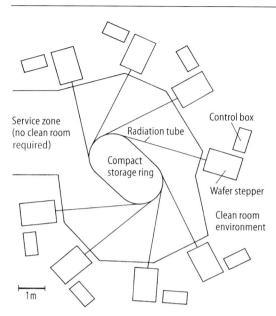

Fig. 4.3.5. Geometrical arrangement of the eight exposure stations (wafer steppers) that can be connected to the compact storage ring COSY

6.2 mm. With an image field dimension of 5 cm exposure of the silicon wafers must therefore involve moving either the wafers or the X-ray beam at a constant velocity in the y-direction (scanning). There are two options for moving the beam, either by tilting an X-ray reflector (X-ray beam at grazing incidence) or sweeping the electron beam in the storage ring (sweep frequency 1 Hz).

The diameter of the electron beam in the storage ring (beam current 0.5 A) and thus the diameter of the X-ray source equals about 0.5 mm. At the large distance between source and mask of 10 m, the effect of source divergence on feature resolution is totally negligible according to the formula above.

The large floor space required by a storage ring such as in BESSY is a major obstacle to the possible use of synchrotron radiation in semiconductor technology. Considerably smaller storage rings can be used, however, if they are designed solely for wafer exposure purposes, and if one uses super-conducting magnets which can produce magnetic fields of 5 T.

The compact storage ring known as COSY (*Compact Synchrotron*) [4.29] has a radius of 38 cm, an electron energy of 0.56 GeV and has a peak radiation at $\lambda = 0.5$ nm. The integral X-ray intensity incident on the mask equals 250 mW/cm^2 and is thus sufficiently large. It takes up a floor area of 100 m^2, and eight radiation tubes can be connected to it, each with a wafer stepper at the end (Fig. 4.3.5). Wafer steppers are required for larger wafers which are able to cover several image fields of 4 cm × 4 cm.

4.3.4
X-ray masks

An X-ray mask consists of the masking patterns and a mask support layer on which the masking patterns are arranged.

The size of the film thicknesses for the mask patterns and the mask support material can be estimated as follows. Around the X-ray wavelength of interest, which is 0.8 nm, gold has an absorption coefficient α of approximately 4 μm^{-1}, and is one of the most highly absorptive materials available (cf. Fig. 4.3.1). If gold is used as the masking material, then the thickness of the gold must be at least 0.8 μm in order to reduce the transmitted intensity to below 4 %. If one allows a 50 % X-ray absorption for the mask support, and one uses a weakly absorptive material such as silicon, silicon nitride, silicon carbide, boron nitride or polyimide[28], which have an absorption coefficient α of around 0.2 μm^{-1} at λ = 0.8 nm (cf. Fig. 4.3.1), then the mask support can be about 3 μm thick. A solid frame made of silicon, glass or metal is provided to hold the thin mask support membrane, making the X-ray mask easier to handle.

Figure 4.3.6 shows an example of the construction of an X-ray mask with silicon as the mask support membrane, and electroplated gold for the mask pattern [4.30]. The thin silicon membrane is produced by exploiting the different etch rates of lightly doped and p$^+$-doped silicon, where the lightly doped silicon has an etch rate that is several orders of magnitude higher in alkaline etching solutions (cf. Table 5.1). If a silicon wafer containing a 3 μm thick p$^+$ doped region on its top side is then etched from the back using ethylene diamine, etching stops when it reaches the p$^+$-doped layer.

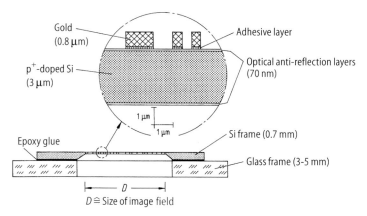

Fig. 4.3.6. X-ray mask construction, with a silicon mask support membrane and gold mask patterns

---

28 Beryllium also has a low absorption coefficient, but cannot be considered if visible light is used for alignment (see Sect. 4.3.5).

Compared with an optical 5 : 1 reticle, an X-ray mask has a very complex design. The main problems are mechanical stability and the stricter demands placed on the line width tolerance and critical defect density arising from 1 : 1 projection.

The specification for mechanical stability can be expressed in terms of the maximum mask distortion that is allowed at the edge of a single image site. For 0.1 µm technology, for instance, the maximum allowable distortion equals just 15 nm for a 4 × 4 cm site[29], which is equivalent to a relative distortion of 10^{-6}. To achieve this figure, the internal stresses within each layer of the X-ray mask must be matched extremely closely to each other. The thermal expansion caused by the energy absorbed by the mask during exposure must also be taken into account.

For a 0.1 µm technology the line width variations must lie within ±5 nm if one allows for the mask half the tolerance of the chip dimensions (see Sect. 4.1). Local pattern defects on the mask with a size of just 30 nm (about 1/3 of the minimum feature) may still be fatal (see Fig. 7.1.3). When one considers that the mask patterns are first produced with an electron beam writing system at a scale of 1 : 1, and then have to be converted to 0.8 µm thick gold patterns, one can see how hard it is to stay within the line width tolerance, which equals just 0.6 % of the mask pattern height. The small size of defects (30 nm) means that repairing them (e.g. with a fine-focused ion beam) is likely to present similar problems.

On the plus side, the problem of particles lying on the mask or on the resist-coated surface is less acute for X-ray exposure than for optical lithography, although it cannot be ignored. The effect of a 1 µm thick flake of resist (e.g. chipped off a resist film) on the line width of a resist pattern is estimated as an example. According to the formula given in Sect. 4.2.6, the intensity gradients at a pattern edge in proximity exposure is:

$$d\left(\frac{I}{I_0}\right)/dx \approx \frac{0.7}{\sqrt{\lambda s}} .$$

This means that for a local change in the exposure intensity $\Delta I/I_0$, the position of the resist edge is shifted by Δx, where

$$\Delta x \approx \frac{\sqrt{\lambda s}}{0.7} \frac{\Delta I}{I_0} .$$

For a resist flake of thickness 1 µm

$$\frac{\Delta I}{I_0} = \left(1 - e^{\frac{-0.15}{\mu m} \cdot 1 \mu m}\right) = 0.14 ,$$

---

29 The value of 15 nm is obtained if one allows the mask distortion half the value of the permitted 3σ overlay error, this value being approximately 30 % of the minimum feature size (see Sect. 4.1).

if one assumes a value of 0.15 µm^{-1} for the resist absorption coefficient α (cf. Fig. 4.3.1). Thus for λ = 0.8 nm, and for a mask-wafer separation s = 10 µm, one obtains for Δx:

$$\Delta x \approx \frac{\sqrt{0.8\,\text{nm} \cdot 10\,\mu\text{m}}}{0.7} \cdot 0.14 \approx 18\,\text{nm}$$

If the flake lies across both edges of a rectangular feature, then the line width changes by 2 × 18 nm = 36 nm, which is already about 36 % of the assumed minimum feature size of 0.1 µm.

4.3.5
Alignment procedure for X-ray lithography

Optical alignment techniques are used for aligning the silicon wafers with the X-ray masks. The various optical alignment principles are illustrated in Fig. 4.2.25. One peculiarity associated with a synchrotron radiation exposure system is that the wafer and mask are arranged vertically. The alignment table then needs a special design [4.31].

An alignment accuracy better than 30 nm can be achieved with a contrast detection method (Fig. 4.2.25 a) using a scanning laser beam.

4.3.6
Radiation damage in X-ray lithography

The question of possible radiation damage arises because the X-rays used for wafer exposure can penetrate the typical layers used in silicon wafers. This could lead to degradation of the electrical performance of the transistors. Results to date suggest that the radiation damage (e.g. traps being created in gate oxide films) can mostly be removed by annealing at temperatures above 400 °C. Whether the long-term stability of the transistors is unaffected still needs to be clarified.

4.3.7
Opportunities for X-ray lithography

As was pointed out in Fig. 4.2.26 optical lithography is likely to be extended down to 0.1 µm feature size or even below.

In the sub-0.1 µm feature region, the problems associated with imaging on a 1 : 1 scale probably cannot be overcome.

Recently an exposure technique was proposed which uses a wavelength in the 11 to 13 nm range for projection at a 4 : 1 reduction scale. The technique is called EUV (Extreme ultraviolet). The projection lens consists of mirrors which are made reflective by depositing e.g. 40 double layers of MoSi and Si with a thickness of λ/4 each. Lenses are being designed using a numerical aperture as large as 0.25. According to the resolution formula b_{min}=0.7 λ/NA (cf.

Sec. 4.2.6) a minimum feature size of 30 nm should be feasible at a depth of focus of 0.2μm

4.4
Electron lithography

Electron beams can be focused and deflected by electric and magnetic fields. They are therefore suitable in principle for both creating patterns (generating masks or direct writing on wafers) and imaging of mask patterns (projection or proximity exposure). These different options are illustrated in Fig. 4.4.1. The current most important application of electron lithography in silicon engineering is generating mask patterns (cf. Fig. 4.2.16). Direct writing onto silicon wafers is used in the manufacture of small batches of customised circuits.

4.4.1
Electron resists

There are numerous positive and negative resists that are sensitive to electron beams [4.32]. Table 4.2 summarizes the data of the most common electron resists. The incident exposure dose and the sensitivity are not specified in mJ/cm^2 as for optical and X-ray resists, but in C/cm^2 (Coulomb/cm^2). The definition of the sensitivity and the contrast is exactly the same as the definition given for optical resists (cf. Fig. 4.2.3).

The chemical mechanism working in negative resists involves cross-linking of chain molecules when exposed to the electron beam. The chain molecules but

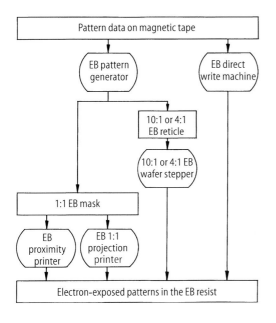

Fig. 4.4.1. Electron beam lithography equipment for converting pattern data stored on magnetic tape into electron-exposed patterns in a resist film (EB = electron beam)

Table 4.2. Properties of a few key electron resists. An exposure energy of 10 keV is assumed. Technology stability refers to the ability of the resist to withstand stressful processes, such as reactive ion etching

Resist	Trade name	Positive/ negative	Sensitivity in µC/cm^2	Contrast	Technology stability
Polyglycidyl-methacrylate -coethylacrylate	COP	negative	0.4	1	poor
Positive photoresist (see Sect. 4.2.1)		positiv	20	3	good
Polymethyl- methacrylate	PMMA	positiv	30	3	medium
Polybutene-sulphone	PBS	positiv	0.7	1	poor

not the cross-linked molecules are soluble in certain organic solvents (developers). Negative resists are generally sensitive, but have poor technology stability (see Table 4.2).

Various mechanisms are employed in positive resists. Positive photoresists can also be used as electron resists, because the conversion of diazonaphthoquinone into carboxylic acid (cf. Fig. 4.2.1) can be triggered not only by UV light but by electrons as well. The same developer is also used. Image reversal using electron beam exposure is also possible with these resists (cf. Fig. 4.2.14). In the resists PMMA and PBS, long-chain molecules are broken up by the electrons. The broken molecules are soluble in suitable solvents (developers).

Although the PBS resist is very sensitive, it is also the least resistant to stressful processes. Like the COP negative resist, it is therefore more often used for producing chromium masks, where only a very thin chromium layer needs to be etched (70 nm). On wafers one either uses a positive photoresist that has high technology stability, or one may apply a tri-level resist technique, although this is rather costly (cf. Fig. 4.2.10).

4.4.2
Resolution capability of electron lithography

In electron lithography equipment, the electrons are accelerated with voltages in the range 5–50 kV. According to the equation

$$\left(\frac{\lambda}{nm}\right) = \sqrt{\frac{1.5}{\left(\frac{U}{V}\right)}},$$

the wavelength λ of the electron beam is of the order of 0.01 nm. Although the numerical aperture of electron-optical imaging systems is about 100 times smaller than in optical projection systems, diffraction effects do not come into play for the feature sizes of interest in silicon technology (cf. Sect. 4.2.6). As explained in the next section, it is possible to generate electron beams with an abrupt rise in the beam current density at the edge of the beam cross-section (although this may be at the expense of beam current).

The resolution capability of electron lithography is not determined by the electron beam, but by electron scattering when the beam is decelerated in the resist or substrate. Figure 4.4.2 shows the result of a Monte-Carlo simulation computing the paths of several electrons incident normal to a PMMA resist surface at the point x = 0, for two different energy levels of 10 and 20 keV respectively [4.33]. One can see that a few electrons are scattered laterally up to a distance of 1 μm at 10 keV, and 3 μm at 20 keV.

Closer study reveals that there are essentially two components contributing to electron scattering [4.34]. One component can be attributed to the forward scattering over a narrow angle of high-energy electrons hitting the resist. The other component is related to the back-scattered electrons from the substrate into the resist. Both components can be easily identified in Fig. 4.4.3, which shows the calculated energy density (dose) absorbed due to electron deceleration in a PMMA resist of thickness 0.4 μm on a silicon substrate[30], when an electron beam of diameter 0.3 μm and energy 20 keV is incident on the resist surface.

The near-effect of electron scattering (lateral range of forward scattering approximately 0.1 μm in Fig. 4.4.3), means that the minimum achievable feature size is about 0.25 μm for the conditions given in Fig. 4.4.3. For thinner resists (e.g. 0.2 μm) and higher electron energies (e.g. 50 keV), or very low energies (below 10 keV), the near-effect is reduced, then allowing a resolution of about 0.1 μm.

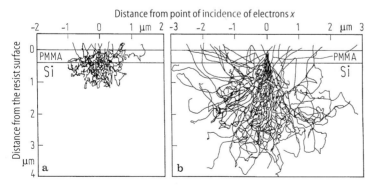

Fig. 4.4.2 a, b. Electron paths computed by Monte-Carlo simulation, for electrons incident at point x = 0 with an energy of 10 keV (**a**) and 20 keV (**b**) onto the surface of a PMMA resist [4.33]

---

30 A silicon substrate is assumed in Fig. 4.4.3. As the electron scattering properties of Si, SiO_2 and glass differ only slightly the diagram also gives a good approximation for glass substrates, commonly used for optical masks and reticles.

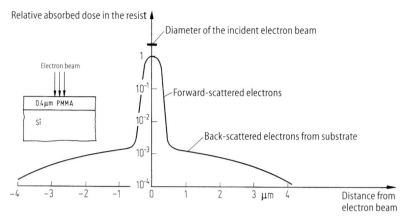

Fig. 4.4.3. The absorbed energy density (dose) due to electron deceleration in a PMMA resist of thickness 0.4 μm, as a function of the distance from an electron beam of diameter 0.3 μm and energy 20 keV incident on the resist surface. The lateral scattering of the electron beam in the resist consists of a high dose component close to the beam, and a lower dose component scattered further laterally

The far-effect of the electron scatter (lateral range of back-scattered electrons is approximately 3 μm in Fig. 4.4.3), appears harmless at first glance, since the dose at the electron beam location only reaches a relative value of about 1 %. The cumulative effect of scattered doses from a surrounding area of some 3 μm in radius can be significant, however (proximity effect). In the extreme case of an isolated unexposed island (e.g. diameter 0.5 μm) surrounded by an exposed area, the island receives a scattered dose of about 10 % for the conditions given in Fig. 4.4.3. Under even less favourable conditions, e.g. higher electron energy (e.g. 50 keV) and a substrate that exhibits higher back-scattering (e.g. gold), the scattered dose can rise to 50 %. Conversely, the back-scattering effect can be considerably reduced by a low electron energy (10 keV and below), and by avoiding substrates with high back-scattering (e.g. tri-level systems with a thick bottom resist layer on a gold substrate).

There is another possible way to compensate for the line width variations caused by electron back-scattering. This is to selectively reduce the electron beam exposure at those regions which receive excessive scattered electrons (proximity correction) [4.35].

4.4.3
Electron beam pattern generators

The design of an electron beam pattern generator is shown schematically in Fig. 4.4.4. A heated tungsten filament is used for the electron source, or an indirectly heated lanthanum hexaboride tip which makes a particularly efficient electron source because of its low work function difference.[31] The electron source is held at a negative potential (10–50 kV), whilst the anode is grounded. The electron beam passes through a hole in the centre of the anode, and is focused by a Wehnelt cylinder, held at a negative potential like the electron source. Once the electron beam has passed through the hole in the anode plate, the electrons are no longer accelerated in the beam direction, but are deflected at right angles to it. An electric field applied to a pair of capacitor plates is used to blank out the electron beam. As the electron beam continues on its path, it is focused into a spot size of 10–100 µm using toroids acting as magnetic lenses. Cylindrical coils are used to deflect the electron beam in the x and y direction. Their longitudinal axes lie in the y direction (for x deflection) and in the x direction (for y deflection), respectively. Finally, the electron beam is focused down to a diameter of 0.01–0.5 µm in the write plane using a magnetic lens. The resist film on the substrate is positioned to lie in the write plane, and the substrate placed on a laser-controlled table that can be moved with a high degree of accuracy in the x and y direction. There must be a high vacuum over the full length of the electron beam, with even the resist and table held in the vacuum. Thus the substrate cannot be held onto the table by suction, as is used in optical and X-ray lithography. Instead the substrate must be held against the support (chuck) either mechanically or electrostatically.

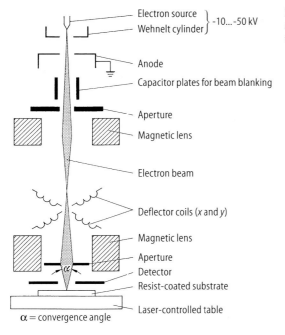

Fig. 4.4.4. Schematic diagram of an electron beam pattern generator

---

31 Field emission cathodes can also be used.

For a given beam current, the beam diameter has a lower limit determined by the spherical and chromatic aberration[32] of the magnetic lenses, and by the brightness of the electron source. For a 1 µA beam current, a minimum beam diameter of about 0.2 µm can be achieved. The corresponding optimum convergence half-angle α/2 of the electron beam (see Fig. 4.4.4) equals about 0.5° in the write plane[33].

The deflector coils don't work perfectly linear. For larger deflections of the electron beam, the errors must be compensated for by correction elements in the electron-optical column, or by appropriately corrected control currents in the deflector coils.

Modern electron beam pattern generators depend on extensive control electronics to convert the pattern data stored on magnetic tape into control currents for the deflector coils or for the table stepper motors. In addition, the electronic control system must ensure that all important parts of the electron-optical column are stable.

There are two possible design options for writing patterns into a resist film: the raster scan approach and the vector scan approach [4.36].

In the raster scan design (Fig. 4.4.5 a), the electron beam is deflected by discrete steps in the x-direction, where the step size equals the beam diameter and the step frequency equals several MHz. The electron beam is blanked out at those locations where exposure of the resist is not required. As soon as the electron beam has been deflected over a full line length, the beam is moved by a distance of one beam diameter in the y direction. This is either achieved by deflecting the electron beam, or by mechanically moving the table.

In contrast to the raster scan approach, the electron beam in the vector scan design is guided selectively from one write pattern to the next by deflecting the beam in both the x and y direction (Fig. 4.4.5 b). A pattern is exposed by deflection of the electron beam in incremental steps equal to the beam diameter. The electron beam can be moved within the rectangular feature in a spiral path for instance (as in Fig. 4.4.5 b). Partial correction of the proximity effect can be achieved by reducing or increasing the write frequency at the edge of the feature. Instead of using a small beam diameter that amounts to 1/4 or 1/10 of the minimum feature size, a square beam cross-section (shaped beam) with side equal to the smallest feature element (e.g. 1 µm × 1 µm) can be useful. Some electron beam pattern generators even have a variable shaped beam [4.37]. This variable beam cross-section is achieved by projecting a square beam onto a square aperture. By varying the deflection of the square beam, the amount of the beam pass-

---

32 Chromatic aberration is mainly caused by electrostatic interaction of closely-spaced electrons in the beam direction. This results in electrons no longer having the same energy, but a certain energy distribution (a spread of a few eV). The beam is also broadened by lateral repulsion between electrons.

33 The convergence half-angle is equivalent to the numerical aperture in optical projection systems.

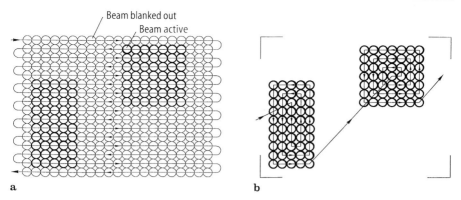

Fig. 4.4.5 a. The raster scan principle and **b** the vector scan principle in electron beam writing, illustrated for the example of two rectangels

ing through the aperture can be adjusted. Thus the beam cross-section can be chosen to take on the shape of all rectangles that fit within the square aperture.

One significant practical difference between raster-scan and vector-scan electron beam pattern generators concerns the freedom to choose which areas to expose. Taking figure 4.4.5 as an example, there may be the requirement to expose the space between the rectangles rather than the rectangles themselves. A raster scan system can easily do this, simply by inverting the blanking commands for the electron beam. This straightforward inversion of the write areas is not possible with the vector-scan approach. Here the inversion of the masked areas can only be achieved by using a negative resist (or a positive resist using an image reversal technique, cf. Fig. 4.2.14) or by using a liftoff technique (see Sect. 3.1.3).

The most widely used electron beam pattern generator called MEBES (*Mask Electron Beam Exposure System*) [4.38] operates on the raster-scan principle (fig. 4.4.6). The electron beam has a diameter that can be adjusted between 0.1 and 1.1 µm. The beam is deflected in the x direction in 1024 steps at a step frequency of up to 100 MHz (raster scan). The step width equals the beam diameter, so that the range of deflection equals 128 µm in the $+x$ direction and 128 µm in the $-x$ direction for a 0.25 µm beam diameter, which corresponds to a beam deflection angle of $\pm 0.5°$. The electron beam can be blanked out during each step, depending on whether the spot concerned needs to be exposed or not. The glass plate, which is coated with a thin chromium film (70 nm) and a film of resist (0.25 µm), lies on the table, which moves continuously in the y direction at the same time as the electron beam is incrementally deflected in the x direction. A high precision laser interferometer controls the table position to an accuracy of 15 nm with respect to the position of the electron beam. The table speed is controlled so that the forward movement of the table equals exactly one beam diameter during the 1024 deflection steps in the x direction. As soon as the table has covered the whole length of the write stripe in the y direction (max. 150 mm), the

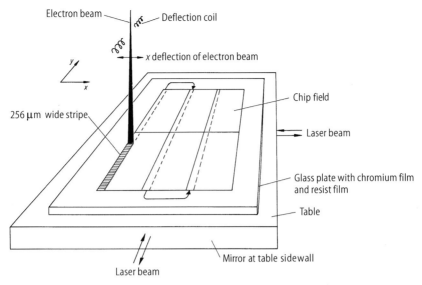

Fig. 4.4.6. Operating mechanism of the MEBES electron beam pattern generator for writing masks and reticle patterns

table is shifted by one whole chip width in the x direction, and the next identical stripe is exposed. Once all identical stripes have been exposed, the pattern data for the next stripe is loaded into the machine memory, and the exposure procedure is repeated in the same way as for the first stripe. The time taken to write over the whole write field (maximum 150 mm × 150 mm) is calculated using the following equation

$$\text{write time} = \frac{\text{write field area}}{(\text{beam diameter})^2 \times \text{write frequency}}.$$

For a beam diameter of 0.1 µm and a write field of 100 mm × 100 mm – that is the reticle write field, for example, for an image field of 20 mm × 20 mm in a wafer stepper with 5:1 image reduction – one obtains a write time of about 170 minutes. The total processing time per reticle is actually slightly longer, because inserting and removing the reticle takes time, as does reading the blocks of pattern data into the machine memory.

Whereas the long processing time can be tolerated for masks and reticles, such long exposure times would be much too expensive for wafer purposes. Electron beam direct write equipment has therefore been developed using the vector scan design and a variable shaped beam (see above) to give considerably reduced write times [4.39]. These designs do, however, require far more complex control and monitoring equipment for shaping and deflecting the electron beam, so that the machine is much more expensive than a wafer stepper. Nevertheless, direct writing using this type of machine can prove economic, for instance when only a few wafers need to be given a certain interconnection pattern (e.g. for ASICs = Application Specific Integrated Circuits).

One factor limiting the speed of the direct write procedure is the sharp temperature rise of the resist during rapid exposure of the individual features (see Sect. 4.4.6). This problem could be overcome using a multi-beam write system, where for instance 1000 separately controlled electron beams work in parallel. This would also further reduce the write time compared with the electron beam direct write equipment of today.

4.4.4
Electron projection equipment

As shown in Fig. 4.4.1, electron beam masks can be projected as a 1 : 1 image or a reduced image using electron beams. Two fundamentally different designs are used for electron beam masks: photo-cathode masks [4.40] and stencil masks that are transparent for the electron beam [4.41].

A photo-cathode mask is based on a chromium mask, such as that used in optical lithography. The entire mask surface is coated with a thin film of material with a low work function (e.g. caesium iodide). If the back of the mask is exposed to ultraviolet light, then electrons are emitted from the caesium iodide film in those regions not covered by chromium. If the caesium iodide film is held at a negative potential (e.g. −10 kV), then the electrons are accelerated away from the mask and focused onto a resist-coated silicon wafer using a magnetic field.

A transmission mask is required where the mask is inserted in the beam path of an electron-optical column (cf. Fig. 4.4.4) [4.41]. The electrons must be absorbed by the mask patterns, but transmitted between the mask patterns without loss of energy. Even an extremely thin film, however, would cause an unacceptably large change in the electron energy (loss of monochromacy). The solution is thus a stencil mask, similar to a X-ray mask, but without a supporting membrane (cf. Fig. 4.3.6). Special measures are required for unconnected mask features[34)] which inevitably involve a reduction in image quality.

Recently a 4 : 1 electron projection technique has been proposed which uses a thin tungsten membrane for electron scattering. This SCALPEL technique (SCattering with Angular Limitation in Electron Projection Lithography) has potential in the 100 nm to 50 nm dimension range. However, the throughput seems to be limited to 5 wafers per hour at 50 nm.

---

34 Two special solutions can be mentioned. One is to use a fine support grid, the other is to expose twice using two masks with complementary support patterns.

4.4.5
Alignment techniques in electron lithography

Whereas alignment is not required when writing mask patterns, the exposure pattern has to be positioned very accurately with respect to an existing pattern on the silicon wafer when writing directly onto the wafer or using electron mask projection.

In electron beam lithography, suitable marks on the silicon wafer are used for alignment. The technique relies on the fact that the intensity of back-scattered and secondary electrons from an incident electron beam is different near the alignment mark [4.42]. An electron detector is placed above the silicon wafer to record the back-scattered and secondary electrons (cf. Fig. 4.4.4). Trenches etched into the monocrystalline silicon may be used as alignment marks. Another solution is tantalum patterns which are created in an extra step at the start of the integrated circuit manufacturing process. Detecting the alignment marks is made more difficult because the marks are covered by at least the actual layer to be patterned (e.g. 1 μm aluminium[35]) and the resist layer[36].

The alignment marks can be placed at the four corners of a chip for instance. Normally alignment marks are not allowed within the chip. In electron beam writing the alignment marks at the chip edge are therefore used to define the locations of the features. The features are then accessed by controlled deflection of the electron beam or by moving the table under laser interferometer control.

4.4.6
Radiation damage in electron lithography

When the electrons are decelerated in direct electron beam writing, the following damage can arise:
– The chemical bonds between atoms in the layers can be modified.
– The exposed areas can become electrostatically charged.
– Rapid severe temperature rises can occur at localized points.

As regards the change in chemical bonds, the processes occurring in the resist and SiO_2 layers are significant. Whereas the breaking of chemical bonds and cross-linking of chain molecules are desirable effects in the resist, as this is precisely what makes the resist work (cf. Sect. 4.4.1), broken bonds in a SiO_2 lattice produce traps, which can become charged e.g. by the injection of hot electrons into the SiO_2 layer, and then contribute to transistor instability. The majority of

---

35 If the liftoff technique is being used (see Sect. 3.1.3), film deposition occurs after the lithographic step. In this case it is easier to detect the alignment marks.
36 In principle one can remove the resist lying over the alignment marks before alignment is performed, by using a relatively coarsely aligned light-optical mask to expose the resist around the alignment marks, and then removing it by development. This procedure adds to the cost however.

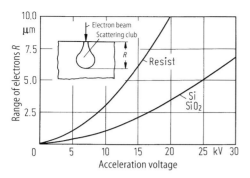

Fig. 4.4.7. Range of the electrons in resist and in silicon or SiO_2 as a function of the electron acceleration voltage

this radiation damage can be repaired by annealing at temperatures around 900 °C.

The electrical charge introduced during electron beam exposure creates stray fields[37] if there are no conducting paths. Within the deceleration region of the electron beam some conductivity can be assumed, even in the insulating layers. For the charge to be removed from the bulk of the scatter region, however, it is necessary for the scatter region to extend into a conducting layer. Where electron beam writing is being used to produce a chromium mask, the scatter beam extends easily through the normally non-conducting resist film into the conducting chromium layer, through which the charge is then removed.

Figure 4.4.7 shows the range of the electrons in resist and in silicon or SiO_2 as a function of the acceleration voltage. From the ranges shown one can conclude that when writing onto a wafer, the scatter region readily extends into the silicon substrate and also the conducting layers on the silicon wafer if the acceleration voltages are above about 20 kV. At an acceleration voltage of 10 kV, however, these conditions may no longer hold, and the electrostatic charges mentioned above can arise.

Finally, the rapid local temperature rise in the resist during electron beam writing shall be estimated. Under real conditions the local exposure time is so short (e.g. 10 ns in the MEBES machine) that the steady-state condition is never reached where equilibrium between the incident power and heat dissipated to the substrate is achieved. Instead one has to consider the ballistic case, where the heat dissipation to the substrate is negligible. In this case the local temperature rise ΔT in the resist during electron beam exposure is given by

$$\Delta T = \frac{Q}{c_v} \frac{dV}{dz},$$

---

37 The stray fields can deflect the electron beam, for instance, so that it hits the resist offset from the required position.

where Q is the exposure charge density (in C/cm^2), c_v the specific heat capacity of the resist and dV/dz the reduction in the acceleration potential of the electrons when decelerated in the resist.

If one inserts $c_v = 1.7$ J/cm^3K and $dV/dx = 3.3$ $kV/\mu m^{38)}$, then for $Q = 0.7$ $\mu C/cm^2$ – which is the value of the sensitivity of the PBS resist (see Table 4.2) – one obtains a temperature rise in the resist of 13.5°C. At $Q = 20$ C/cm^2 – which is the sensitivity of a typical positive photoresist (see Table 4.2) – the resist is rapidly heated to a temperature of 390 °C. This latter temperature rise can degrade the resist properties.

4.5
Ion lithography

Since ions, like electrons, can be focused and deflected using electrical and magnetic fields, they are in principle also suitable for direct pattern writing and for the projection of mask patterns. Pattern generation is either achieved using the conventional resist process, or using one of the other techniques shown in Fig. 4.5.1.

In the traditional resist technique (Fig. 4.5.1 a), the solubility of a suitable resist is modified by exposing the resist to an ion beam. As in the lithography techniques described in Sects. 4.2 through 4.4, both positive and negative resists can be used.

Figure 4.5.1 b shows another option for creating a pattern. In this case a mask for a subsequent anisotropic etch process is created in the resist by ion implantation. Inorganic resists such as $Ag_2S/GeSe$ can also be used in ion lithography (Fig. 4.5.1 c). Ion bombardment causes the Ag atoms to migrate out of the Ag_2S film into the GeSe layer (see also Fig. 4.2.12). This reduces the solubility of the GeSe, so that only the unexposed area is removed by the developer solution.

As outlined in Fig. 4.5.1 d, the ion beam can also induce chemical etching or deposition (see Chaps. 3 and 5). The ions can even etch away material directly (Fig. 4.5.1 e and chapter 5). Ion lithography is also suitable for selective ion implantation (Fig. 4.5.1 f and Chap. 6).

Almost without exception, the pattern generation techniques described here using ion lithography are currently at the research and development stage.

4.5.1
Ion resists

The impact of the ions entering the resist produces secondary electrons. In the positive resists these break up the long chain molecules, and in negative resists

---

[38] The value of 3.3 kV/µm is obtained by taking the mean of the potential gradient for an acceleration voltage of 10 kV. From Fig. 4.4.7 one can deduce a range in the resist of 3 µm, i.e. a mean potential drop of 10 kV/3 µm = 3.3 kV/µm.

4.5 Ion lithography

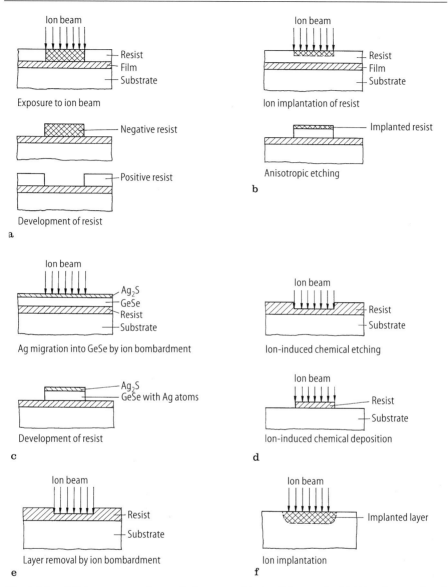

Fig. 4.5.1 a–f. Pattern generation techniques used in ion lithography [4.43]. **a** Conventional resist technique; **b** resist implantation; **c** inorganic resist; **d** ion-induced etching and deposition; **e** ion beam etching; **f** ion implantation

they cause cross-linkage of molecules. In a similar to way to X-ray radiation, it is therefore the electrons that are responsible for creating the pattern. This means that all electron-sensitive resists are also suitable for ion lithography. As can be

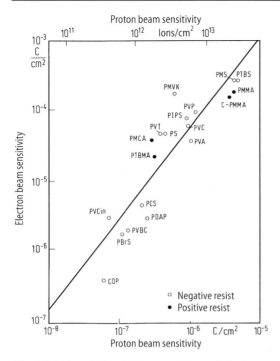

Fig. 4.5.2. Correlation between proton and electron sensitivity for various resists (electron energy = 20 keV, proton energy = 100 keV) [4.47, 4.58]. Further data on ion resists can be found in [4.43]. The resists have been labelled with the abbreviations given in the relevant literature

seen in Fig. 4.5.2, there is actually a strong correlation between proton and electron sensitivity, where the protons can be taken to represent the ions. In ion resists the sensitivity is specified in C/cm^2 or as the number of ions/cm^2. The definition of the sensitivity is the same as for photoresists (Sect. 4.2.1). As Fig. 4.5.2 shows, the resists are far more sensitive to ions than to electrons.

The most sensitive resists, such as COP (Polyglycidyl-methacrylate-coethyl-acrylate), have a low etch resistance, as was mentioned in Sect. 4.4.1. They are therefore only suitable for patterning very thin films.

4.5.2
Ion beam writing

In ion beam writing, a focused ion beam (FIB) is scanned over the substrate to be patterned. The pattern is generated serially, and so speeds tend to be as low as in electron beam writing. The advantages of the technique are its particularly high resolving power and the high degree of flexibility for generating patterns.

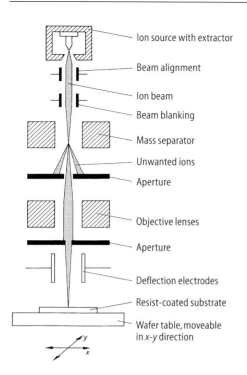

Fig. 4.5.3. Main elements of an ion beam write system [4.46]

Ion beam writing is also suitable for repairing masks for X-ray and photolithography. Track shorts can be removed by focused ion etching, and breaks in interconnects can be bridged using ion-induced deposition. It might also be possible to use ion beam writing for producing X-ray masks.

Figure 4.5.3 shows the schematic design of an ion beam write machine. The major components are listed below.

- *Ion source with extractor.* Plasma, field ionization or liquid metal sources are suitable for generating the ions [4.43]. The extractor pulls the ions out of the source and concentrates them into a beam.
- *Beam alignment.* Voltages applied to the plates of a capacitor are used to align the ion beam
- *Beam blanking.* Depending on the pattern to be written, the ion beam is either blanked out or allowed to pass, by selectively applying a voltage to capacitor plates as required.
- *Mass separator.* The unwanted ions are deflected using intersecting electric and magnetic fields so that they cannot pass through the subsequent aperture.
- *Objective lenses.* The ion beam is focused onto the write plane using the objective lenses and subsequent aperture.

- *Deflection electrodes.* The ion beam is scanned line-by-line by applying sawtooth voltages to the deflection electrodes. The required patterns are selectively exposed by blanking the ion beam in and out.
- *Wafer table.* The wafer table carrying the resist-coated substrate can be adjusted to a high degree of accuracy in the x and y direction. This means that each exposure sub-region can be aligned with its neighbour.

A high vacuum must be maintained over the whole length of the beam path to avoid ion collisions. When a liquid gallium source is used the ions have an energy of typically 60 keV.

Complex control electronics are used to convert the pattern data held on magnetic tape into electrical voltages to control ion beam blanking and deflection, and to control the movement of the wafer table in the x and y direction.

4.5.3
Ion beam projection

In ion beam projection a mask containing holes (stencil mask) is projected onto the surface to take the pattern (MIBL, *M*asked *I*on *B*eam *L*ithography). Pattern generation is performed in parallel, and is therefore faster than in ion beam writing.

Figure 4.5.4 shows the schematic design of an ion beam projection device. The major components are listed below.
- *Ion source with extractor.* The duoplasmatron [4.44; 4.45] is suitable for generating a stable ion beam with the high beam current density required. The anode aperture and the extraction electrode are designed so that the ion beam leaves the source with a beam angle of 4°. The ion current is typically greater than 200 µA. The ion source is coupled to a mass separator which blanks out the unwanted ions. The duoplasmatron can be used to generate ions of hydrogen, nitrogen, helium, neon, argon and xenon.
- *Immersion lens.* The ions hit the ion mask with an energy of about 5 keV. The ions that pass through the mask holes are accelerated in the immersion lens to an energy of more than 60 keV.
- *Beam alignment.* An eight-pole configuration is particularly suitable for aligning the beam [4.44].
- *Projection lenses.* The projection lenses produce a scaled image of the ion mask on the substrate to be patterned. Reduction factors of 5 or 10 are typical.
- *Wafer table.* As in light-optical projection equipment (Sect. 4.2.5), the wafer table can move in the x and y direction, so that again each exposure area can be placed side-by-side when used in "step and repeat" mode.

Geometrical and chromatic aberrations are minimized using a suitable combination of immersion and projection lenses. Chromatic aberrations arise because of the variation in the energies of the ions exiting the source, which are of the order of 1–10 eV. The ion-optical system works with a very small numerical ap-

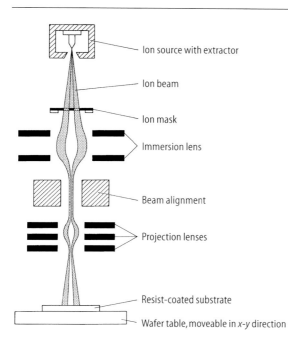

Fig. 4.5.4. Main elements of an ion projection lithography system (MIPL, *Masked Ion Projection Lithography*) [4.44; 4.45]

erture, and therefore has a relatively large focal depth (of the order of several 100 μm).

The exposure time depends on the necessary exposure dose and the ion current density. The ion current density should be as high as possible for a short exposure time, as can be seen in Fig. 4.5.5. Values of up to 1 mA/cm² are currently being achieved. Figure 4.5.5 also shows that significantly longer exposure times are required for ion-induced chemical etching and deposition than for resist-based ion lithography or for ion implantation.

Figure 4.5.6 shows the construction of an ion mask[39]. It essentially consists of a mask film, a frame and a mask support. The mask film contains holes that correspond to the patterns to be produced.

The optimum thickness of the mask film lies between 2 and 5 μm. This is thick enough for only parallel, unscattered ions to pass through the mask holes [4.46]. At the same time, the mask film is still thin enough for the holes to meet dimensional tolerances even in the sub-micron region. Using this technology, an image-reduction ion projection system can produce patterns with dimensions below 0.1 μm.

---

39 This type of mask contains holes for patterns. This means that isolated mask features cannot be included. This drawback can be overcome by performing two exposures with two complementary masks.

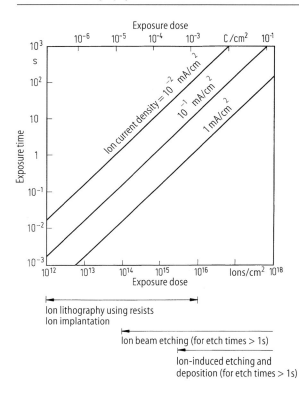

Fig. 4.5.5. Exposure time as a function of the exposure dose and the ion current density [4.44] based on the equation: dose = current density × exposure time

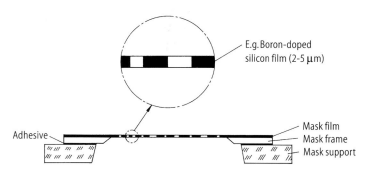

Fig. 4.5.6. Ion mask construction [4.44; 4.47; 4.60]

4.5.4
Resolution capability of ion lithography

There are several factors that can limit the resolution in ion lithography. One of these is the lateral scatter of ions in the resist. Back-scattering and secondary electrons, both of which affect the resolution capability of electron beam litho-

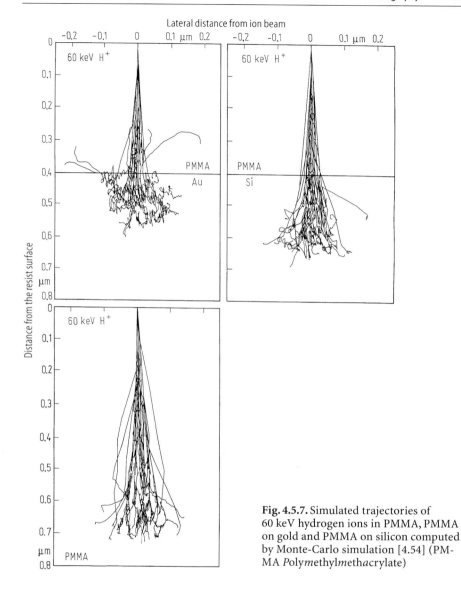

Fig. 4.5.7. Simulated trajectories of 60 keV hydrogen ions in PMMA, PMMA on gold and PMMA on silicon computed by Monte-Carlo simulation [4.54] (PMMA Polymethylmethacrylate)

graphy, are usually negligible here. Thanks to their relatively high mass, the incident ions only produce low-energy secondary electrons in the range 5–50 eV. This corresponds to a range of less than 10 nm, which is why secondary electrons do not start to affect the resolution until one gets down to the region below 0.02 μm.

Figure 4.5.7 shows simulated trajectories of several incident ions in PMMA, PMMA on gold and PMMA on silicon, computed using a Monte-Carlo simula-

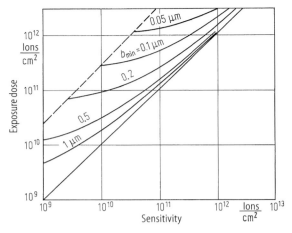

Fig. 4.5.8. Minimum pattern dimension b_{min} that can be obtained according to the resist sensitivity and the exposure dose, determined on the basis of the statistical distribution of ion emissions. The dashed line shows the resolution limit set by the resist sensitivity. At this point an area of b_{min}^2 will still just intercept one ion [4.55; 4.61]

tion. The ion beam diameter was assumed to be infinitely small in this case. A comparison with the results in Fig. 4.4.2 shows that the ions are scattered far less widely than the electrons. Back-scatter of the ions only arises with the gold film. Another factor limiting the resolving power is the statistical distribution of the emitted ions from the ion source. The smaller the pattern area, the lower the probability that it will intercept an ion. This is why smaller patterns require a higher exposure dose. As Fig. 4.5.8 shows, however, a resist with a lower sensitivity can then be used.

The resolution capability of ion lithography also depends on the lithography system being used. For instance the beam diameter constitutes an absolute resolution limit in ion beam writing, whereas in ion beam projection the smallest pattern that can be resolved is mainly determined by the lens aberration (cf. Sect. 4.4.3, in particular footnote 32).

Figure 4.5.9 shows the resolution limit b_{min} imposed by the lens aberration for the ion projection system described in [4.56], as a function of the diameter d_f of the image site.

The absolute limit of resolution in an aberration-free imaging system is determined by diffraction effects (Fig. 4.5.10). The smallest resolvable pattern is then given by the same equation as in light optical projection (Sect. 4.2):

$$b_{min} = 0.5 \lambda / NA \ .$$

The dimension b_{min} is the minimum distance between two points that can still just be resolved, λ is the de Broglie wavelength of the ions, and NA the numerical aperture of the projection system.

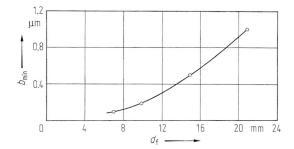

Fig. 4.5.9. Resolution capability b_{min} imposed by the lens aberration, as a function of the image field diameter d_f for the ion projection system described in [4.56]

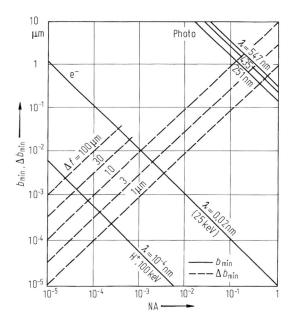

Fig. 4.5.10. Absolute limit of resolution b_{min} (continuous lines) and Δb_{min} (dashed lines) for an aberration-free ion projection system ($\lambda = 10^{-4}$ nm, H$^+$, 100 keV) as a function of the numerical aperture NA, showing comparative curves for photo- and electron lithography [4.57]. b_{min} is the minimum pattern dimension that can still just be reproduced before resolution is lost because of diffraction effects. Δb_{min} is the deterioration in the resolving power because of defocusing, where Δf is the distance between the image and focal plane and λ is the de Broglie wavelength of the radiation in question

If the substrate surface on which the pattern is to be produced (image plane) does not lie in the focal plane of the ion projection system, then a worse resolution must be expected, as in light optical projection (Sect. 4.2). The deterioration in the resolution limit caused by defocusing is given by:

$$\Delta b_{min} \approx \Delta f \cdot \mathrm{NA} \; ,$$

where Δf is the distance between the image and focal plane, and NA is the numerical aperture.

Looking at Fig. 4.5.10, one can conclude that even for a defocusing distance of 10 μm, ion projection lithography still has a potential resolving power of 1 nm. The main challenge of ion projection lithography is the 4× stencil mask. If the stability and cost issues of the stencil mask can be solved, ion lithography is a favourite candidate to replace optical lithography in the sub-0.1μm range.

4.6
Pattern generation without using lithography

The principle of lithography outlined in Fig. 4.1 is not the only way of creating required patterns. For example, lithography is not used to produce the lightly doped drain (Fig. 3.5.2 a) or to define the channel length of the DMOS transistor (see Sect. 8.3.2). The size of the area to be doped is defined by the thickness of a spacer or by the different penetration depths of two dopants, i.e. not by a mask geometry. The advantage of this type of pattern generation technique is mainly that pattern dimensions can be produced below the minimum achievable with

- Fabrication of a SiO_2-step

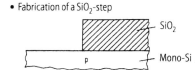

- Gate oxidation
- Fabrication of a n^+-doped poly-Si-spacer

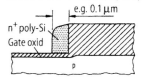

- Anisotropic SiO_2 etching
- Thermal oxidation
- Source/Drain-implantation

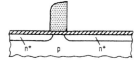

Fig. 4.6.1. Process sequence for the manufacture of MOS transistors. The length of the gate is defined by the thickness of poly-Si-spacers

- Trench etching in mono crystal silicon

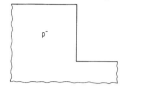

Fig. 4.6.2. Process sequence for the manufacture of vertical MOS transistors. The length of the gate is defined by the etch depth of a trench

- Source/Drain-implantation (vertical)
- Channel implantation (slanting)

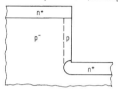

- Gate oxidation
- n^+-doped poly-Si-spacer

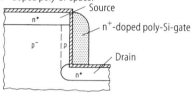

lithography. The fundamental limitation, however, is that the linewidth cannot be varied within the integrated circuit.

Figures 4.6.1 and 4.6.2 show two process sequences for MOS transistors. In Fig. 4.6.1 the gate length is defined by the thickness of a poly-Si spacer, whilst in Fig. 4.6.2 it is defined by the etch depth of a trench [3.40]. Vertical transistors of the type shown in Fig. 4.6.2 are particularly useful because they consume little space by making use of vertical surfaces as well.

A final mention should be made here of the trench capacitor in a DRAM memory cell, for which the detailed process sequence is shown in Sect. 8.6.4. The features produced in the trench wall to create the buried plate, collar and buried strap are not produced lithographically here, but involve processing steps such as spacer formation, defined resist removal, defined etch-back and oblique implantation.

5
Etching technology

Etching technology is used either to remove complete surface layers, or to transfer lithographically generated mask patterns into the underlying layer. The quality of pattern transfer depends on the type of etch process used. Typical etch profiles are shown in Fig. 5.1. In an isotropic etch process the same degree of attack occurs in all directions, so that undercutting of the layer occurs beneath the mask (Fig. 5.1 a). An anisotropic etch process is directional in character. Thus if etching only occurs normal to the wafer surface, the mask pattern is accurately transferred to the underlying layer without a change in dimensions (Fig. 5.1 b). The etch profile shown in Fig. 5.1 c is the most likely outcome of an etch process. A measure of the degree of anisotropy is the anisotropy factor f, defined as:

$$f = \frac{\text{vertical etch rate }(r_v) - \text{horizontal etch rate }(r_h)}{\text{vertical etch rate }(r_v)}$$

where the etch rate is given by

$$r = \frac{\text{etch depth }(\Delta z)}{\text{etch time }(\Delta t)}.$$

In an isotropic etch process the vertical etch rate r_v equals the horizontal etch rate r_h, and the anisotropy factor f is then equal to zero. In a purely anisotropic etch process f has a value of one.

In addition to the degree of anisotropy and the etch rate, the selectivity S is very important in an etching process. It is defined as follows:

$$\text{Selectivity } S_{12} \text{ between material 1 and material 2} = \frac{\text{etch rate of material 1 }(r_1)}{\text{etch rate of material 2 }(r_2)}$$

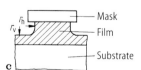

Fig. 5.1 a–c. Characteristic etch profiles. **a** Isotropic; **b** anisotropic; **c** common etch profile. r_v vertical etch rate; r_h horizontal etch rate; $f = (r_v - r_h) / r_v$ = anisotropy factor; $f = 0$ for isotropic etch process; $f = 1$ for anisotropic etch process

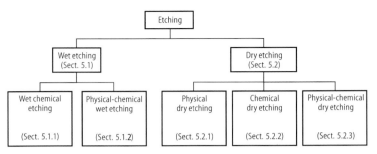

Fig. 5.2. Classification of the various etching processes used in the manufacture of integrated circuits

An etch process used in the manufacture of VLSI circuits should ideally meet the following requirements:
- low undercutting (anisotropy factor f as close to 1 as possible)[1]
- high selectivity to the material lying below, so that is it etched as little as possible[2]
- low contamination of, and damage to, the etched surface
- high etch rate (for commercial reasons)
- highly uniform etch rate over the semiconductor wafer
- high degree of repeatability in order to meet production tolerances
- minimum heating of the wafer to be etched, since the photoresist starts to flow at a temperature of about 100 °C
- low accumulation of charge to avoid oxide breakdown.

Figure 5.2 classifies the various etching techniques under wet and dry processes, and indicates which section in this chapter deals with each technique.

5.1
Wet etching

Wet etching refers to the removal of solid material by dissolving it in a chemical solution. Wet etching is classified as wet chemical etching if the material is dissolved away by purely chemical means, or as physical-chemical wet etching if a mechanical factor is also involved.

5.1.1
Wet chemical etching

Wet chemical etching of semiconductor wafers is performed in practice either by immersing in chemical baths or by spraying with etching solutions. Etching is

---
1 Sometimes sloping etch edges are required (see Sect. 5.3.5).
2 Does not apply for planarization (see Sect. 5.3.7).

generally isotropic. There are, however, a range of etching solutions such as KOH or NaOH, which etch monocrystalline silicon anisotropically and at rates that depend on the doping level. Both effects are exploited in micromechanics to produce fine patterns [5.2, 5.54].

When monocrystalline silicon is etched with KOH, an $n-p^+$ junction acts as an etch stop, because the etch rate of the heavily p-doped (p^+) silicon is more than 100 times lower than that for the n-doped region [5.53].

Isotropic wet chemical etching causes undercutting beneath the etch mask (see Fig. 5.1 a). This is why isotropic wet chemical etching is only rarely used for the fine patterns in VLSI circuits. An exception is the edge bevelling of contact holes shown in Fig. 8.5.2, which is performed by combined isotropic and anisotropic etching.

Wet chemical etch techniques have the general characteristics below:
– high selectivity
– very low contamination of, and damage to, the etched surface
– high degree of uniformity and repeatability, and
– an adjustable etch rate by changing the ratio of components in the etching solution

Wet chemical etching techniques are mainly used for full-surface etching of films, and for removal of unwanted thin insulating films.

Table 5.1 catalogues the major etching solutions used in integrated circuit technology. When these etching solutions are formulated, small percentages of wetting agents must also be added to ensure that good wetting of the etch surface is achieved.

5.1.2
Chemical-mechanical polishing

Chemical-mechanical polishing (CMP) is a planarizing technique that meets the more demanding requirements of sub 0.5 μm technology. It can either be thought of as chemically enhanced mechanical polishing, or as a mechanically enhanced wet chemical etch process.

Figure 5.1.1 shows the schematic of chemical-mechanical polishing.

A flexible, perforated pad containing polishing slurry lies on a revolving polishing table. The semiconductor wafer to be polished is pressed onto the pad by the chuck, with the wafer rotating in the opposite direction to the polishing table.

Instead of performing purely mechanical polishing, which is a well-established technique [5.2], the slurry contains active chemical components in addition to abrasive particles. They enable selective removal of the films on the semiconductor wafers. The abrasive particles have diameters ranging from 20 to 500 nm, and are usually made of quartz, aluminium oxide or cerium oxide. The chemical additives are chosen to suit the film material to be removed. For in-

Table 5.1. Etching solutions used in integrated circuit manufacture. CVD = Chemical Vapour Deposition. LTO = Low Temperature Oxide, LP = Low Pressure, PSG = Phosphosilicate Glass, thermal = thermal oxidation. The etching solution data is mainly obtained from [5.1]. This reference contains a comprehensive list of wet chemical etching materials. The percentage figures in brackets indicate the concentration of the solution

Etching solution	Material to be etched	Etch rate nm/min	Film	Etch temp. °C	Comments
20 H_3PO_4 (85 %) 1 HNO_3 (65 %) 5 H_2O	aluminium	220	sputtered	40	selective to SiO_2
76 H_3PO_4 (85 %) 3 HNO_3 (65 %) 15 CH_3COOH (100 %) 5 H_2O and small percentage of NH_4F (40 %) at 1 vol% NH_4F at 5 vol% NH_4F	aluminium	 160 100	sputtered	40	selective to SiO_2
7 NH_4 (40 %) 1 HF (49 %)	SiO_2 PSG	130 240–800	thermal PSG	30	buffered hydrofluoric acid (BHF); selective to Si; etch rate depends on SiO_2 doping
3 HF (49 %) 2 NHO_3 (65 %) 640 H_2O	SiO_2 PSG	1.9 3–4	thermal PSG	25	"PSG etch", "P– etch" for etching PSG and very thin SiO_2 films
20 H_3PO_4 (85 %) 1 NHO_3 (65 %) 4 H_2O	SiO_2 PSG	5.8 7–41 (depends on phosphorus content)	thermal PSG	25	"backdoor etch" for etching PSG and very thin SiO_2 films
H_3PO_4 (85 %)	SiN_4	6.0	LP-CVD	160	selective to SiO_2, SiO_2 etch rate 0.3–0.4 nm/min
2 HF (49 %) 15 HNO_3 (65 %) 5 CH_3COOH (100 %)	Si	depends on doping	CVD poly-Si; monocrystalline Si	25	"planar etch", selective to SiO_2, isotropic etch

Table 5.1. (continued)

Etching solution	Material to be etched	Etch rate nm/min	Film	Etch temp. °C	Comments
KOH (3-50 %)	Si [100] orientation	20	monocrystalline Si or poly-Si	70–90	anisotropic etch with respect to crystallographic orientation for V trenches (NB attacks positive photoresist) Heavily p-doped region acts as etch stop
	Si [111] orientation	≈ 0			
1 HF (49 %) 10 NH$_4$F (40 %)	TaSi$_2$	20	sputtered		silicide on polysilicon
1 HF (49 %) 10 NH$_4$ F (40 %)	TiSi$_2$	≥ 150	sputtered		silicide on polysilicon

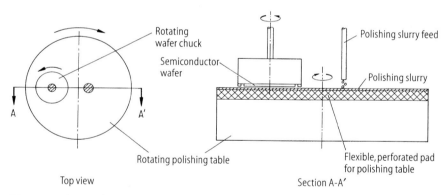

Fig. 5.1.1. Chemical-mechanical polishing equipment [5.55]

stance a wolfram film can be selectively planarized down to the SiO$_2$ using a ferricyanide phosphate additive [5.55].

Chemical-mechanical polishing is mainly used in the following applications, illustrated in Fig. 8.5.3:
- planarization of trench fillings
- planarization of metal plugs (e.g. wolfram plugs) in contact holes and vias
- planarization of intermediate oxides and intermetal dielectrics

The following features of chemical-mechanical polishing are important in addition to its planarizing action:
- increased yield thanks to reduced defect density and simpler processing
- smooth substrate makes pattern transfer easier during subsequent dry etching
- increased integration density because a smaller metallization pattern can be used on a smooth substrate.

The selectivity of the thinning process to the underlying layer is not infinitely high, so endpoint detection is still required in chemical-mechanical polishing. The following two techniques are particularly suitable:
- capacitance measurement to determine the thickness of the insulating layers
- measurement of the power consumption of the rotating chuck; where the power changes at the point where the film materials change.

5.2
Dry etching

In dry etching the material is etched away by atoms and molecules in a gas and/or by bombarding the surface to be etched with ions, photons or electrons. The etch process can be physical or chemical in nature, or a combination of the two. The various forms of dry etching are shown schematically in Fig. 5.2.1.

5.2.1
Physical dry etching

In physical dry etching the surface to be etched is bombarded with ions, electrons or photons.

Electron or photon bombardment causes vaporization of the material to be etched, and the associated techniques are called electron beam and laser vaporization respectively. The laser technique is of more interest in semiconductor technology, where it is used for repairing photomasks, isolating faulty regions in integrated circuits and for labelling wafers.

In ion etching, atoms are knocked out of the etching surface by ion bombardment. It can be used to etch all materials that occur in semiconductor technology. Ion etching processes do have the following serious disadvantages however:
- the etch rates are low
- there is low selectivity between different materials, one consequence often being severe etching of the resist masks
- the etch profile usually has an edge angle of less than 90 °C
- reflection of the incident ions at the sloping etch sides creates trenches around the edges of the film to be etched, or in the substrate (trench effect)
- the high energy of the ions often causes damage to the etching surface
- the atoms knocked out by ion bombardment are often redeposited at the mask edge, and produce unwanted ridges after the mask is removed (redeposition)
- pattern edges are removed at a faster etch rate.

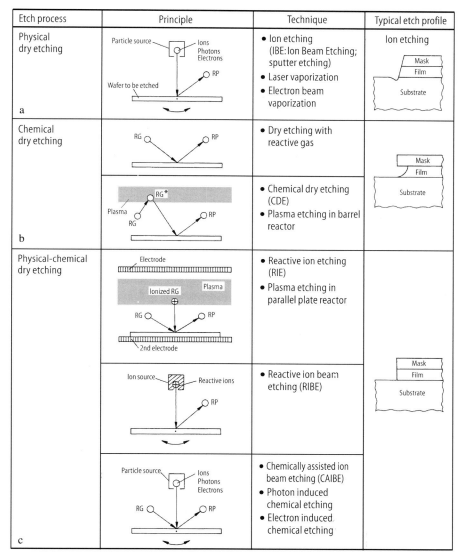

Fig. 5.2.1 a–c. Various forms of dry etching: **a** physical; **b** chemical; **c** physical-chemical; RG = reaction gas, RG* = excited reaction gas, RP = reaction product

The last effect is exploited in the so-called Dep./Etch process for planarizing film surfaces. In this technique, PECVD deposition is performed at the same time as ion etching in a parallel plate reactor (see Fig. 3.1.7 b). Apart from this application, ion etching is hardly used nowadays in the manufacture of integrated circuits.

5.2.2
Chemical dry etching

In chemical dry etching, a chemical etching reaction takes place between neutral particles in a gas and atoms in the surface to be etched. For an etching process to occur, a gaseous, volatile reaction product must be formed. For example, excited fluorine atoms from a plasma form the volatile reaction product SiF_4 with atoms from the silicon surface. Since the velocity of the neutral particles is normally evenly distributed in all directions, an isotropic etch profile is usually obtained with this process (Fig. 5.1 a). Etching techniques that are purely chemical are therefore rarely suited to producing very fine features. In ULSI circuit production they are therefore used almost exclusively for etching away complete layers of films.

The barrel reactor, also known as the tunnel reactor, was the first commercial reactor for chemical dry etching. Originally it was used solely for stripping photoresist. The main components of photoresist (C, H, N, O) combine with oxygen to produce volatile reaction products, so it is easily etched off in an oxygen plasma. Later it was discovered that silicon, SiO_2 and Si_3N_4 could also be etched in a barrel reactor, although not by oxygen but by fluorine. A typical barrel reactor design is shown in Fig. 5.2.2.

The barrel reactor consists of a vacuum chamber 4, electrodes 5 for the rf voltage, an inlet 2 for the etching gases and an outlet 3 to the vacuum pump. A holder 7 (boat) carrying the wafers for etching is located inside the vacuum chamber. A perforated metal cylinder 6 is placed between the vacuum chamber walls and the wafers to screen the etching area from electromagnetic fields.

A suitable gas for etching is fed into the previously evacuated reactor. At a pressure of around 100 Pa and with a constant gas flow, a plasma is ignited by applying an rf voltage to the electrodes.

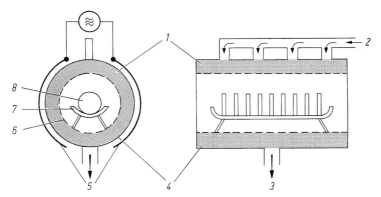

Fig. 5.2.2. Basic design of a barrel reactor [5.1] *1* plasma; *2* gas; *3* vacuum pump; *4* vacuum chamber; *5* electrodes; *6* perforated metal cylinder (tunnel); *7* holder; *8* wafers

5.2 Dry etching   177

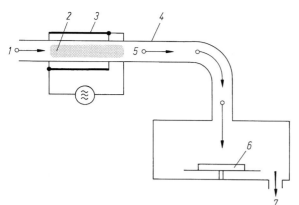

Fig. 5.2.3. Schematic diagram of chemical dry etching where plasma and etching are physically separated. *1* gas inlet; *2* plasma; *3* electrodes for generating the plasma; *4* gas pipe; *5* excited gas atoms/molecules (radicals); *6* semiconductor wafer to be etched; *7* outlet to vacuum pump

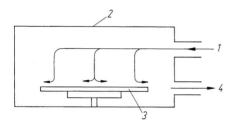

Fig. 5.2.4. Basic design of a chamber for dry etching with a neutral unexcited gas. *1* etching gas; *2* etching chamber; *3* semiconductor wafer; *4* outlet to vacuum pump

Excited neutral atoms or molecules (radicals) diffuse out of the plasma towards the wafers, and react chemically with the atoms on the wafer surface. For etching to occur, volatile reaction products must be formed, which can be extracted by the vacuum pump. This is therefore the main factor when selecting the etching gases. Etching occurs in every direction, and thus produces an isotropic etch profile, similar to that usually obtained with wet chemical etching (Fig. 5.1 a). Selectivity is generally high.

A specific version of chemical dry etching (CDE) is shown in Fig. 5.2.3, where plasma generation is physically separated from the etching space.

A plasma *2* is ignited in a tube *4* through which gas is flowing, either by applying an rf voltage to the electrodes *3*, or by coupling in microwave energy. The radicals *5* generated in the plasma, flow towards the surface for etching. There they form gaseous reaction products which are extracted by the vacuum pump *7*. In this method only neutral gas particles reach the wafers, and no ions, which results in a purely chemical etching process.

Chemical dry etching can also occur with an unexcited gas. Figure 5.2.4 shows the basic design of a processing chamber where direct etching of the semiconductor wafer surface occurs by the gas introduced to the chamber.

SiO_2 films can be etched very homogeneously using a mixture of HF, N_2 and water vapour fed into the etching chamber. This etching process can be used to etch unannealed TEOS SiO$_2$ films (Sect. 3.5.1) with a selectivity of more than 10:1 over thermal SiO$_2$ [5.56].

This dry etch technique is also suitable for cleaning silicon surfaces [5.57], where other gases such as HCl are used in addition to HF/H$_2$O.

5.2.3
Physical-chemical dry etching

Physical-chemical dry etching is of major importance in ULSI technology, because it can be used to produce very fine features. A chemical etching reaction is triggered on the wafer surface by bombarding with ions, electrons or photons. If particle bombardment is normal to the surface, then the mask pattern can be transferred to the layer below without loss of dimensional accuracy (Fig. 5.1 b). The selectivity of the etching process lies between that of purely physical and that of purely chemical dry etching. Once again a volatile reaction product must be produced for the etching process to take place.

The reactors currently used in semiconductor engineering for physical-chemical dry etching employ ion bombardment to trigger the chemical etching reaction. The most widely used equipment is the parallel plate reactor shown schematically in Fig. 5.2.5. It meets the primary requirement for an anisotropic etching process, since the ions bombard the etch surface perpendicularly.

In a similar design to the sputter chamber (Sect. 3.1.4), the reactor basically comprises a vacuum chamber with an inlet for the etching gas, an outlet to the vacuum pump, and two parallel electrodes. The wafers to be etched are placed on one of the two electrodes. The reactor is first evacuated, then a suitable etching gas introduced. Electronic control is used to maintain a constant gas pressure and flow rate. An rf voltage applied to the electrodes sets up a gas glow discharge between them. The result is a low pressure, low temperature plasma containing ions, electrons and excited neutral particles (radicals). The electrons, which are lighter than the ions, can track the rf field between the electrodes, whilst the ions cannot. This means that far more electrons than ions reach the electrodes during each rf half-phase. The electrodes thus become negatively charged, and attract the positive ions out of the plasma. The negative charge of the electrodes settles at a level where the electron and ion currents to the electrodes exactly cancel out when averaged over time. Under suitable processing

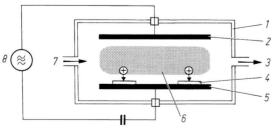

Fig. 5.2.5. Design of a parallel plate reactor for physical-chemical dry etching [5.2]. *1* vacuum chamber; *2* upper electrode; *3* vacuum pump; *4* semiconductor wafers; *5* lower electrode; *6* plasma *7* etching gas inlet; *8* rf generator

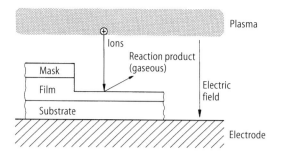

Fig. 5.2.6. Formation of the anisotropic etch profile in physical-chemical dry etching in the parallel plate reactor. The chemical etch reaction is triggered by ion bombardment

conditions, the ions hitting the wafer surface can trigger a chemical etching reaction. The reactive particles that are required for this either come from the surrounding plasma, or directly from the incident ions. If the etching reaction is solely triggered by the incident ions, which arrive normal to the surface, then the result is an anisotropic etch profile, as shown in Fig. 5.2.6. The etch mask is then transferred accurately to the layer below, which means that very fine patterns can be produced with this technique.

It is therefore important when developing etching processes for ULSI circuit manufacture that chemical etching reaction solely occurs in the region of ion bombardment.

The most important etching processes performed with the parallel plate reactor are summarized in Fig. 5.2.7.

In reactive ion etching, also referred to as reactive sputter etching, the rf voltage is capacitively coupled to the lower electrode. The upper electrode is connected to the vacuum chamber, which is earthed. Because it is connected to the chamber, the upper electrode has a larger surface area than the lower electrode. This results in the lower electrode being more negatively charged than the upper one (cf. Sect. 3.1.4). As they travel towards the wafer, the ions from the plasma attain a sufficiently high kinetic energy ($W > 100$ eV) to trigger a chemical reaction. The gas pressure in the etching chamber is also relatively low (0.1–10 Pa), so that the ions suffer few collisions. This means that they arrive normal to the etching surface, enabling the mask pattern to be accurately transferred to the layer below.

Anodically coupled plasma etching differs from reactive ion etching by the coupling of the electrodes: the lower electrode is now earthed, whilst the rf voltage is capacitively coupled to the upper electrode. The lower electrode thus has only a relatively low negative charge, and the positively charged ions reach the etching surface with only a low kinetic energy ($W < 100$ eV). The relatively high gas pressure (10–1000 Pa) means that the ions collide with atoms and molecules on their way to the etching surface, further reducing their energy. Often it is too low for a chemical etching reaction to be triggered. Furthermore, the ions are scattered by collisions and thus do not arrive normal to the etching surface. This means that the etch mask is rarely transferred accurately to size to the layer below.

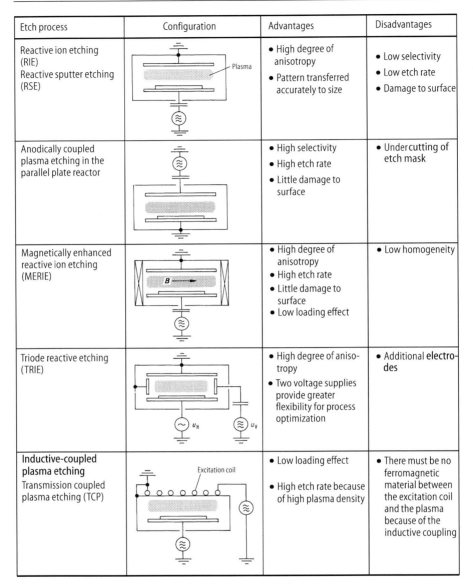

Etch process	Configuration	Advantages	Disadvantages
Reactive ion etching (RIE) Reactive sputter etching (RSE)		• High degree of anisotropy • Pattern transferred accurately to size	• Low selectivity • Low etch rate • Damage to surface
Anodically coupled plasma etching in the parallel plate reactor		• High selectivity • High etch rate • Little damage to surface	• Undercutting of etch mask
Magnetically enhanced reactive ion etching (MERIE)		• High degree of anisotropy • High etch rate • Little damage to surface • Low loading effect	• Low homogeneity
Triode reactive etching (TRIE)		• High degree of anisotropy • Two voltage supplies provide greater flexibility for process optimization	• Additional electrodes
Inductive-coupled plasma etching Transmission coupled plasma etching (TCP)		• Low loading effect • High etch rate because of high plasma density	• There must be no ferromagnetic material between the excitation coil and the plasma because of the inductive coupling

Fig. 5.2.7. Etching processes performed in the parallel plate reactor

The reactor for magnetically enhanced reactive ion etching [5.31, 5.37, 5.63] differs from that for reactive ion etching by having a built-in magnet. The magnetic field it produces increases the density of the plasma over the wafers to be etched. Thus considerably more ions and reactive species are available for etching. One can then afford to reduce the pressure compared with reactive ion etching (0.01–10 Pa). Since the ions suffer fewer collisions when the pressure is low, the degree of anisotropy increases. The etch rate is also higher than that of reactive ion etching because of the high density of ions and reactive species. Another advantage of the magnetically enhanced technique is the relatively low kinetic energy of the ions (typically < 100 eV). It is normally high enough to trigger the chemical etching reaction, yet is also low enough to cause almost no damage to the surface being etched, and to keep wafer heating during etching within limits. A further increase in the etch rate can be achieved by exposure to light (e.g. using a laser) [5.32]. These photon-enhanced techniques are still under development.

The reactor used for triode reactive ion etching contains two vertical electrodes in addition to the electrodes lying horizontally. An rf voltage u_V applied to the vertical electrodes generates a plasma in the reactor. The excitation frequency is typically 13.56 MHz. The voltage u_H across the horizontal electrodes attracts the ions out of the plasma to the surface for etching. Either a dc voltage is used, or an ac voltage with normally so low a frequency (< 500 kHz) that the ions can follow the alternating field. In another option, a combined dc and ac voltage is applied.

A certain degree of independent control can be exercised in triode reactors over the ion energy and the ion current density. Similar results to those obtained with the excitation described above can also be achieved if both the rf voltage and the low frequency or dc voltage are applied to just the two horizontal electrodes.

Unlike the techniques described so far, inductive-coupled plasma etching uses inductive coupling rather than capacitive coupling for the rf power [5.58]. This highly effective coupling method increases the plasma density. An additional rf source, capacitively coupled to the lower electrode, attracts the positive ions out of the plasma and accelerates them towards the etching surface. Here they trigger a chemical reaction. Thanks to the low gas pressure (0.1–10 Pa), the ions hardly suffer any collisions, and impact the wafer normal to its surface. Thus the etching mask is transferred exactly to size to the layer being etched.

In the etching reactors shown in Fig. 5.2.7, it is not normally possible to vary the plasma density independently of the ion energy. Thus one cannot reduce the ion energy, say, to reduce surface damage whilst simultaneously increasing the plasma density to maintain the etch rate. New types of etching reactors have been developed to provide this flexibility, where plasma generation and ion extraction mainly occur independently.

Key examples of these new etching reactors are summarized in Fig. 5.2.8.

A reactor for reactive ion beam etching (RIBE) [5.38] basically consists of an ion source and a vacuum chamber containing a neutralizer, a wafer table, an aperture and a vacuum pump feed. RIBE equipment currently available on the market mainly uses Kaufmann ion sources. In these sources electrons are emitted from a heated filament, and accelerated towards the anode by a voltage applied between the filament and the anode. They collide with and ionize gas atoms in their path, creating a low pressure, low temperature plasma. By applying

5 Etching technology

Etching technique	Etching reactor design	Source	Advantages	Disadvantages
Reactive ion beam etching (RIBE) / Chemically assisted ion beam etching (CAIBE)		• Kaufmann source	• Low gas pressure • High degree of anisotropy • Etch profile adjustable by setting table angle	• Low etch rate • Source sensitive to reactive gases • Inhomogeneous etch rate • Complex reactor compared with parallel plate reactor
Electron cyclotron resonance (ECR) etching / Reactive ion stream etching (RISE)		• ECR source	• Low gas pressure • High plasma density • Little surface damage • High degree of anisotropy	• Complex reactor compared with parallel plate reactor
Etching with inductively coupled plasma source (ICP) / High density plasma (HDP)		• Inductively coupled source	• High etch rate because of high plasma density • Low gas pressure • High degree of anisotropy	• Complex reactor compared with parallel plate reactor
Etching with helicon source Mode M = 0 resonant induction (MORI) / Resonant inductive plasma etching (RIPE)		• Helicon source	• Low gas pressure • High plasma density • Little surface damage • High degree of anisotropy • Uniform etch rate	• Complex reactor compared with parallel plate reactor

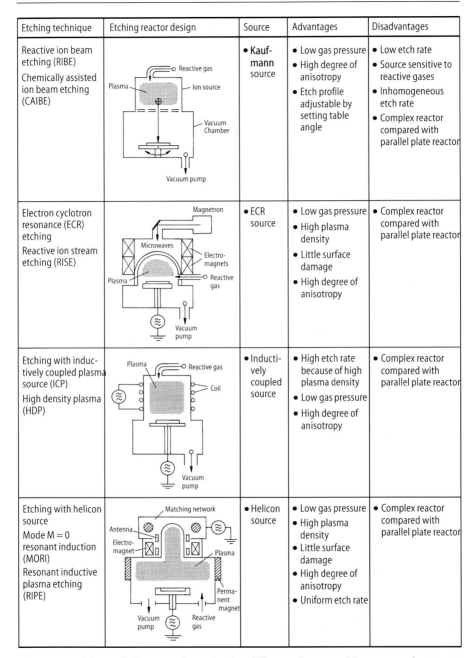

Fig. 5.2.8. Etching techniques performed using different plasma and ion sources [5.42, 5.58, 5.59, 5.62]

a positive voltage U to the extraction grid, positive ions are attracted from the ion source and accelerated to a kinetic energy of about 1 keV. The ion source works at a pressure of around 10^{-2} Pa.

The neutralizer, usually a heated metal filament, emits electrons to neutralize the positively charged ions.

There are two options for performing reactive ion beam etching:
1. using ions from a reactive gas (e.g. CF_3^+) (standard RIBE)
2. using ions from an inert gas (e.g. Ar^+) together with a reactive gas near the wafer being etched (chemically assisted ion beam etching, CAIBE).

In both cases a chemical etching reaction is triggered by ion bombardment.

Reactive ion beam etching offers the following benefits compared with etching in a parallel plate reactor:
1. The kinetic energy and the current density of the ions can be set independently of the other process parameters.
2. The working pressure range is very low, which means that the isotropic etch component is also very small. Furthermore, the reaction products are volatile even at low temperatures, which is not the case in reactive ion etching (particularly important when etching metals)
3. The table can be angled to adjust the etch profile (from sloping to vertical edges: important in contact hole etching, Sect. 5.3.5).

Reactive ion beam etching does, however, suffer from the following disadvantages:
1. Parts of the reactor are sensitive to reactive gases
2. Inhomogeneity in the density of the ion beam produces a non-uniform etch rate
3. The ion beam tends to be divergent
4. The etch rate is usually lower than for etching in a parallel plate reactor
5. It tends to be harder to detect the etching endpoint than in a parallel plate reactor.

The drawback of using the commercial source in RIBE equipment is that the heated filament does not last long in the reactive gases [5.36]. Alternative sources have therefore been developed. Of these, the plasma source is particularly interesting [5.58]. It does not generate an ion beam, but produces a plasma beam containing ions and electrons. The main advantage of this is that the wafers being etched do not build up an average charge over time. Furthermore, plasma sources work in the ion saturation region, so they generally provide far higher current densities than ion sources. The plasma is generated by rf or microwave fields.

Figure 5.2.8 shows three different types of etching reactor fitted with plasma sources. Each uses a different method of plasma generation: ECR (*Electron Cyclotron Resonance*), inductive coupling of an rf voltage or helicon mode plasma excitation.

5 Etching technology

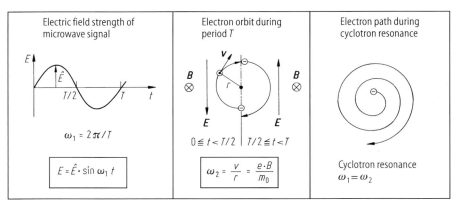

Fig. 5.2.9. Principle of electron cyclotron resonance in an ECR plasma source. The angular velocity ω_2 of the orbiting electrons is determined by the magnetic flux density B and the electron mass m_0. The period T of the microwave signal is selected so that the electrons are constantly accelerated by the electric field of field strength E. Since ω_2 is constant, both the radius of orbit r and the radial velocity increase. The electron describes a spiral path with increasing kinetic energy. (Frequently used figures: $B = 87.5$ mT, $f = 1/T = 2.45$ GHz)

The basic components of an ECR plasma source include a magnetron for generating microwave energy, and a magnet which deflects moving electrons in a circular path within its field. The period T of the microwave energy and the magnetic flux density B are designed so that the electrons reach cyclotron resonance. As shown in Fig. 5.2.9, the electrons then follow a spiral path, their kinetic energy constantly increasing. The result is that the ionization efficiency of the electron collisions increases, and thus the plasma density increases.

In the ECR reactor shown in Fig. 5.2.8, two electromagnets are positioned around the plasma source. The first magnet produces an inhomogeneous field, in which the conditions for cyclotron resonance are met within a certain region. The second magnet is used to produce a magnetic field close to the wafers, in order to increase the etch rate and the etch homogeneity. The energy of the extracted ions can be controlled by the rf voltage capacitively coupled to the electrode. ECR sources usually work in the pressure range 10^{-2}–1 Pa, at a microwave frequency of 2.45 GHz. The magnetic flux density required for cyclotron resonance is then $B = 87.5$ mT.

A high plasma density is also a feature of inductively coupled plasma sources. Whereas in capacitively coupled systems the rf power is shared between electrons and ions, in inductively coupled equipment it only benefits the electrons. Increased energy to the electrons then leads to a higher plasma density.

Figure 5.2.8 shows the basic design of an etch reactor with an inductively coupled source. The rf energy is inductively coupled via a coil positioned around the plasma source. A second rf voltage is capacitively coupled through the wafer table, and this is used to control the kinetic energy of the ions impacting the wafer.

In helicon sources, helicon waves are excited within a plasma. These are right-hand circularly polarized electromagnetic waves which can be set up in a highly conductive medium, such as a metal at low temperatures, or a gas plasma in which a stationary magnetic field exists [5.60]. In gas plasmas, this phenomenon is also referred to as "whistler" waves. The term "whistler" comes from the whistling sound that can be heard when listening to a short wave receiver. This sound is produced by atmospheric whistler waves, which are generated at very high energies by lightning, and propagate along the Earth's magnetic field.

Whistler waves are a very efficient means of generating plasmas. Electrons having a velocity equal to the phase velocity of the wave are accelerated because of Landau decay, extracting energy from the wave in the process. The accelerated electrons then produce new electrons and ions in ionizing collisions. This mechanism produces plasma densities of up to $n_e = 10^{13}$ cm^{-3} [5.58]. In helicon sources the whistler waves are excited by an inductively coupled rf field. The design of the coupling antenna is a critical factor in determining the excitation mode of the wave, defined by the mode no M ($M = 0, 1, 2...$). The mode $M = 0$ is particularly important here, because it can produce very homogeneous plasmas.

MORI ($M = 0$, Resonant Induction [5.59]) and RIPE (*Resonant Inductive Plasma Etching* [5.60]) are alternative names for helicon source techniques.

Figure 5.2.8 shows the basic design of an etching system using a helicon source, in which the plasma is excited in mode $M = 0$. An rf field, usually at the industry standard frequency $f = 13.56$ MHz, is coupled into the system through an antenna placed around the source. An electromagnet encircles the antenna, generating a perpendicular field within the helicon source, which is necessary for the excitation of whistler waves. The etching chamber lies below the helicon source and contains the wafer table and the pipes to the vacuum pump and gas supply. A multi-pole permanent magnet is placed around the walls of the upper section of the etching chamber to prevent plasma losses. The ions are extracted from the plasma by an rf voltage coupled to the wafer table.

Etching processes using helicon sources have the following characteristics:
- excellent plasma homogeneity giving a very uniform etch rate
- high plasma density producing a high etch rate
- low voltage between plasma and wafer reducing the damage to the surface being etched
- thin dark space region between plasma and wafer reducing collisions of incident ions, and thus producing a high degree of anisotropy for the etch.

The quality of the dry etch process often suffers because of wafer heating caused by the process. As the temperature rises, the isotropic etch component increases and the photoresist suffers increased erosion. Both effects impair the quality of pattern transfer. The wafer therefore needs to be cooled. Simply cooling the wafer chuck is not enough, however, because the thermal conductivity between chuck and wafer is too low when the wafer just rests on the chuck. Thermal conductivity can be increased if the wafer is pressed onto the chuck, which can be

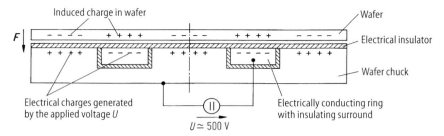

Fig. 5.2.10. Principle of electrostatic wafer clamping. Applying a dc voltage U sets up an electrostatic attractive force F between the wafer and chuck because of charge transfer

done either mechanically or electrostatically. Electrostatic wafer clamping has proved the most effective option in practice [5.51, 5.52].

Optimum cooling is achieved if, in addition, a cooling gas such as helium is passed over the back of the wafer.

Figure 5.2.10 shows the basic principle of electrostatic wafer clamping. An electrically conducting ring surrounded by an insulating layer is set into the chuck. By applying a voltage of about 500 V between the chuck and the ring, electric charges of different polarity are set up. These in turn induce opposite charges in the wafer. An electrostatic attractive force F results, which pulls the wafer onto the chuck.

5.2.4
Chemical etching reactions

As already stated, the formation of a gaseous, volatile reaction product is necessary for a chemical etching reaction to take place. This is therefore the main criterion for selecting etching gases. Table 5.2 summarizes the common etching reactions. Since the temperature at which a substance boils depends on the pressure, Table 5.2 specifies both the temperature and pressure range for the gaseous state of aggregation of the reaction products. The vapour pressure curves for the reaction products listed in Table 5.2, as well as many others, can be found in [5.6].

One can see from Table 5.2 that silicon can be etched with fluorine, chlorine bromine and iodine, although higher process temperatures are required when using iodine. Si, SiO_2 and Si_3N_4 can be etched with fluorine, having reaction products that are gaseous down to very low temperatures. Aluminium only forms a volatile reaction product with fluorine at very high temperatures, which means that aluminium cannot be etched with fluorine. Chlorine, bromine and iodine are suitable alternatives.

The refractory metals molybdenum, tantalum, titanium and wolfram can all be etched with fluorine, chlorine, bromine and iodine, although mostly only at a raised temperature. Titanium does, however, form a reaction product with flu-

Table 5.2. Chemical etching reactions [5.6]

material	etching reaction	reactions products are gaseous for	
		temperatures greater than (°C)	gas pressures less than (Pa)
Si	Si+4F→SiF$_4$	−130	10^3
	Si+4Cl→SiCl$_4$	−40	10^3
	Si+4Br→SiBr$_4$	+25	10^3
	Si+4J→SiJ$_4$	+140	10^3
SiO$_2$	SiO$_2$+4F→SiF$_4$+O$_2$	−130	10^3
Si$_3$N$_4$	Si$_3$N$_4$+12F→3SiF$_4$+2N$_2$	−130	10^3
Al	Al+3F→AlF$_3$	880	10
	Al+3Cl→AlCl$_3$	60	2
	2Al+6Br→Al$_2$Br$_6$	40	2
	2Al+6J→Al$_2$J$_6$	+140	10
Mo	Mo+6F→MoF$_6$	−50	10^3
	Mo+6Cl→MoCl$_6$	+50	2
	Mo+5Br→MoBr$_5$	85	2
	Mo+4J→MoJ$_4$	220	10
Ta	Ta+6F→TaF$_6$	40	2
	Ta+6Cl→TaCl$_6$	70	2
	Ta+5Br→TaBr$_5$	125	2
	Ta+5J→TaJ$_5$	180	2
Ti	Ti+6F→TiF$_6$	170	10^3
	Ti+6Cl→TiCl$_6$	20	10^3
	Ti+4Br→TiBr$_4$	20	20
	Ti+4J→TiJ$_4$	185	10^3
W	W+6F→WF$_6$	−50	10^3
	W+6Cl→WCl$_6$	90	2
	W+4Br→WBr$_4$	125	2
	W+4J→WJ$_4$	240	10
C	C+O→CO	−191	10^5
	C+2O→CO$_2$	−78	10^5

orine and oxygen that is volatile under typical etching process conditions ($p < 100$ Pa, $T \approx 300$ K).

Carbon can be etched very easily with oxygen. The main reaction products are CO and CO$_2$, which are both gaseous at very low temperatures. Since the main components of organic polymers are C, H, N, O, F and Cl, these polymers can also be etched very easily with oxygen.

When etching semiconductor and insulating films, it is crucial that the doping elements boron, gallium, phosphorus, arsenic and antimony also form volatile reaction products at standard etching temperatures with fluorine, chlorine, bromine and iodine. Otherwise the etching reactors would rapidly become contaminated by the dopants, which could later return indiscriminately to the semiconductor wafer, diffusing back into the wafer during the next high temperature process.

The etching reactions listed in Table 5.2 mostly only occur spontaneously, i.e. without external supply of energy, if the gases (F, Cl, Br, I or O) are present in excited atomic form. As this is not their naturally occurring state, these atoms must either be generated directly in a plasma, or at the surface being etched by ion bombardment. Etching is isotropic if the atoms are generated in the plasma and have a sufficiently long lifetime to diffuse to the etching surface. If the atoms are not produced until ion bombardment occurs at the etching surface, then this constitutes the classic case of a chemical reaction triggered by ion bombardment. The resultant etching process is anisotropic.

5.2.5
Etching gases

Compounds of the halogen gases F, Cl, Br and I are used to etch inorganic compounds, whilst organic films are etched almost exclusively by oxygen. Table 5.3 contains a list of those etching gases frequently used in integrated circuit technology. In all, some hundred etching gases and etching gas mixtures are known of from the literature. [5.1] lists many of these in alphabetical order according to the gases and materials to be etched. Numerous new gases and gas mixtures are continually being added to the list. Table 5.3 is therefore just a snapshot, and cannot claim to be complete or up to date in the future.

5.2.6
Process optimization

The characteristics of an etch process such as etch rate, selectivity and anisotropy are normally optimized using the following process parameters, which can be set independently of each other:
1. Pressure, flow and composition of the etch gas,
2. Power of the rf field or the electrical potential of the electrodes
3. Frequency of the rf voltage
4. Temperature of the electrode on which the wafer for etching is lying.

When an etching process is being developed for the manufacture of ULSI circuits, the parameters above are varied to achieve the maximum degree of anisotropy and selectivity, and as far as possible a maximum etch rate. The effect of each process parameter on the etching process is discussed below.

Gas flow

The gas flow determines the maximum amount of reactants available. The actual quantity depends on the generation and recombination processes of the reactive

Table 5.3. Selection of gases for etching Si, SiO$_2$, Si$_3$N$_4$, metals, metal silicides and photoresist

Gas	Etched material	Etching process selective to	Etch profile	Literature
BCl$_3$/Cl$_2$	Si, Poly-Si	SiO$_2$	anisotropically	[5.1]
BCl$_3$/CF$_4$	Si, Poly-Si	SiO$_2$	anisotropically	[5.44]
BCl$_3$/CHF$_3$	Si, Poly-Si	SiO$_2$	anisotropically	[5.44]
Cl$_2$/CF$_4$	Si, Poly-Si	SiO$_2$	anisotropically	[5.44]
Cl$_2$/He	Si, Poly-Si	SiO$_2$	anisotropically	[5.42]
Cl$_2$/CHF$_3$	Si, Poly-Si	SiO$_2$	anisotropically	[5.44]
HBr	Si, Poly-Si	SiO$_2$	anisotropically	[5.44]
HBr/Cl$_2$/He, O$_2$	Si, Poly-Si	SiO$_2$	anisotropically	[5.42]
HBr/NF$_3$/He, O$_2$	Si, Poly-Si	SiO$_2$	anisotropically	[5.49; 5.50]
HBr/SiF$_4$/NF$_3$	Si, Poly-Si	SiO$_2$	anisotropically	[5.44]
HCl	Si, Poly-Si	SiO$_2$	anisotropically	[5.42]
CF$_4$	Si, Poly-Si	SiO$_2$	anisotropically, isotropically	[5.1]
CF$_4$/O$_2$	Si, Poly-Si	SiO$_2$	isotropically	[5.42]
SF$_6$/He	Si, Poly-Si	SiO$_2$	isotropically	[5.42]
CF$_4$/H$_2$	SiO$_2$	Si	anisotropically	[5.1]
C$_2$F$_6$	SiO$_2$	Si	anisotropically	[5.44]
C$_3$F$_8$	SiO$_2$	Si	anisotropically	[5.44]
CHF$_3$	SiO$_2$	Si	anisotropically	[5.1]
CHF$_3$/O$_2$	SiO$_2$	Si	anisotropically	[5.42; 5.44]
CHF$_3$/CF$_4$	SiO$_2$	Si	anisotropically	[5.42; 5.44]
CF$_4$/O$_2$	SiO$_2$	Al	isotropically	[5.42]
CF$_4$/H$_2$	Si$_3$N$_4$	Si	anisotropically	[5.1; 5.44]
CF$_4$/CHF$_3$/He	Si$_3$N$_4$	Si, SiO$_2$	anisotropically	[5.42]
CHF$_3$	Si$_3$N$_4$	Si, SiO$_2$	anisotropically	[5.1; 5.44]
C$_2$F$_6$	Si$_3$N$_4$	Si, SiO$_2$	anisotropically	[5.42]
SF$_6$	Si$_3$N$_4$	SiO$_2$	anisotropically	[5.44]
CF$_4$/O$_2$	Si$_3$N$_4$	SiO$_2$	isotropically	[5.42]
SF$_6$/He	Si$_3$N$_4$	SiO$_2$	isotropically	[5.42]
BCl$_3$	Al	SiO$_2$	anisotropically	[5.1]
BCl$_3$/Cl$_2$	Al	SiO$_2$	anisotropically	[5.42; 5.43; 5.44]
BCl$_3$/Cl$_2$/He	Al	SiO$_2$	anisotropically	
BCl$_3$/Cl$_2$/CHF$_3$/O$_2$	Al	SiO$_2$	anisotropically	[5.42]
HBr	Al	SiO$_2$	anisotropically	[5.42]
HBr/Cl$_2$	Al	SiO$_2$	anisotropically	[5.44]
HJ	Al	SiO$_2$	anisotropically	[5.42]
SiCl$_4$	Al	SiO$_2$	anisotropically	[5.1; 5.42; 5.44]
SiCl$_4$/Cl$_2$	Al	SiO$_2$	anisotropically	[5.44]

Table 5.3. (continued)

Gas	Etched material	Etching process selective to	Etch profile	Literature
Cl_2/He	Al	SiO_2	isotropically	[5.42]
BCl_3/N_2	Cu	SiO_2	anisotropically	[5.39]
SF_6	W	SiO_2, TiN	anisotropically, isotropically	[5.42; 5.44; 5.48]
SF_6/Ar	W	SiO_2, TiN	anisotropically	
SF_6/Cl_2/CCl_4	W	SiO_2	anisotropically	[5.48]
NF_3/Cl_2	W	TiW, TiN	anisotropically	[5.44] [5.44]
Cl_2/O_2/Ar	Cr	SiO_2	anisotropically, isotropically	[5.46]
CF_4/Cl_2	$CoSi_2$	SiO_2	anisotropically	[5.44]
CCl_2,F_2	$CoSi_2$	SiO_2	anisotropically	[5.44]
CCl_2,F_2/NF_3	$CoSi_2$	SiO_2	anisotropically	[5.44]
Cl_2/O_2/He	$MoSi_2$	SiO_2	anisotropically	[5.45]
SF_6/Cl_2	$MoSi_2$	SiO_2	anisotropically	[5.44]
SF_6/HBr/O_2	$MoSi_2$	SiO_2	anisotropically	[5.45]
BCl_3/Cl_2	$TaSi_2$	SiO_2	anisotropically	[5.1]
Cl_2/Ar	$TaSi_2$	SiO_2	anisotropically	[5.1]
SF_6/Cl_2	$TaSi_2$	SiO_2	anisotropically	[5.1]
CF_4/Cl_2	$TiSi_2$	SiO_2	anisotropically	[5.44]
CCl_2,F_2	$TiSi_2$	SiO_2	anisotropically	[5.44]
CCl_2,F_2/NF_3	$TiSi_2$	SiO_2	anisotropically	[5.44]
CF_4/Cl_2	WSi_2	SiO_2	anisotropically	[5.44]
CCl_2,F_2	WSi_2	SiO_2	anisotropically	[5.44]
CCl_2,F_2/NF_3	WSi_2	SiO_2	anisotropically	[5.44]
O_2	photoresist	Si, SiO_2, Si_3N_4, metals	isotropically, anisotropically	[5.1; 5.42]
O_2/CF_4	photoresist	Si, SiO_2, Si_3N_4, metals	isotropically	[5.1; 5.42]

species. These species are primarily generated by oscillating electrons in the rf field, which collide with the incoming gas molecules and thus generate excited particles, electrons and ions. The reactive species are lost through the etch reaction, recombination with other particles and extraction by the vacuum pump. The rate of loss depends heavily here on the mean residence time of the reactive species in the etch space. The mean residence time Δt is given by:

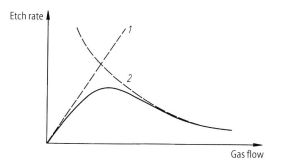

Fig. 5.2.11. Theoretical curve of the etch rate with respect to the gas flow $\dot{m} = dm/dt$. *1* etch rate ~ gas flow; *2* etch rate limited by the residence time of the reactive species

$$\Delta t = \frac{pV}{RT\dot{m}} \tag{5.1}$$

where p is the gas pressure, V is the etching chamber volume, R is the gas constant, T is the absolute temperature and $\dot{m} = dm/dt$ is the gas flow (mass flow) in the etching chamber.

Figure 5.2.11 shows the theoretical curve of the etch rate with respect to the gas flow. For low gas flow rates, the etch rate is defined by the incoming gas flow. In this region the residence time Δt of the reactive species is much greater than their life time τ. The situation reverses when the frequency of etch reactions of the reactive species is limited by their residence time Δt. The etch rate is then proportional to the residence time Δt and is thus proportional to the reciprocal of the gas flow $\dot{m}$ according to Eq. (5.1). Increasing the gas flow then causes the etch rate to fall.

Fluorine atoms generated in the plasma have a long lifetime, so that over a large gas flow range the etch rate is set by the residence time. On the other hand chlorine atoms in the plasma have a very short life time, so in this case the etch rate is proportional to the gas flow rate for most of its range.

Gas pressure

Both the etch profile and the etch rate depend on the gas pressure. As the gas pressure decreases, the degree of anisotropy of the etching process increases. This is because the incident ions are scattered less as a result of the longer mean free paths, and can thus acquire a higher kinetic energy.

The active species (excited neutral particles, ions) are generated in the plasma by collisions between the gas molecules or atoms and electrons. The energy needed by the electrons to generate the etching species is derived from the electric field of the plasma. As the gas pressure rises, there is an increased probability that a gas molecule or atom will be hit by an electron. This is why the etch rate initially rises with increasing gas pressure, as shown in Fig. 5.2.12. As the gas pressure continues to rise, an opposing process takes over: the reduction in the mean free path of the electrons means that the kinetic energy they have derived

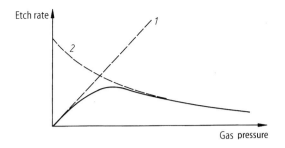

Fig. 5.2.12. Theoretical curve of the etch rate with respect to the gas pressure. *1* etch rate ~ gas pressure; *2* etch rate restricted by the mean free path of the colliding electrons

from the electric field reduces. This in turn causes a fall in the number of etch species that are generated, and hence a fall in the etch rate.

Another reason for the decrease in etch rate with rising gas pressure, is that the etching surface can become increasingly covered by polymers, etch residues or adsorbed reaction products.

Coupled rf power and frequency

When etching is performed in a plate reactor (e.g. RIE), the etch rate usually increases linearly with rf power. It is therefore very important to ensure that the reflected rf power from the etching reactor is minimized, i.e. the system must be matched. Any mismatch has a significant effect on the etch rate. The upper limit for the rf power is set by the maximum temperature that the wafers are permitted to reach. Normally they must not exceed 100 °C, because otherwise the photoresist mask starts to deform. In etching using plasma or ion sources, the etch rate generally increases as the energy and current density of the incident ions increase. The kinetic energy of the ions depends strongly on the frequency of the applied voltage, but in most etching systems the frequency is fixed.

Temperature

If the etching process is a purely chemical reaction, then the temperature dependence of the etch rate r can usually be expressed by the Arrhenius relationship below [5.8]:

$$r = r_0 \exp(-W_A / kT) \tag{5.2}$$

where W_A is the activation energy, T the absolute temperature and k is Boltzmann's constant. According to (5.2), the etch rate increases with rising temperature, although inverse temperature relationships have been measured in isolated cases. An increase in thermal desorption as the temperature rises might be the cause here, since the thermally desorbed species would no longer be available for the etching process. In practice, one tries to maintain a constant temperature for the wafers being etched, in order to achieve as uniform and consistent an etch rate as possible. For this reason the electrode carrying the wafers is usually cooled and thermostatically controlled. One must, however, ensure that there is good thermal contact between the wafer and electrode (see Fig. 5.2.10).

Loading effect

Apart from the parameters already discussed, the etch rate also often depends on the size of the surface to be etched. This is called the "loading effect". For the case where there is only one reactive species responsible for the etching process, the etch rate r is related to the loading effect as follows:

$$r = \frac{\beta \tau G}{1 + k\beta \tau A} \tag{5.3}$$

where β is a constant of the reaction, τ is the lifetime of the active species, G their generation rate, k a constant that depends on the material and reactor geometry, and A is the area of the surface to be etched. It can be seen from equation (5.3) that the etch rate falls as the etch surface area A increases.

5.2.7
Endpoint detection

Since the selectivity of most etching processes to the underlying layer is not sufficiently high, it is necessary to use instrumentation to detect when etching is complete. This is either performed directly by measuring the film thickness, or indirectly by measuring the plasma parameters.

At the point where the layer to be etched changes to the layer below, there is often a change in the plasma characteristics, such as a change in the density of the species. The variation in these parameters over time can therefore indicate the endpoint. The density of individual species can be measured using either emission spectroscopy or mass spectroscopy, although the former is most common for reasons of cost. Emission spectroscopy can be used to detect the emission lines over a specific wavelength region. The measured signal is proportional to the light intensity and therefore to the density of the species associated with that emission line. The spectral line that changes most in intensity at the layer transition is selected for endpoint detection purposes.

Figure 5.2.13 shows the variation in intensity of a nitrogen emission line during endpoint detection of an Si_3N_4 film. As soon as the Si_3N_4 film has been etched away, no more nitrogen is generated and the signal for the nitrogen emission line falls to the background intensity level.

Table 5.4 shows some of the emission lines used in end point detection.

Emission spectroscopy detects the integral endpoint, i.e. the endpoint is indicated when the film being etched has been removed from all the wafers in the reactor. The sharpness with which the intensity reading falls is here a measure of the uniformity of the etch process. Simple photodiodes are often used for endpoint detection. This is possible because the intensities of nearly all emission

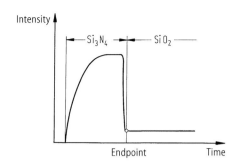

Fig. 5.2.13. Intensity of a nitrogen emission line over time during etching of an Si$_3$N$_4$ film (λ = 337.1 nm, N$_2^+$, 2nd positive) [5.9]

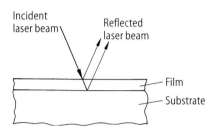

Fig. 5.2.14. Principle of laser interferometry for endpoint detection and measurement of the etch rate [5.41]

lines change together at the layer transition. Since the change in the intensity reading is proportional to the surface area being etched, emission spectroscopy is only of very limited use where small surface areas are being etched, such as in contact hole etching (Sect. 5.3.5).

Laser interferometry can detect the endpoint at a particular location. It can also detect the etch rate "in situ", i.e. during the etch process, when transparent films are being etched. In laser interferometry, a laser beam is directed at the film being etched (Fig. 5.2.14). For transparent films the reflected laser beam is made up of two components, a reflected beam from the film surface, and one from the interface between film and substrate. The phase difference between the two reflected beams depends on the thickness of the film being etched. Depending on this phase difference, the two beams either interfere constructively or destructively. The overall intensity of the reflected laser beam therefore varies periodically with the film thickness, and thus with the etch time t (Fig. 5.2.15).

The etch rate r is obtained from the period T of the reflected beam by the following equation:

$$r = \frac{\lambda}{2nT} \tag{5.4}$$

where λ is the wavelength of the laser beam and n is the refractive index of the film being etched. The reflectivity of the surface usually changes at the transition between the film and substrate. This produces a knee in the intensity curve of the

Table 5.4. Emission lines for endpoint detection

Layer	etch gas	emission lines from reactant	etch product	wavelength nm	literature
silicon silicides	containing F	F		704	[5.10; 5.40]
			Si	252	
			SiF	778	[5.40]
			SiF	440	
	containing F and O		CO	298	[5.11]
			CO	484	[5.11]
			CO	520	[5.11]
	containing Cl		SiCl	287	[5.11; 5.40]
			Si	288	
SiO$_2$	containing F and C		CO	484; 451; 482; 520 184	[5.17] [5.40]
SiO$_2$ doped with P	containing F		P	254	
Si$_3$N$_4$	containing F	F		704	[5.10]
			N$_2$	337; 362	[5.9; 5.40]
			N	674	[5.10; 5.40]
	containing F and C		CN	387	[5.10; 5.40]
Al	containing Cl		Al	396	[5.14; 5.40]
			Al	394	[5.40]
			Al	391	
			Al	309	[5.40]
			Al	266	
			Al	257	
Al	containing Cl		AlCl	309	
			AlCl	308	
			AlCl	261	
			AlCl	522	
			CCl	306	[5.17; 5.40] [5.40]
refractory metals	containing F			704	[5.10]
Cu	Ar		Cu	325	
Cr	Ar		Cr	358	

5 Etching technology

Table 5.4. Emission lines for endpoint detection

Layer	etch gas	emission lines from reactant	emission lines from etch product	wavelength nm	literature
organic polymers	O_2		CO	298	[5.15]
				308	[5.40]
				484	[5.16]
				283	
				520	
	O_2		OH	309	
	O_2		H	656	
	O_2	O		777	[5.17]
				843	
				616	

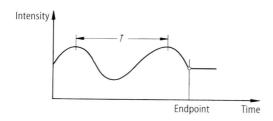

Fig. 5.2.15. Typical intensity curve over time for the reflected laser beam when etching a thin transparent film. The etch rate can be found from the period T

reflected laser beam over time, as shown in Fig. 5.2.15. The technique is therefore suitable for detecting both the endpoint and the etch rate for transparent films, but can only detect the endpoint if the film is not transparent.

At the transition from the etched film to the underlying substrate, the composition of the plasma changes. This produces slight fluctuations in the gas pressure, gas flow rate and the plasma impedance. Any of these parameters can therefore be used for endpoint detection as well. Ellipsometry, a technique that is being used more widely, can measure the etch rate in situ in a similar way to laser interferometry.

Some endpoint detection techniques also perform the special function of plasma diagnostics, detecting important plasma parameters during the etching process. Apart from optical emission spectroscopy which has already been described, other suitable options are mass spectroscopy, laser-induced fluorescence spectroscopy, microwave interferometry and the Langmuir probe method.

5.3
Dry etch processes

Dry etch processes are predominantly used in the manufacture of integrated circuits for creating fine patterns. The process involves transferring an etch mask, usually in the form of a photoresist pattern, to the layer beneath the mask with the maximum possible dimensional accuracy. Etching must not normally continue into the following layer. In order to meet these requirements, the etch process that is used must have a high degree of anisotropy (Fig. 5.1 b) and high selectivity to the underlying layer and to the etch mask. The layers to be etched can be made of silicon (monocrystalline, polycrystalline or amorphous), SiO_2, Si_3N_4, metals (usually aluminium with small additions of silicon, copper or titanium), metal silicides and organic polymer films [5.7, 5.9].

5.3.1
Dry etching of silicon nitride

Silicon nitride films are used for local oxidation (LOCOS nitride, Table 2.1, process numbers 7–9 and Sect. 3.7.2), and for passivation of integrated circuits (Table 2.1, process number 20 and Sect. 3.7.4). These Si_3N_4 layers are normally produced by the CVD technique (Sect. 3.7). For local oxidation, the nitride films are between 100 and 200 nm thick. These films must be etched very precisely to size because they define the active areas in which the electronic components will later exist. Beneath the nitride film there is an SiO_2 layer which is about 20–50 nm thick. It does not matter if etching proceeds into this layer, so a selectivity to SiO_2 of between 5 : 1 and 10 : 1 is adequate for the Si_3N_4 etching process, assuming that this process has good uniformity over the whole wafer. Table 5.3 includes some of the gases suitable for etching the LOCOS nitride. When these gases are used under optimum processing conditions, the selectivity between Si_3N_4 and SiO_2 lies in the required range of 5 : 1 to 10 : 1.

Sometimes a polysilicon film also lies between the LOCOS nitride and SiO_2 layers, designed to shorten the bird's beak feature (transition between thick and thin oxide region, Sect. 3.4.2). If this is the case, then the Si_3N_4 must be etched selectively to polysilicon. Most of the gases used for SiO_2 etching are suitable here (e.g. CHF_3 with added O_2, H_2 etc.).

The etching process that is used to make holes for the pads in the passivation layer is uncritical (Table 2.1, process number 20). These pads are so large that dimensional accuracy is not a factor here. Furthermore, since the holes go down to Al, selectivity is not a problem. This is because Si_3N_4 is etched by fluorine-only gases, and aluminium does not produce a volatile reaction product with fluorine at the prevailing etching temperatures.

5.3.2
Dry etching of polysilicon

Polysilicon is used as a gate and interconnection material in MOS technology, and as a contact material and dopant source in bipolar technology (Sect. 3.8). In MOS transistors the channel length is defined by the polysilicon gate (Fig. 8.3.1). Since the electrical properties of the transistor are highly dependent on this length, the polysilicon gate must be etched accurately to size. The polysilicon gate also lies on a very thin layer of gate oxide (Table 2.1, process number 12, gate oxide thickness: 5-50 nm), so both the anisotropy and the SiO_2 selectivity must satisfy very high demands. Both requirements, however, are very hard to achieve at the same time.

High selectivity is obtained when the chemical etching component dominates. In this case neutral etching species react chemically with the atoms in the surface being etched, producing a volatile reaction product. Since the velocity of the neutral etching species is isotropically distributed, the etching process is also mostly isotropic (Fig. 5.1 a). Where the physical component predominates, i.e. the ions have a high kinetic energy, the etching tends to be anisotropic. The selectivity is low, however, because in purely physical etching processes (e.g. ion etching with Ar^+) nearly all materials are removed at the same rate.

Polysilicon can be etched with fluorine-based gases such as CF_4 and SF_6. In anodically coupled plasma etching (Fig. 5.2.7), i.e. at a relatively high gas pressure, etching is mainly isotropic, however, because of the long lifetime of the fluorine-based etching species.

Anisotropic etch profiles are obtained in RIE operation because of the higher ion energy and the lower gas pressure (Fig. 5.2.7), although the selectivity of the etch process to SiO_2 is too low here for gases containing fluorine. The requirements for anisotropy and selectivity are not normally met with fluorine-based gases.

Etching species (radicals or ion) containing chlorine, bromine or iodine usually have a much shorter lifetime compared with fluorine species. The etching species reaching the semiconductor wafer either react rapidly with silicon or with suitable recombination partners. In comparison with fluorine, these etching species therefore remain only a very short time on the etching surface, which promotes anisotropic etching. Etch gases containing Cl, Br or I are thus more likely to meet the anisotropy and selectivity requirements than fluorine-based gases. Frequently used etch gases include Cl_2, Cl_2/He, Cl_2/BCl_3, Br and HCl (see Table 5.3). It has also be found that both the etch rate and the anisotropy depend on the doping of the polysilicon. Heavily n-doped layers are etched at more than ten times the rate of heavily p-doped silicon. Whilst undoped layers are usually etched anisotropically, heavily n-doped layers often exhibit undercutting.

The dependence of the etch rate on the doping can be explained by the position of the Fermi level. In n-doped silicon the Fermi level lies near the bottom of the conduction band, and in p-doped silicon it lies near the top of the valence band. Electrons for the Si-halogen atom bond thus have a lower energy barrier to overcome in n-doped silicon than in p-doped silicon. This produces the higher etch rate in n-doped silicon.

The degree of anisotropy can be increased by adding recombination partners such as CF_4.

In chlorine and bromine processes the loading effect is small compared with that of fluorine processes.

As the level of integration increases, the component dimensions get smaller and the gate oxides become thinner (Fig. 8.3.3). This means that the demands

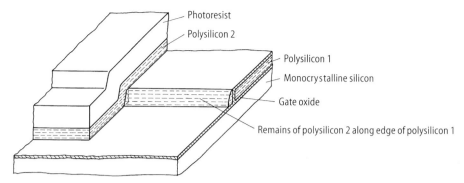

Fig. 5.3.1. Illustration of the problems associated with residues of polysilicon 2 at the edges of polysilicon 1

placed on polysilicon etching processes also increase. The ideal process would have an anisotropy factor of 1 and a selectivity to SiO_2 of more than 50 : 1.

In the double or triple polysilicon gate process, the different polysilicon planes often overlap (Fig. 5.3.1). Since the thickness of polysilicon film 2 is greater at the edges of polysilicon 1, some of polysilicon film 2 is normally left at these points. Such polysilicon residues must always be removed, otherwise they can short-circuit adjacent components. This requires a very high selectivity to SiO_2, since the thin gate oxide film is exposed over wide areas.

5.3.3
Dry etching of monocrystalline silicon

One application requiring etching of monocrystalline silicon is the manufacture of dynamic memory cells (Table 8.14).

Silicon etching is illustrated below using the example of a trench capacitor in a dynamic memory cell, since particularly strict requirements need to be met here. As shown in Fig. 5.3.2, a trench capacitor consists of a polysilicon electrode, a heavily n-doped monocrystalline silicon electrode (n^+) and a thin SiO_2 or ONO film as the dielectric. When used in a 256 Mbit DRAM, the trench shown in Fig. 5.3.2 would be about 10 µm deep and have a cross-sectional area of less than 0.3 µm².

The etching process should ideally meet the following requirements:
- the trench walls should be smooth and without contamination after etching, so that the subsequent oxidation process can produce a perfect dielectric;
- the trench floor should have rounded edges to avoid too high an electric field strength, which could cause breakdown of the dielectric at these points;
- the trench walls should be vertical, to enable optimum filling with polysilicon.

The following gases are just a selection of those suitable for etching silicon trenches (Table 5.3):
- $HBr/Cl_2/He$; $HBr/NF_3/He$, O_2.

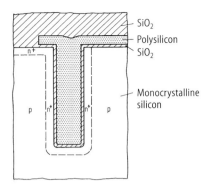

Fig. 5.3.2. Structure of a trench capacitor in a dynamic memory cell. A thin SiO_2 film acts as the dielectric between the two electrodes (polysilicon and n^+-doped silicon) [5.65]

SiO_2 is often used as the etch mask, because the organic resists that are normally employed are removed too rapidly for etching deep trenches.

5.3.4
Dry etching of metal silicides and refractory metals

The refractory metals molybdenum, tantalum, titanium, wolfram and their silicide compounds $MoSi_2$, $TaSi_2$, $TiSi_2$ and WSi_2 have a much lower resistivity than polysilicon. Since the switching times of integrated circuit components are shorter for a lower resistance, polysilicon is often replaced or combined with a metal silicide or refractory metal (Sect. 3.9 and 3.10).

A requirement for plasma induced etching of silicides and refractory metals is that the reaction products are volatile under etching process conditions. One can see from Table 5.2 that some of the reaction products of refractory metals with fluorine, chlorine, bromine and iodine are only gaseous at relatively high temperatures, and this should be taken into account when selecting etching gases. Furthermore, the etching chamber often has to be heated to avoid the formation of etch residues. Heating must be kept below fairly low limits, however, since the photoresist normally cannot be heated above 100 °C. This is why $TiSi_2$, for instance, should not be etched using a gas containing only fluorine. Adding oxygen to the fluorine/titanium reaction allows a gaseous reaction product to be obtained at lower temperatures.

For various reasons the polysilicon is usually not completely replaced by a metal silicide or refractory metal. Normally one uses a double layer of polysilicon and silicide (polycide: see Sect. 3.9.2), which places greater demands on the etching process. Both the silicide and the polysilicon need to be etched anisotropically, otherwise the situation shown in Fig. 5.3.3 can arise under unfavourable conditions. This profile presents incredible problems for subsequent film deposition.

The following gases are just some of the gases used for anisotropic etching of polycides and refractory metals: SF_6/Cl_2, BCl_3/Cl_2, Cl_2/Ar (Table 5.3).

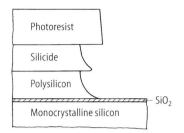

Fig. 5.3.3. Edge profile of an etched polycide layer. The isotropic etch component is here greater for polysilicon than for the silicide

The problem discussed in Sect. 5.3.2 regarding the residues of polysilicon 2 along polysilicon 1 edges, is heightened in polycide applications. This is because the anisotropic etching required to avoid the etch profile shown in Fig. 5.3.3 is usually achieved at the expense of selectivity. Thus the selectivity to SiO_2 is normally not great enough to remove the remaining polysilicon. In fact the ideal etching process would again be anisotropic with a selectivity to SiO_2 of more than 50 : 1.

Of the refractory metals, wolfram has acquired a particular significance, as it is ideally suited to providing high-quality electrical contacts (Sect. 8.5.2). In this application wolfram needs to be selectively etched over Ti/TiN. Suitable etching gases include SF_6 and SF_6/Ar [5.48], which can achieve selectivities of more than 50 : 1.

5.3.5
Dry etching of silicon dioxide

Silicon dioxide is used for isolating the electronic components and the various interconnection planes (see Table 2.1 and Sects. 3.4.1 and 3.5.2).

The devices in an integrated circuit are electrically connected to create the required circuit operation. In the simplest case just one metallization layer is used to connect them, but more often there are several interconnection levels with up to four polysilicon or polycide layers plus up to six metal layers. Contact holes are etched through the SiO_2 insulating layers to link the various interconnection levels. These holes can lead to monocrystalline silicon, polysilicon, a metal silicide or a metal.

Since they are relatively thin, minimal etching of the interconnections should occur when the contact holes are being etched. The SiO_2 etch process must therefore meet very high selectivity requirements. This is not a problem when the contact holes lead to aluminium, as SiO_2 is usually etched with a fluoride-based gas, and aluminium does not produce a gaseous reaction product with fluorine. Silicon, however, is etched by fluorine, forming volatile SiF_4. High selectivity between SiO_2 and silicon can only be achieved by optimizing the gas composition and process parameters.

SiO_2 films can be etched using CF_4, C_2F_6, C_3F_8 and CHF_3, either alone or combined, and/or with added H_2, O_2, CH_4, C_2H_4, C_2H_2 etc. Experiments have

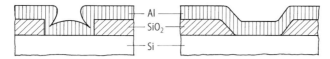

Fig. 5.3.4. Sputtered metal contacts on contact holes with vertical and sloping sides

shown that the CF_x^+ molecules with $x \leq 3$ are the most likely etch species, and that the etch reaction is triggered by incident ions. This means that the etch profile is mostly anisotropic (Fig. 5.1 b), irrespective of the selected etching mode.

The SiO_2 etching process is usually accompanied by polymeric deposition. One factor determining this polymerization tendency is the F/H ratio in the plasma. The etch rate increases with the fluorine concentration, whilst the polymerization rate increases with the hydrogen concentration. One obtains the highest selectivity between SiO_2 and Si at the cross-over point between etching and polymerisation. In order to achieve high selectivity, the process parameters (pressure, gas flow rate, rf power) and the gas composition are adjusted so that a polymer film is just forming on the Si surface, but the SiO_2 is still being etched. One reason for the different polymerisation behaviour of Si and SiO_2 is that the oxygen released during etching of SiO_2 forms volatile reaction products (CO, CO_2) with the carbon in the polymer film.

Apart from high selectivity, etching of contact holes must meet several other demands. For instance contamination must be kept to a minimum. Heavy metal contaminants such as iron or copper are particularly harmful here, because they diffuse into the monocrystalline silicon during subsequent high temperature processing (see Sect. 6.4) creating defects in the crystal. The result is a drastic deterioration in the semiconductor properties. Heavy metal contamination can arise, for example, from atoms sputtered off a stainless steel chamber.

After etching, small contact holes are often left with polymer linings. These must be removed before deposition of the next interconnection layer to prevent excessive contact resistances. Usually such polymer coatings can be removed in a barrel reactor with an oxygen plasma.

During contact hole etching, the wafers are exposed to radiation from the plasma. This can lead to radiation damage, which can only be repaired by annealing at higher temperatures.

As already mentioned, the SiO_2 etching process is usually anisotropic. This often causes edge coverage problems for subsequent film deposition (Fig. 5.3.4). This is particularly true for sputtered metal films, where narrowing of the film frequently occurs at edges. The increased current density at these points heightens material transport effects (electromigration, Sect. 3.11.3) and leads ultimately to cracks in the interconnection. Where contact holes have sloping sides, this problem does not arise, or at least only to a lesser degree.

Contact holes with sloping sides can be produced using the techniques listed below (Fig. 8.5.2).

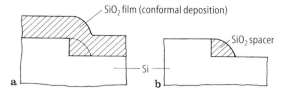

Fig. 5.3.5 a, b. Formation of an SiO$_2$ spacer. **a** Cross-section through structure after conformal SiO$_2$ deposition; **b** cross-section after anisotropic etching of the SiO$_2$ film

- *Reflow technique* (see also Sect. 3.6.2). The SiO$_2$ film is heavily doped with phosphorus and/or boron. This produces a silica glass (flow glass) with a relatively low melting point (800–1100 °C). Heating to temperatures above the melting point causes the silica glass around the contact holes to flow, rounding off the SiO$_2$ sides. The silica glass that has flowed into the contact holes has to be etched away again afterwards however. This is usually done by a full-surface wet chemical overetch.
- *Combination of wet and dry etching.* In this technique the contact holes are first etched by a wet chemical technique, and then completed using a dry etching process. Since the wet chemical etch is isotropic and the dry etching process is anisotropic, this sequence produces sloping contact hole sides.
- *Reactive ion etching using variable plate separation.* When SiO$_2$ is etched by the reactive ion etching process, the degree of anisotropy can be adjusted by varying the plate separation. Sloping SiO$_2$ sides are then obtained if the plate separation is varied appropriately during the etching process [5.27].

The anisotropic nature of SiO$_2$ etching is exploited to produce "spacers". These are bands of material that form along steps after anisotropic etching. Figure 5.3.5 shows the principle behind SiO$_2$ spacer formation. As already explained in Sect. 5.3.2, anisotropic etching of conformal films leaves behind part of the film along steps. When this residual material is polysilicon it can cause faults and must be removed, but for SiO$_2$ this effect can be put to good use for spacer technology (see Sect. 3.5.3).

5.3.6
Dry etching of aluminium

Aluminium is used as interconnection material in integrated circuits (Table 2.1, process numbers 17–19). In order to improve its technological properties, small quantities of silicon, copper or titanium are added to the aluminium (see Sect. 3.11). Aluminium films with thicknesses in the range 0.3–1.5 µm are mostly produced by sputtering. They normally lie on an insulating layer of SiO$_2$, so Al etching must be selective to SiO$_2$.

Aluminium cannot be etched with fluorine-based gases, because the reaction product AlF$_3$ is only volatile at temperatures above 800 °C. Gases containing chlorine, iodine or bromine are suitable as etching gases, however, although their reaction products still require a temperature of more than 50 °C before they become volatile (Table 5.2). Equipment for aluminium etching must therefore be heated to at least these temperatures. Commonly used etching gases are Cl_2/He, $SiCl_4$, HBr, BCl_3 and BCl_3/Cl_2. When process parameters are optimized, anisotropic etching of aluminium can be achieved with these gases.

Several different problems need to be overcome when etching aluminium. For instance, a thin Al_2O_3 film is usually present on the Al surface, and this cannot be etched by Cl or Cl_2. It therefore has to be removed before the actual Al etching process, either by sputtering it off, or by chemical reduction, e.g. in an H_2 plasma.

Another problem arises because of the hygroscopic nature of the reaction product $AlCl_3$. Some of the $AlCl_3$ precipitates onto the internal walls of the etching reactor, and when the reactor is opened up it absorbs moisture from the air to form $Al(OH)_3$ and HCl. These reaction products are desorbed during the next etching process, so that Al_2O_3 forms again on the surface of the aluminium, which can slow down or even inhibit etching progress. This sequence of events is often the cause of poor reproducibility, and is one reason why Al etching equipment uses interlocks for loading the wafers for etching, as this can prevent air entering the etching area.

The problems outlined above can be practically eliminated by using BCl_3 as the etch gas. It has the following useful properties:
– low selectivity between Al and Al_2O_3
– able to bind moisture and oxygen in the etching chamber
– low level of polymerization

The main advantage of BCl_3 is obviously its low selectivity to Al_2O_3, which means that the Al_2O_3 film can also be etched from the Al surface.

In addition, few polymer deposits are formed, either on the wafers being etched or the reactor walls. This means that the etching chamber does not need to be cleaned so frequently.

The low etch rate of BCl_3 must be one of its main drawbacks. The etch rate can be increased by adding Cl_2, although the amount that can be added is limited, otherwise the anisotropy of the process is lost. When process parameters have been optimized, a selectivity between Al and SiO_2 of up to 50 : 1 is achieved, and 10:1 between Al and a photoresist. Selectivity to silicon is low, however, because chlorine produces a volatile reaction product with both aluminium and silicon. It is therefore essential that the Al interconnections completely cover contact holes leading to polysilicon or monocrystalline silicon (Table 2.1, process numbers 17 and 19).

The addition of silicon and titanium to the aluminium does not create a problem, since both elements form volatile reaction products with chlorine. Suitable Cl-based gases can also etch away the copper frequently added to aluminium (Table 5.3).

When the etched wafers are removed from the reactor, another problem often arises. On the wafer are residues containing chlorine. These react with the moisture in the air to form HCl, which corrodes the aluminium. Most of the Cl residues lie in the photoresist, so it is essential that the photoresist is removed directly after Al etching. The best solution would be to actually use the same equipment to etch off the photoresist, whilst ensuring that no air entered between processes.

Another option for avoiding corrosion is to subject the etched wafers to post-processing in a fluorine-based plasma. This converts the chlorine residues into non-reactive fluorides.

When chlorine-based gases are used for etching, problems generally arise because of the aggressivity of many of the reaction products, such as HCl. All parts of the etching system coming into contact with these materials must be resistant to acids and alkalis. Furthermore, all parts of the system must be heated to at least 50 °C, otherwise they will become coated in residues of the reaction products.

Carcinogenic chlorinated hydrocarbons can occur in the plasma and in the oil used in the vacuum pumps. Extreme safety precautions must therefore be taken when ventilating and cleaning the etching equipment, and when changing the pump oil.

5.3.7
Dry etching of polymers

Polymers are used in semiconductor technology for photoresists, electron and X-ray resists, for sacrificial films in multi-layer lithography, in planarization techniques and for electrical insulation (Sect. 3.12.2). Organic polymer films can be etched using oxygen. The polymer elements (C, H, F, Cl, O etc.) are either gaseous in themselves (H_2, F_2, Cl_2, O_2 etc.) or they form volatile reaction products with oxygen (CO, CO_2 etc.).

Full-surface etching of polymer films (stripping) is usually performed in barrel reactors (see Sect. 5.2.2). Pure oxygen is used as the etch gas, or else oxygen with small amounts of other gases added. CF_4 gas is a typical additive, which can be used to prevent alkaline residues. Microwave reactors are also used for stripping polymer films, having particularly short etch times.

Reactive ion etching (RIE) and plasma source or ion source etching techniques are suitable for anisotropic etching (Fig. 5.2.8). In these processes an inert gas (usually argon) is often added to the oxygen to stabilize the plasma. Anisotropic etching only occurs once the gas pressure is below about 1 Pa (Fig. 5.1 b; [5.38]). Any inorganic film that does not form volatile reaction products with oxygen can be used as an etching mask (e.g. SiO_2, Si_3N_4, spin-on glass, Al etc.).

As the complexity of integrated circuits increases, the surface starts to look more and more like mountainous terrain (Fig. 8.5.1). This causes severe prob-

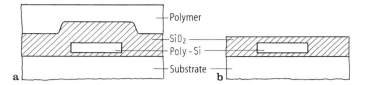

Fig. 5.3.6 a, b. Principle of layer planarization. **a** Before etching; **b** after etching

lems for film deposition and lithography unless planarization techniques are employed (Sect. 8.5.1). One of these techniques is used for planarizing SiO_2 films. This involves depositing a polymer film on the wafer surface to level off the unevenness in the underlying layer. This is followed by an etching process where the selectivity between the polymer and the underlying layer equals 1 : 1, i.e. the polymer film and the layer beneath are etched at the same rate. A flat surface is obtained once the polymer layer has been completely etched through (Fig. 5.3.6). There are a few problems in the practical implementation of the procedure, however. First, it is difficult to achieve complete planarization of the structured surface with the polymer coating, second it is very hard to achieve an etching process selectivity of exactly 1 : 1. Nevertheless, this planarization technique can at least take the edge off many problems.

6
Doping technology

The fourth key process in silicon technology apart from film production, lithography and film patterning, is the doping of the p- and n-type regions in monocrystalline and polycrystalline silicon. Boron (for p-type regions) and arsenic, phosphorus and antimony (for n-type regions) are the preferred dopant atoms that are currently used.

Sections 3.3, 3.6 and 3.8 have already dealt with bulk doping of monocrystalline silicon wafers and in-situ doping of epitaxial layers, polysilicon wafers and phosphorus glass films. Whilst these doping processes are blanket techniques, this chapter is concerned with the selective doping of geometrically defined areas. The regions are doped according to the principle, that a uniform supply of dopant to the wafer surface is restricted to the required areas using a mask. Figure 6.1 shows the three techniques that are used for doping specific regions. In Fig. 6.1 a, the dopant diffuses from a dopant gas atmosphere into the silicon, whilst in Fig. 6.1 b it is implanted into the silicon by ion implantation. In Fig. 6.1 c the dopant diffuses into the silicon from a covering layer, the dopant having been introduced into the layer itself either by diffusion from the gas phase or by ion implantation. Alternatively the dopant can be introduced in the layer during deposition. Although ion implantation is the most frequently used technique, it has not fully replaced the other methods. This is partly for commercial reasons, partly because the other techniques are more suited to certain applications than ion implantation.

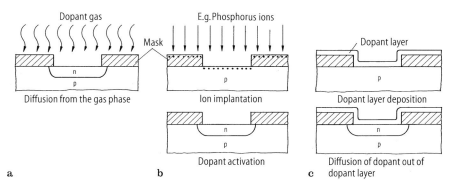

Fig. 6.1 a–c. Basic principles of selective doping of silicon. **a** Diffusion from the gas phase; **b** ion implantation; **c** diffusion from a dopant layer

6.1
Thermal doping

Thermal doping means dopant diffusion out of the gas phase into the surface being doped [6.1]. Thermal doping is either performed in the same type of tube furnace as is used for thermal oxidation (Fig. 3.1.14), or in a CVD reactor (Fig. 3.1.4). In the CVD reactor, doping occurs at the same time as film deposition (polysilicon or SiO_2). Since many wafers can be processed at once in both a tube furnace and a CVD reactor, thermal doping is cheaper than ion implantation. It is also the preferred option if the typical side-effects of ion implantation – crystal damage, channelling and anisotropic doping – are a problem or unwanted. For instance, anisotropic doping, where the supply of dopant atoms to the wafer surface is directional, means that the back of the wafer cannot be doped at the same time. Thermal doping is e.g. preferred for n^+ phosphorus doping and for trench doping. On the other hand, ion implantation is used when exact doses (in particular low doses) or buried doping profiles are required, or when photoresist is employed for masking.

Table 6.1 lists the most popular thermal doping techniques. A detailed description can be found in [6.1]. The phosphorus doping technique using a PH_3 or $POCl_3$ source is described in more detail in Sect. 3.6.1 (Fig. 3.6.1). In this technique it is important that the surface of both the silicon and the SiO_2 mask is converted into phosphorus glass.

Table 6.1. The most popular thermal doping techniques

	Boron	Phosphorus	Arsenic	Antimony
Doping source	B_2H_6 (gaseous) BBr$_3$ (liquid) BN (solid, as wafers)	PH_3 (gaseous, see Sect. 3.6.1) $POCl_3$ (liquid, see Sect. 3.6.1) In-situ phosphorus-doped films (see Sect. 3.6.1)	AsH_3 (gaseous) As (solid, box technique) In-situ arsenic doped films (see Sect. 3.6.1) Dopant film (see Sect. 3.12.1)	Dopant resist film (see Sect. 3.12.1)
Temperature	800–1200 °C	800–1200 °C for PH_3 and $POCl_3$ 400–700 °C for in-situ doped films	800–1200 °C for AsH_3 and As 400–700 °C for in-situ doped films 200–400 °C for dopant film	200–400 °C for glass formation

6.2
Doping by ion implantation

In ion implantation [6.2] the dopant is initially ionized in a plasma. The charged particles are accelerated towards the silicon surface by voltages of typically 100 kV, and usually penetrate some 0.1 μm deep into the silicon. The silicon lattice is damaged in the process, and even the dopant atoms themselves do not occupy lattice sites. Annealing at between 500 and 1000 °C is necessary to restore the monocrystalline state and to locate the dopant atoms at lattice sites (electrical activation).

The main advantages of ion implantation over thermal doping are the excellent accuracy and uniformity of the implanted dose (particularly important for small doses), the opportunity to set the peak doping concentration below the surface into the substrate, the photoresist masking capability and the ability to dope materials such as silicide films. Furthermore, the process does not create dopant films such as boron glass, phosphorus glass or arsenic glass, that can cause problems. It is these benefits that have made ion implantation the leading doping technology, even though the process is more expensive. Although typical ion implantation side-effects, such as crystal damage, channelling[1], asymmetries[2] or substrate charging, may be unwelcome, they can usually be minimized by suitable adjustments to the implantation process.

6.2.1
Ion implantation machines

Today's industrial ion implanters [6.3] are classified into two types: low-/medium-current machines (beam current up to 1 mA) and high-current machines (typical beam current 10 mA). The main design difference between these two types of machines relates to the manner in which the ion beam is scanned over the wafer in order to supply an equal dose over the whole surface (Figs. 6.2.1 and 6.2.2). In the medium-current machines the ion beam (diameter approx. 1 mm) is deflected electrostatically in the x and y direction, so that it scans across the whole wafer surface, whilst in the majority of high-current machines (beam a few cm in diameter), the ion beam is stationary and the wafers are moved in both directions perpendicular to the beam.[3]

Most ion sources supplying boron, phosphorus and arsenic use the gases BF_3, PH_3 and AsH_3. The plasmas of these gases not only contain the singly charged

---

1   Channelling refers to the long range attained by ions scattered in the preferred crystallographic orientation of the silicon crystal lattice.
2   In order to avoid channelling during ion implantation, the ion beam is usually tilted by 7° with respect to the wafer normal. This can lead to asymmetries at steep mask edges because of shadowing.
3   There are also machines in which the ion beam is deflected magnetically in one direction, and the wafers move just in one direction perpendicular to the beam.

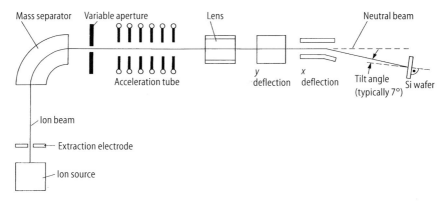

Fig. 6.2.1. Diagram of a medium-current ion implantation machine. The silicon wafers are fed successively from a cassette through a vacuum interlock into the ion irradiation position. In order to avoid the channelling effect, the ion beam is not incident normal to the wafers (tilt angle typically 7°). The ion beam is also deflected slightly before hitting the wafers, in order to avoid the silicon wafers being contaminated by neutral particles, for instance metal atoms sputtered off the aperture shield. The beam current is adjusted using the variable aperture

ions B^+, P^+ and As^+, but also multiply charged ions, although these are generated in a far lower concentration. In addition all possible molecular ions occur as well, for instance high concentrations of BF_2^+ ions are produced when BF_3 gas is used.

The various ions are spatially separated in the mass separator. An aperture is positioned in such a way that only the required ion type can pass through. If a molecular ion is chosen for ion implantation, there is a risk that some ions may break down under molecular impact. For instance if BF_2^+ is used, then some BF_2+ ions will break down into B^+ and F_2. In the type of implantation machine shown in Fig. 6.2.1, where ion acceleration occurs after the mass separator, this kind of breakdown can mean that B^+ ions as well as BF_2^+ ions are implanted in the silicon wafer, both ions having passed through the same acceleration voltage. Being the lighter ion, the B^+ ion hits the wafer at a higher speed, and therefore penetrates further. The result is a deeper doping profile than one would expect with just BF_2^+ ions.[4]

The implanted dose can be set extremely accurately (1 %) in the implantation machine by measuring the charge. The uniformity of the dose across the wafer surface can achieve the same degree of accuracy when the scanning mechanism is set up precisely.

The acceleration voltages in high-current implantation machines typically equal 80 kV, and range from 25–180 kV in medium-current machines. Certain

---

4   In ion implantation machines where ion acceleration occurs before the mass separator, this effect does not arise.

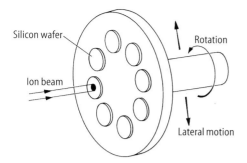

Fig. 6.2.2. Schematic diagram of wafer movements in a high-current machine during ion implantation. Since the ion beam sweeps over an area that is much larger than a single silicon wafer – which is not the case in medium-current machines – the wafers are not overheated

applications also require acceleration voltages of less than 25 kV (for a shallow profile), and some require voltages exceeding 180 kV, e.g. for retrograde wells in CMOS technology (see Sect. 8.3.1), or for pedestal collectors (see Fig. 8.3.8). Specialized equipment of this type is available on the market.

6.2.2
Implanted doping profiles

Using the so-called LSS theory [6.4], the distribution of the implanted dopant in (amorphous) solids can be calculated to a good degree of accuracy. The result is a Gaussian profile, characterized by the projected range R_P and the standard deviation ΔR_P (Fig. 6.2.3).

Figure 6.2.4 shows the projected range R_P in silicon of implanted boron, phosphorus and arsenic ions with respect to the acceleration voltage, and Fig. 6.2.5 shows the corresponding standard deviation ΔR_P. There is no significant difference between the R_P and ΔR_P values for silicon and those for SiO_2, Si_3N_4 and Al. With photoresists, however, values of 20–30 % higher should be expected.

Agreement between theoretical and measured values for R_P and ΔR_P is not only good for amorphous silicon, but also for monocrystalline and polycrystalline silicon under real processing conditions[5]. There is one region of the curve, however, where measurement may differ significantly from the Gaussian profile for monocrystalline and polycrystalline silicon. This is in the far tail region of the doping profile, i.e. where concentrations are < 10 % of the maximum con-

---

5 "Real conditions" include, for instance, the implantation through a thin SiO_2 film (screen oxide) into silicon. Apart from spreading the angle of the implanted ions, the SiO_2 film also traps the heavy metal atoms during ion implantation, and prevents the implanted atoms from vaporizing during subsequent annealing.

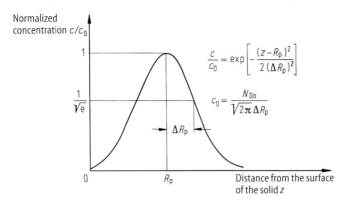

Fig. 6.2.3. The characteristic parameters R_P (projected range) and ΔR_P (standard deviation) of the implanted doping profiles in amorphous solids. $c(z)$ is the concentration of the dopant atoms at position z, c_0 is the concentration at position $z = R_P$ and N_{Do} the total implanted dose

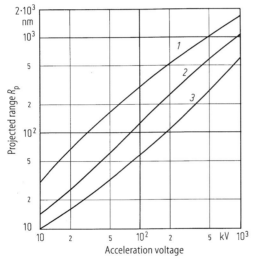

Fig. 6.2.4. Projected range R_P in silicon of implanted singly charged ions of *1* boron, *2* phosphorus and *3* arsenic, as a function of the acceleration voltage. For doubly charged ions, the correct R_P value is obtained by reading off at twice the acceleration voltage. Curve 1 is also valid for BF_2^+ ions, but the effective acceleration voltage must be reduced by a factor of 4.6 (mass ratio BF_2:B = 4.6)

centration. Here the foreign atom concentration is higher than that predicted by the LSS theory. This phenomenon is caused by channelling, where ions that are travelling in the direction of a crystallographic axis in the silicon suffer less deceleration [6.5]. The effect is most pronounced when the direction of ion bom-

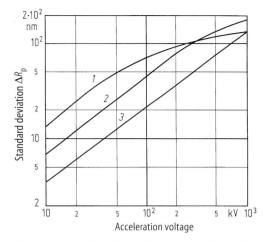

Fig. 6.2.5. Standard deviation ΔR_P in silicon of implanted singly charged ions of *1* boron, *2* phosphorus and *3* arsenic, as a function of the acceleration voltage. For doubly charged ions and BF_2^+ ions the same conversion applies to the acceleration voltage as in Fig. 6.2.4

bardment exactly coincides with the orientation of the crystallographic axis, and can be a problem if flat doping profiles are required. On the other hand, what initially might seem a useful technique for producing profiles deep into the silicon, is in fact of no practical use. This is because only slight angle deviations of 1–2° can significantly reduce the channelling effect, so that the reproducibility of such a method is doubtful. Modern implantation practice therefore aims to reduce any channelling effect.

Two measures are commonly used to inhibit channelling. The first is to tilt the silicon wafers in the ion implantation machine by about 7° (see Fig. 6.2.1), i.e. to set up a 7° angle between the ion beam direction and the wafer normal, which may lie in a crystallographic ⟨100⟩ direction. The second measure is to cover the monocrystalline silicon surface with a thin amorphous film (usually SiO_2). After passing through this film the ions acquire a certain angular spread.

Yet even these measures cannot totally prevent channelling, since a small fraction of the ions will still be scattered down the "open channels". Figure 6.2.6 shows a measured boron implantation profile with the crystallographic ⟨100⟩ orientation tilted by 7° with respect to the direction of ion bombardment [6.6]. Although deviation from the Gaussian profile is actually only significant for small boron concentrations, in practical cases this deviation can still be significant. For example, for a substrate doping of 10^{16} cm^{-3} (n-type doping) the pn-junction should theoretically lie at a depth of 0.23 μm for the Gaussian profile, but in practice it would lie at 0.31 μm.

If in the example shown in Fig. 6.2.6, the silicon substrate is made amorphous to a depth of about 0.2 μm using silicon implantation (dose approx. 10^{15} cm^{-2}),

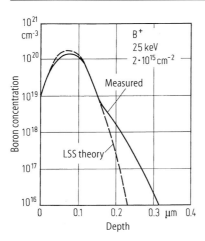

Fig. 6.2.6. Measured implanted boron profile in monocrystalline silicon with a $\langle 100 \rangle$ surface. The incident ion beam is tilted at 7° to the surface normal

then the conditions for channelling are effectively removed. One then obtains an almost ideal Gaussian profile as given by the LSS theory [6.7].

The extreme tail of a doping profile is also relevant when masking against ion implantation. Take for example a MOS silicon gate process where boron is implanted in the source and drain at an energy of 80 keV and a dose of $5 \cdot 10^{15}$ cm^{-2}. In order not to affect the doping concentration in the transistor channel, less than e.g. $5 \cdot 10^{10}$ boron atoms per cm^{-2} are allowed to penetrate into the channel region which is equivalent to a transmission factor of 10^{-5}. The typical materials used for implantation masking (SiO$_2$, Si$_3$N$_4$, amorphous and fine-grained polysilicon[6] and photoresist) can be assumed to follow a Gaussian profile[7]. The minimum film thickness can then be found from Fig. 6.2.7: for a transmission factor of 10^{-5}, a depth of $z = R_P + 4.25 \Delta R_P$ would be required. Using Figs. 6.2.4 and 6.2.5, one obtains for the implantation data in the example above, the values $R_P = 280$ nm and $\Delta R_P = 63$ nm. Thus $z = 550$ nm. This means that the total thickness of the SiO$_2$ and polysilicon films lying over the transistor channel must equal at least 550 nm. Using similar reasoning one can evaluate the minimum thickness of a field oxide (see Sect. 3.4.2).

Just as vertical doping profile tails can be important, so can the lateral range of doping concentrations. Figure 6.2.8 shows the scatter of implanted dopant atoms around a mask edge [6.8]. If one refers again to the above example of boron source/drain implantation, then the mask edge in this case is the edge of the

---

[6] For coarse-grained polysilicon, where the grains can extend through the whole thickness of the film (see Fig. 3.8.1), channelling effects may arise again as in monocrystalline silicon.

[7] A Gaussian profile can at least be assumed for the far tail region of the doping profile (i.e. extending into the substrate), which is the area of interest here. In the front (i.e. towards the substrate surface) region higher concentrations are measured.

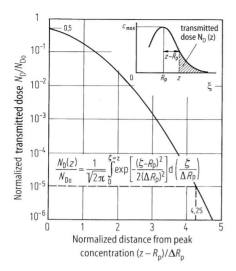

Fig. 6.2.7. Transmission factor of masking films in ion implantation. $N_D(z)$ is the dose that is still transmitted at a distance z from the surface. N_{Do} is the total implanted dose

$$\frac{N_D(z)}{N_{Do}} = \frac{1}{\sqrt{2\pi}} \int_0^{\zeta=z} \exp\left[-\frac{(\zeta-R_p)^2}{2(\Delta R_p)^2}\right] d\left(\frac{\zeta}{\Delta R_p}\right)$$

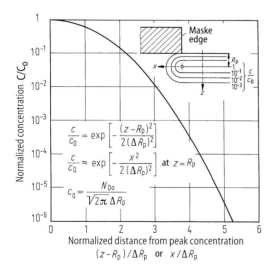

Fig. 6.2.8. Lateral and vertical scatter of implanted atoms around a mask edge

$$\frac{C}{C_0} = \exp\left[-\frac{(z-R_p)^2}{2(\Delta R_p)^2}\right]$$

$$\frac{C}{C_0} \approx \exp\left[-\frac{x^2}{2(\Delta R_p)^2}\right] \text{ at } z = R_p$$

$$C_0 = \frac{N_{Do}}{\sqrt{2\pi}\,\Delta R_p}$$

gate electrode. With a dose of $5 \cdot 10^{15}$ cm^{-2} and an energy of 80 keV, the maximum concentration c_0 at a depth $z = R_P$ is then $3.2 \cdot 10^{20}$ cm^{-3}. If, for instance, the concentration of the n-type doping in the transistor channel region is $3 \cdot 10^{15}$ cm^{-3}, then according to Fig. 6.2.8, laterally scattered boron atoms cancel out this concentration over a distance of 300 nm from the edge of the gate electrode.

For MOS transistors with channel lengths below 1 μm, this lateral effect is not negligible. For this reason, and also because of the vertical distribution (thick masking layers required, doping profile tail see Fig. 6.2.6), either BF_2^+ implantation (see Sect. 6.2.1) or amorphization of the silicon prior to implantation (see above) are the preferred options in p^+ source/drain implantation.

The deceleration of the implanted ions in the masking and screen oxide films as well as in the monocrystalline silicon causes a series of unwanted side-effects.

The extent of the problem caused by dopant atoms trapped in the masking films is considered first, according to the function of these films. Where these films are going to be removed, for instance in photoresist masking, the dopant atoms can be removed at the same time. Removing the film can, however, be difficult because of the changes to the photoresist film brought about during implantation. (see Sect. 7.3).

When an SiO_2 mask is acting as an implantation mask that shall remain on the silicon wafer as part of the integrated circuit (e.g. the LOCOS oxide), then charges can arise in the SiO_2 film, particularly at high implantation doses. These charges are bound to traps in the oxide[8] created by the deceleration of the implanted ions [6.9]. Here they can reduce the thick oxide threshold voltage, for instance, which is critical for the isolation of neighbouring transistors. One measure that is often employed to avoid this damaging effect is to etch off in dilute hydrofluoric acid the implanted part of the SiO_2 mask close to the surface. This also removes any heavy metal contaminants that might be present. These can get onto the wafer surface during ion implantation, for instance, by being sputtered off the aperture shield[9] in the implantation machine.

Like the SiO_2 masking films, the screen oxide films are also damaged by the penetration of the implanted ions. If the screen oxide is not removed after implantation this can effect device characteristics. For instance at the edge of a polysilicon gate electrode the damaged oxide can degrade the breakdown performance (early gate-drain breakdown, Fig. 6.2.9). It has been shown that even relatively low implanted doses such as those used for MOS transistor channel implantations will create permanent traps in the gate oxide. It is therefore preferable to perform channel implantation before gate oxidation and to remove the screen oxide prior to gate oxidation.

During ion implantation, some of the silicon atoms in the monocrystalline silicon lattice are knocked out of their lattice sites, and even the dopant atoms do not normally come to occupy lattice sites. For high implanted doses, a dense network of dislocations often remains even after annealing. Fortunately, the maximum crystal damage lies nearer to the silicon surface than the peak of the doping profile, so that crystal damage often only has a limited effect on the leakage

---

8   At high implanted doses (e.g. 10^{15} cm^{-2} of arsenic), the SiO_2 (or even a polysilicon film) is so badly damaged that the etch rate for isotropic etching increases. This effect is sometimes exploited to produce bevelled edges. Annealing at high temperature may not fully repair the damage to the film.
9   Reducing the beam current produces a more focused beam, which reduces the risk of sputtered contaminants from the aperture shield.

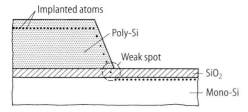

Fig. 6.2.9. Formation of a weak spot in the SiO$_2$ layer during source/drain implantation. Early electrical breakdown between gate and drain can arise at the weak spot

of pn junctions. Crystal damage still poses a risk to pn junctions lying near the surface, however, particularly since post-implantation annealing often cannot completely remove the defects (see Sect. 6.3) [6.10].

If implantation occurs through an SiO$_2$ screening film, then another damaging effect occurs: the implanted ions collide with numerous oxygen atoms as they pass through the SiO$_2$ film, and some of these atoms are scattered into the silicon [6.11] (recoil or knock-on implantation). Here they can act as dislocation centres. At the implantation energies used in practice, the silicon penetration depth of the oxygen atoms is less than 100 nm. This means that the risk they pose to the electrical performance is similar to that posed by for silicon lattice damage, which was discussed above.

Apart from implantation damage, the other unwanted side-effect of ion implantation that should be mentioned is contamination of the silicon wafers by heavy metal atoms, organic residues and particles. The first two contaminants can be greatly reduced in modern ion implantation systems, and can also be removed from the silicon wafers by the overetching technique mentioned earlier. Particulate contamination, however, can present a more severe problem, particularly if the silicon surface is contaminated before implantation. This can happen, for instance, during loading and automatic transportation of the wafers in the implantation facility, or if splinters of photoresist flake off under mechanical impact. In such cases a single particle can cause a chip to fail, because the particle may mask the implantation process (see Sect. 7.1).

An "electron shower" is used in implantation facilities to cancel out the positive charge induced on the silicon wafers during ion implantation. If compensation is incomplete, electrostatic charges can be set up (e.g. on isolated gate electrodes) which may lead to gate oxide breakdowns.

The masking films used in ion implantation are mainly SiO$_2$ and photoresist. The incident electric charge is carried through these insulating layers and into the substrate once the breakdown field strength E$_{bd}$ has been reached (approximately 10^6 V cm^{-1} for oxide and resist). According to the formula

$$N_D = \frac{\varepsilon_0 \cdot \varepsilon_r \cdot E_{bd}}{e}$$

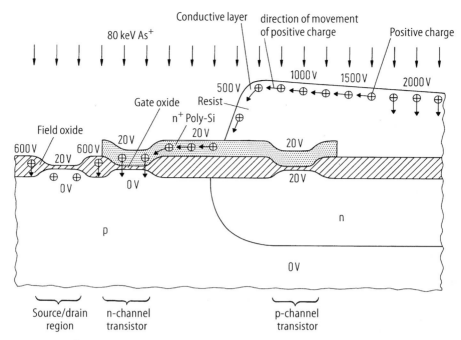

Fig. 6.2.10. Illustration of the charge flow off the wafer surface into the Si substrate during source/drain implantation of arsenic in a CMOS process. Because of the conductivity of the poly-Si and the resist surface, the charge collected over an extended area flows through the small gate oxide region of the n-channel transistor. This can degrade the gate oxide until it reaches breakdown

an implanted dose of just $N_D = 2.5 \cdot 10^{12}$ cm^{-2} is sufficient for the breakdown field strength to be reached in the masking oxide or resist films, where ε_r is the relative dielectric constant and e is the elementary charge.

The flow of charge through the oxide might then be damaging if the total charge passed were to approach the charge–to–breakdown Q_{bd} (approx. 1 Ccm^{-2} for oxide, see Fig. 3.4.13). This should not occur, however, even for high-dose implantation without a flood gun. For example, a charge of only $8 \cdot 10^{-4}$ Ccm^{-2} flows through the oxide for an implanted dose of $5 \cdot 10^{15}$ cm^{-2}.

Nevertheless, it can be a problem if the charge does not flow uniformly down to the substrate in a vertical direction, but instead flows e.g. through gate oxide regions. Figure 6.2.10 illustrates how such conditions can arise during source/drain implantation in a CMOS process. The lateral flow of charge occurs along the poly-Si tracks ("antenna effect") and the resist surfaces, which are conductive during ion bombardment. The collected charge ultimately flows off through the gate oxide of a MOS transistor into the Si substrate. If the charge capture surface area is 1000 times larger, say, than the gate oxide area, then a charge of 0.8 Ccm^{-2} would then flow through the gate oxide for an implanted dose of $5 \cdot 10^{15}$ cm^{-2} (without flood gun). This amount of charge is already enough to damage the gate oxide, and weak spots in the gate oxide (see Fig. 3.4.14) can be degraded to breakdown in this way.

The following measures can be taken to reduce the charge flowing through the gate oxide regions:
- optimum adjustment of the flood gun installed in modern implanters;
- reduction of the beam current (to reduce the conductivity of the resist)
- designing poly-Si tracks to provide the smallest possible field oxide : gate oxide surface area ratio (e.g. < 100:1)
- avoiding resist masking patterns extending unnecessarily over field oxide regions
- designing poly-Si tracks not to cross resist masking patterns.

6.3
Activation and diffusion of dopant atoms

Sections 6.1 and 6.2 looked at thermal doping and ion implantation as the two main techniques for introducing dopant atoms into silicon. In this section the effects of temperature treatments on the doped films are considered. These effects are mainly the activation of dopant atoms, repair of crystal defects and diffusion of dopant atoms [6.12].

6.3.1
Activating implanted dopant atoms

Ion implantation of dopant atoms into a monocrystalline silicon substrate initially leaves the silicon crystal lattice in a damaged state, the extent of the damage depending on the implantation dose. As a result of collisions as the implanted ions decelerate, numerous silicon atoms are displaced from lattice sites, creating deep-level traps for electrons and holes. Even the dopant atoms do not occupy lattice sites, so they are unable to generate free charges (electrons or holes). The severe shortage of free charges and the presence of the traps means that the implanted layer has an extremely high resistance. Temperature treatment is needed to activate the dopant atoms, i.e. bring them to silicon lattice sites, and also to remove the traps. Annealing is thus referred to as a repair process.

At low implanted concentrations (up to about 10^{17} dopant atoms per cm^3), which is a typical figure for e.g. the channel doping of MOS transistors, the main damage involves just point defects in the silicon lattice. These can be completely annealed out at relatively low temperatures of between 600 and 800 °C in just a few minutes. After annealing, the dopant atoms occupy silicon lattice sites, and are thereby "activated", i.e. they then act as acceptors (if boron) or donors (if phosphorus or arsenic). As the following section shows, only marginal diffusion of the dopant atoms occurs in the process (diffusion length less than about 10 nm), so that the acceptor or donor profile is practically identical to the implanted doping profile.

For medium implanted concentrations (about 10^{17}–10^{20} dopant atoms per cm^3), severe crystal defects (dislocations) occur above 500 °C, which only partially disappear even at temperatures as high as 1000 °C. Below about 900 °C some of the dopant atoms are trapped in the dislocation region, and are then electrically inactive.

Since the dislocations act as traps for the free charges, the electrical conductivity in the concentration range of 10^{17}–10^{20} dopant atoms per cm^3, is not increased a proportional to the dopant atom concentration, but rises at a lower rate (see Sect. 6.3.3 as well).

If annealing to activate dopant atoms is performed in a standard tube furnace (Fig. 3.1.14), then annealing times inevitably take several minutes because of the thermal inertia of the quartz boats. At temperatures above 900 °C the diffusion of dopant atoms is no longer negligible during the annealing period. If annealing is carried out in a rapid annealing process (Fig. 3.1.25), however, e.g. for 5 s at 1150 °C, then complete dopant atom activation can be achieved with almost negligible diffusion of the dopants.

At high implanted concentrations of arsenic and phosphorus (more than about 10^{20} cm^{-3}), practically all the silicon atoms are displaced from their original lattice sites because of the numerous collisions that occur as the dopant atoms are decelerated. The implanted layer is therefore amorphous where the dopant atom concentration is high. An amorphous silicon layer on a monocrystalline silicon substrate can then be annealed at relatively low temperatures (about 600 °C), because the conversion of the amorphous layer into monocrystalline silicon with the same crystallographic orientation as the substrate (solid phase epitaxy) requires a relatively low activation energy. Nevertheless, the doping profile of a high implant dose inevitably involves a region with dopant atom concentrations of 10^{17}–10^{20} cm^{-3}, which requires temperatures of above 900 °C as discussed above.

6.3.2
Intrinsic diffusion of dopant atoms

At low dopant atom concentrations (less than about 10^{18} cm^{-3}), interaction between the dopant atoms is negligible. If such dopant atoms diffuse into monocrystalline silicon, then this is referred to as intrinsic diffusion. The diffusion mechanism is based on the interaction between the dopant atom and a charged or uncharged vacancy in the silicon lattice.

The temporal and spatial change in the dopant atom concentration c can be defined for intrinsic diffusion by the diffusion law:

$$\frac{\partial c}{\partial t} = D \frac{\partial^2 c}{\partial z^2},$$

where D is the diffusion constant. Figure 6.3.1 shows the diffusion constants as a function of temperature for boron, phosphorus, arsenic and antimony in lightly doped silicon.

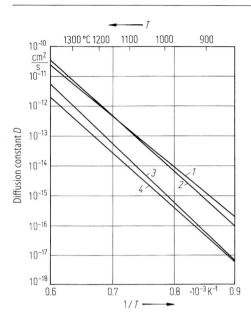

Fig. 6.3.1. Diffusion constants as a function of the temperature for *1* boron, *2* phosphorus, *3* arsenic and *4* antimony in lightly doped silicon

The diffusion equation has explicit solutions for the two separate boundary conditions of "dopant atom concentration at $z = 0$ constant over time" and "dose constant over time".

The first condition of a constant dopant atom concentration over time at $z = 0$ is, for instance, approximately met when the dopant atoms are diffusing out of the environmental gas or out of a deposited layer into the silicon, where the supply of dopant to the silicon surface is supposedly constant during diffusion into the silicon. The solution to the diffusion equation in this case is then:

$$\frac{c(z)}{c_0} = 1 - \frac{2}{\sqrt{\pi}} \int_{\zeta=0}^{\zeta=z} e^{-\frac{\zeta^2}{\sigma^2}} d\left(\frac{\zeta}{\sigma}\right), \text{ for } c(z=0,t) = c_0.$$

The normalized concentration $c(z)/c_0$ is shown in Fig. 6.3.2 as a function of the normalized depth z/σ.[10] The normalization parameter σ is called the diffusion length, and is given by

$$\sigma = 2\sqrt{Dt}.$$

It is a measure of how far the dopant atoms diffuse in time t.

The second condition is that the dose, i.e. the number of dopant atoms per cm² of surface area, does not change over time in the substrate (drive-in condi-

---

10 The function $c(z)/c_0$ is also referred to as erfc (z), where erfc stands for the *c*omplementary *error function*.

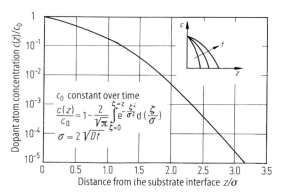

Fig. 6.3.2. Doping profile $c(z)/c_0$ arising from diffusion, for the situation where the dopant atom concentration c_0 at the substrate interface $z = 0$, is constant over time. σ is the diffusion length, D is the diffusion constant and t is the period of time for which the substrate is held at the diffusion temperature

tion). This condition is approximately met when dopant atoms have been implanted with a dose N_{Do} in a silicon substrate, and then diffuse into the silicon at a raised temperature. The solution to the diffusion equation is in this case a Gaussian distribution (Fig. 6.3.3):

$$\frac{c(z)}{\left(\dfrac{N_{Do}}{\sqrt{\pi}\sigma}\right)} = e^{-\dfrac{(z-z_0)^2}{\sigma^2}}, \text{ for } \int_{z=-\infty}^{z=\infty} c(z,t)\,dz = N_{Do},$$

where z_0 is the distance from the substrate surface $z = 0$ at which the peak dopant atom concentration occurs. For ion implantation, $z_0 = R_P$ (R_P = projected range, see Fig. 6.2.3). The Gaussian doping profile with standard deviation ΔR_P that already exists after ion implantation and before actual diffusion (time $t = 0$), can be accounted for by setting σ equal to the following expression:

$$\sigma = 2\sqrt{\frac{1}{2}(\Delta R_P)^2 + Dt}.$$

Multi-stage diffusions, where successive diffusion processes take place at different temperatures, can be treated as follows

$$\sigma = 2\sqrt{\frac{1}{2}(\Delta R_P)^2 + D_1 t_1 + D_2 t_2 + \ldots}.$$

The doping profiles shown in Figs. 6.3.2 and 6.3.3 can actually describe practical situations very closely if the dopant atom concentrations are sufficiently

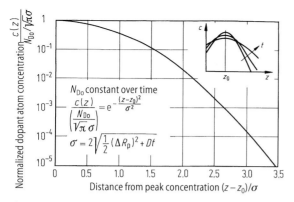

Fig. 6.3.3. Doping profile arising from diffusion, for the situation where at time $t = 0$, an implanted Gaussian profile already exists with the maximum at $z = z_0$, dose N_{Do}, and standard deviation ΔR_P

small ($< 10^{18}$ cm^{-3}) and if processes at the silicon surface such as oxidation enhanced diffusion (Sect. 6.3.4) or segregation (Sect. 6.3.5) play a secondary role. Examples of where the doping profiles of Figs. 6.3.2 and 6.3.3 can be used are diffusion from a buried layer into an epitaxial layer (see Sect. 3.3.2), the (one-dimensional) doping profile in the channel region of a MOS transistor, and the well profile in CMOS technology.

6.3.3
Diffusion for high concentrations of dopant atoms

At high dopant atom concentrations (above about 10^{19} cm^{-3}, which is equivalent to an implanted dose of more than some 10^{14} cm^{-2}), deviations from the standard diffusion behaviour occur. The degree of anomaly is different for boron, phosphorus and arsenic, because of the specific diffusion mechanisms in each case. The complex mechanisms involved in dopant atom diffusion at high concentrations is still the subject of intensive studies. Nevertheless, in most cases models are able to simulate fairly accurately the change in doping profiles under given conditions. [6.13, 6.14].

Figure 6.3.4 shows typical diffusion differences for boron, phosphorus and arsenic at high concentration levels.

For boron, one obtains an activated boron concentration at 900-1000 °C of about 10^{20} cm^{-3} maximum. Where the total boron concentration exceeds this value, the excess concentration is not electrically active. Furthermore, there is almost no boron diffusion in this region. At the other extreme, in the moderate to low concentration region of the profile, boron diffusion is enhanced because of interstitial atoms which diffuse out of the high concentration region into the low concentration region (diffusion tail) [6.16].

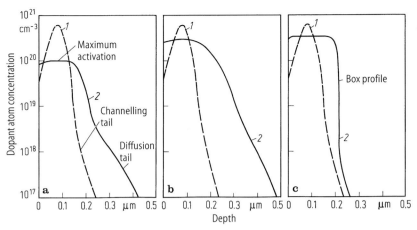

Fig. 6.3.4 a–c. Typical diffusion profiles (activated dopant atom concentration), obtained when annealing a high concentration implanted layer (dashed curves). Annealing at about 900-1000 °C. **a** Boron; **b** phosphorus; **c** arsenic

Phosphorus also exhibits a diffusion tail, for the same reason as boron.[11]

Arsenic exhibits a different behaviour, however. In this case diffusion is enhanced at high concentrations. This produces a steep fall in the concentration at the far end of the doping profile. Since the maximum concentration of dopant atoms that can be activated equals about $3 \cdot 10^{20}$ arsenic atoms per cm^3 at 900–1000 °C, the top section of the profile is flattened off at high implantation doses. This flattening and the steep fall-off give the profile a rectangular shape ("box profile").

6.3.4
Oxidation enhanced diffusion

Boron, phosphorus and arsenic diffuse more rapidly under a silicon surface being oxidized than when thermal oxidation is not occurring (Fig. 6.3.5). This phenomenon can be explained by silicon interstitial atoms, which are produced at the oxidizing silicon surface, diffusing away from the surface and then interacting with the dopant atoms.

The diffusion enhancement factor is different for boron, phosphorus and arsenic, and depends on the oxidation temperature, the oxide growth rate and the crystal orientation. It can reach values of 2 and above.

---

11 The so-called emitter push effect (which refers to the enhanced diffusion of boron beneath a high concentration phosphorus emitter region) can be explained by the same mechanism.

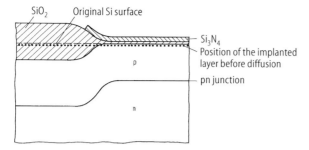

Fig. 6.3.5. Oxidation enhanced diffusion of boron. Beneath the oxidizing silicon surface the boron atoms diffuse more rapidly than under the SiO_2/Si_3N_4 film where no thermal oxidation is taking place

6.3.5
Diffusion of dopant atoms at interfaces

Close to the interface between monocrystalline silicon and other layers, the diffusion doping profile is not just determined by the different diffusion constants in silicon and in the layer concerned, but also by what is known as dopant atom segregation.

The segregation coefficient specifies the ratio of the dopant atom concentration in silicon to the concentration in the relevant layer under thermal equilibrium at the boundary.

For an Si/SiO_2 boundary, the segregation coefficient for boron is < 1 (varies between 0.2 and 0.9 depending on oxidation conditions), whilst for arsenic and phosphorus it has a value of about 10. Figure 6.3.6 shows typical profiles that arise close to the SiO_2/Si boundary during thermal oxidation. The depletion of boron at the silicon surface is referred to as "pile-down", and the increase in concentration exhibited by arsenic and phosphorus is called "pile-up". In the latter case, the advancing SiO_2/Si interface during thermal oxidation acts like a snowplough, with most of the dopant atoms caught by the advancing interface and pushed along in front of it.[12]

The depletion or enhancement of dopant atoms at the Si surface during thermal oxidation is a particularly important factor in LOCOS technology. If the silicon substrate is doped with boron, for instance, then the boron depletion must be compensated for by boron implantation, to ensure that adjacent active regions are electrically isolated (Sect. 3.4.2).

---

12 At very low oxidation temperatures (e.g. in high pressure oxidation at 800 °C), the oxidizing interface can advance at too fast a rate for the arsenic atoms to follow. The result is that the snowplough effect is practically eliminated, and the original arsenic profile remains almost unchanged.

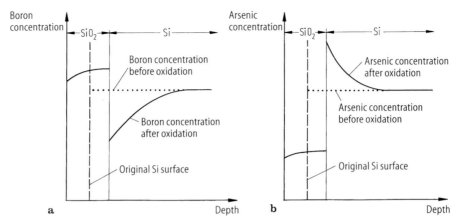

Fig. 6.3.6 a, b. Effect of dopant atom segregation at the SiO$_2$/Si interface on boron and arsenic profiles during thermal oxidation. **a** For boron, depletion (pile-down) occurs at the silicon surface; **b** for arsenic or phosphorus, enhancement occurs (pile-up)

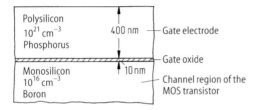

Fig. 6.3.7. Illustration of the doping ratios and the different thicknesses in the gate region of a MOS transistor. The thin gate oxide film must transmit less than 10^{-5} of the phosphorus dose to the gate electrode at the prevailing process temperatures (900–1000 °C)

If the SiO$_2$/Si interface is stationary during high temperature treatment (e.g. the gate oxide interfaces in Fig. 6.3.7), then segregation has a far less pronounced effect on the doping profile. The reason for this is the low diffusion rate of boron, phosphorus and arsenic in SiO$_2$ (see Sect. 6.3.6). In this case the stationary Si/SiO$_2$ interface is effectively acting as a diffusion barrier, with only a relatively small amount of dopant diffusion over the boundary.

As shown in Fig. 6.3.8, there are no significant segregation effects at the polysilicon/monosilicon interface. On the other hand, there can be considerable segregation at a silicide/silicon boundary. For instance the boron concentration in the silicide increases significantly at a TiSi$_2$/Si interface (cf. Fig. 3.9.1).

At the interface between a doped layer and the surrounding atmosphere, the dopant atoms diffuse out of the layer at elevated temperatures. This out-diffusion is normally unwelcome, not just because of the loss of dopant, but also because the dopant can be transferred to other wafers in an undefined manner. For

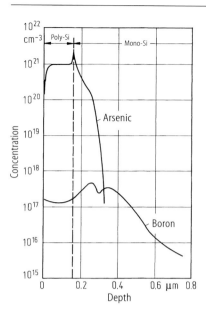

Fig. 6.3.8. Arsenic and boron profiles (total concentrations) in a bipolar transistor with a polysilicon emitter (see Fig. 8.3.8), measured by secondary ion mass spectroscopy (SIMS). The arsenic was implanted in the polysilicon, whereas the boron was implanted in the monosilicon prior to polysilicon deposition. The diffusion temperature was 900 °C

this reason, doped regions (including backs of wafers) are usually covered by a thin SiO$_2$ film before temperature treatment. This is also the reason why an oxidizing atmosphere is already established when the silicon wafers are fed into the tube furnace for thermal oxidation (see Fig. 3.1.15). This technique creates a thin sealing SiO$_2$ film before dopant atoms can diffuse out or in. Section 3.3.2 dealt with the problem of dopant atoms diffusing out during the growth of an epitaxial layer.

6.3.6
Diffusion of dopant atoms in films

The diffusion rate of boron, phosphorus and arsenic in films lying on monosilicon is just as relevant to the shape of the doping profiles as the diffusion rate in the monosilicon itself. Diffusion behaviour in SiO$_2$ films is one of the most important factors. After all, the whole concept of silicon planar technology relies on the fact that SiO$_2$ acts as a practical diffusion barrier. On the other hand, a high diffusion rate is desirable in those films which are supposed to supply dopant to the silicon. Such films include phosphorus glass, polysilicon and silicide films.

The highest demands are placed on the diffusion barrier performance of SiO$_2$ films when they are used as the gate oxide layer with a polysilicon gate electrode. Figure 6.3.7 illustrates the huge difference in dopant concentration: 5 orders of magnitude for the example given here of a phosphorus-doped polysilicon electrode. A phosphorus dose of only 10^{11} cm^{-2} – that is less than 1/100 000 of the dose in the polysilicon – diffusing through the thin gate oxide would be enough to significantly affect the dopant ratios in the channel region of the MOS transistor, and thus its electrical behaviour. Practical experience has shown, however, that even a gate oxide just 10 nm thick constitutes an adequate diffusion barrier to phosphorus at process temperatures of 900–1000 °C. The same appears to be true for arsenic. On the other hand, a low level of diffusion has been found to occur for boron-doped polysilicon gate electrodes under certain conditions. The boron atoms in the gate oxide also seem to be responsible for threshold voltage instabilities under thermal/voltage stress.

The diffusion rate of phosphorus in phosphorus glass layers, i.e. SiO_2 films with a few percent by weight of phosphorus, is several orders of magnitude higher than in pure SiO_2 films. This provides another means of doping silicon with phosphorus, namely by out-diffusion of phosphorus from a phosphorus glass film (see Fig. 3.6.1). If, however, the phosphorus glass film lies on a pure SiO_2 layer, then the SiO_2 again acts as an excellent diffusion barrier. This is useful where, for instance, the phosphorus glass is being used solely as a planarizing and gettering film, but not as a dopant film for silicon doping. In this case, a thin SiO_2 film lying between the phosphorus glass and the silicon prevents the phosphorus diffusing out into the silicon (see Fig. 8.5.3).

The diffusion rate of boron, phosphorus and arsenic in polycrystalline silicon films is higher than in monosilicon, because diffusion along the grain boundaries is up to 100 times faster (cf. Sect. 3.8.3). Thus doped polysilicon in direct contact with monosilicon provides an ideal doping source for monosilicon doping. Well-known applications for this include the buried contact (Fig. 3.8.6), as well as the polysilicon emitter and the polysilicon base connection in bipolar transistors (Fig. 8.3.8). Figure 6.3.8 shows the measured arsenic profile of a polysilicon emitter, which was produced by arsenic implantation into the polysilicon and subsequent diffusion of the arsenic at 900 °C. The relatively high diffusion rate of the arsenic in the polysilicon can be seen by the predominantly homogeneous arsenic distribution in the polysilicon.

In silicides the diffusion constants of boron, phosphorus and arsenic appear to be much higher than in monosilicon and even higher than in polysilicon. On the one hand this is a welcome characteristic. For instance, one can implant dopant atoms into the silicide film, and then drive these dopants into the underlying silicon at a raised temperature. This can be used to produce low-resistance films with small junction depths (see Sect. 6.3.7). Furthermore, this form of doping avoids implantation damage to the monosilicon.[13] On the other hand, the high lateral diffusion rate of dopant atoms in the silicide film can be a problem e.g. when using p- and n-type polysilicon in a CMOS circuit (see Table 8.8). In this situation, unwanted mixing of the p- and n-type doping can occur in the polysilicon (Fig. 6.3.9).

---

13 The same applies to implantation in an SiO_2 layer or polysilicon film, and subsequent drive-in of the dopant into the monosilicon, as in a polysilicon emitter for instance.

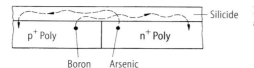

Fig. 6.3.9. Lateral diffusion of boron and arsenic in a silicide film

6.3.7
Sheet resistance of doped layers

The relationship shown in Fig. 3.2.2 between the electrical resistivity and the dopant atom concentration applies to homogeneously doped monocrystalline silicon where the dopant was added at the melt stage. This relationship still holds for doped layers that have been implanted and then annealed, if the concentrations are low to moderate. For high concentrations (greater than about 10^{19} cm^{-3}), the resistivity is actually larger than in the case of a homogeneously doped silicon wafer, because the unrepaired crystal defects reduce both the level of activation and the charge carrier mobility (see Sect. 6.3.1).

It is common practice to specify the electrical resistance characteristic of a near-surface doped region by its sheet resistance R_s:

$$R_s = 1 / \int_0^{z_j} \frac{dz}{\rho(z)}$$

$$R = R_s \frac{l}{w}$$

where $\rho(z)$ is the resistivity at distance z from the silicon surface, and R is the resistance of the doped region of length l and width w for current flowing in the longitudinal direction, and where z_j represents the distance from the pn junction to the surface. The dimension of R_S is written as $\Omega/\square$, to indicate that it represents the resistance of a doped region with a square surface ($l = w$).

The limited number of dopant atoms that can be activated means that the sheet resistance has a lower limit. Figure 6.3.10 shows the minimum sheet resistance that can be achieved for a given junction depth z_j of the pn junction[14]. For example, a sheet resistance of 20 $\Omega/\square$ is obtained for a furnace annealed n$^+$ doped region with a 0.3 µm junction depth.

Figure 6.3.10, curve *4* illustrates the interesting effect obtained when a layer implanted with a high concentration of boron atoms is activated by rapid annealing (e.g. 5 s at 1150 °C, see Sect. 3.1.8). A higher level of activation is achieved because of the high temperature, without significant diffusion occurring during the short annealing time.

---

14 The sheet resistence of doped polysilicon films has already been discussed in section 3.8.3.

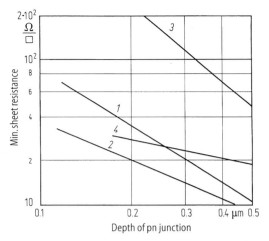

Fig. 6.3.10. Minimum sheet resistance of p- and n-type near-surface layers for a given pn junction depth. It is assumed that the dopant atoms have been implanted and then annealed in a furnace or by rapid annealing. *1* n^+ doping, furnace annealing; *2* n^+ doping, rapid annealing; *3* p^+ doping, furnace annealing; *4* p^+ doping, rapid annealing

The rapid annealing effect is not so pronounced for phosphorus or arsenic doping as it is for boron (curve *2* in Fig. 6.3.10).

In order to obtain low sheet resistances even for very thin doped layers, metal silicide films are often used, which are deposited on top of the doped layers, and act as low resistance "shunts". If the doped silicon layer is polysilicon, the polysilicon/silicide double layer is called polycide (see Sect. 3.9.2). Where monosilicon/silicide double layers are used, these are referred to as salicide films (*self-ali*gned sili*cide*, see Sect. 3.9.3).

6.3.8
Diffusion at the edge of doped regions

If one implants dopant atoms in a region defined by the implantation mask, the dopant atoms diffuse into the substrate not only vertically but also laterally (Fig. 6.3.11). The lateral distance of the pn junction from the mask edge is usually about 30 % shorter than the depth of the pn junction, because the dopant atoms implanted close to the mask edge are diffusing into a larger volume.
Lateral diffusion beneath the mask edges may be harmful. For instance it will increase the gate/drain capacitance of a MOS transistor (Miller capacitance), and degrade the electrical isolation of closely spaced doped regions. Possible countermeasures, apart from reducing the penetration depth of the dopant atoms, include creating a spacer at the mask edge (see Sect. 3.5.3) or to extend the mask edge into the underlying silicon, as is done for instance in some versions of the LOCOS technique (see Fig. 3.4.7) or in trench isolation (see Sect. 3.5.4).

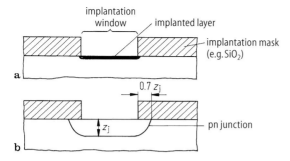

Fig. 6.3.11 a, b. When an implanted layer diffuses into the monosilicon, the lateral distance of the pn junction from the mask edge equals about $0.7\,z_j$, where z_j is the depth of the pn junction. **a** After implantation; **b** after diffusion

6.4
Diffusion of non-doping materials

Apart from dopant materials, the diffusion of heavy metals, sodium, oxygen and hydrogen is of interest. Figures 6.4.1 and 6.4.2 show the diffusion constants for these materials in Si and SiO_2.

Gold, copper, iron and other heavy metals have such a high diffusion rate[15] in silicon that they can diffuse throughout the whole wafer thickness at processing temperatures of around 900 °C. This means that no matter where they enter the silicon, heavy metal atoms can reach critical areas of the wafer and cause serious damage. Some notable effects include heavy metal atoms acting as centres for the generation of charge carriers, as "lifetime killers" (i.e. they reduce the lifetime of minority charge carriers) and as nuclei for oxidation-induced stacking faults and weak spots in thin oxide films.

Avoiding heavy metal contamination is therefore crucial in wafer processing. For instance the heavy metal content of any gases, liquids and solids coming into contact with the wafers during manufacture must be kept as low as possible (see Chap. 7). Other measures include adding HCl during thermal oxidation and when cleaning the tube furnace, as well as providing gettering layers. The gettering layers act as heavy metal sinks, and are produced in the silicon substrate itself or placed in direct contact with the silicon. Phosphorus doped films and phosphorus doped polysilicon in direct contact with the monosilicon wafer have also proved effective. Creating extensive crystal damage over the back of the wafer, e.g. by argon implantation or mechanical stress, is another useful gettering technique for reducing contamination.

---

15 The diffusion length $2\sqrt{Dt}$ (see Sect. 3.3.2) is the mean distance that the diffusing particles travel in time t.

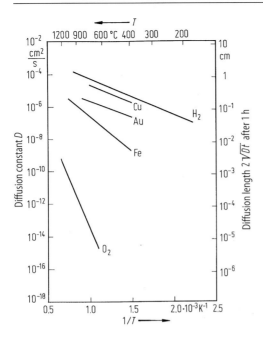

Fig. 6.4.1. Diffusion constants of important non-doping materials in silicon

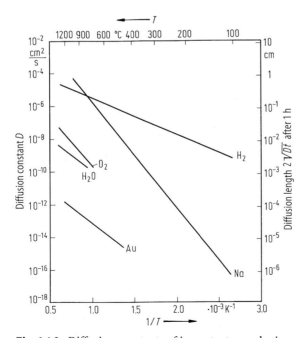

Fig. 6.4.2. Diffusion constants of important non-doping materials in SiO_2

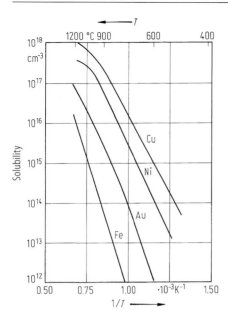

Fig. 6.4.3. Solubility of some heavy metals in silicon

If heavy metal contaminants land on the silicon wafers once temperatures are below about 500 °C (this is the maximum temperature that can be tolerated once aluminium metallization is present), then these will cause almost no damage, since the solubility of heavy metals at these temperatures is greatly reduced [6.15] (Fig. 6.4.3).

Sodium has a particularly high mobility in SiO_2 films (Fig. 6.4.2) when it exists in the form of positively charged ions. In an electric field these ions move in the direction of the field and in turn induce opposite charges in the silicon substrate. The effect of sodium ions is particularly harmful if the direction of the field is such that the sodium ions migrate to the Si-SiO_2 interface, for instance when a positive voltage is applied to a gate electrode. The threshold voltage V_T of a MOS transistor is then changed by an amount

$$\Delta V_T = -\frac{eN}{C_{ox}},$$

where N is the number of sodium ions per unit area, e is the elementary charge and C_{ox} is the oxide capacitance per unit area.

The sodium can get onto the semiconductor surface either during manufacture of the integrated circuit or even afterwards. In modern silicon technology the sodium problem has basically been solved by taking two precautionary approaches. The first is to ensure low sodium contamination by following clean working practices (e.g. using gloves to avoid transfer of salt from hand perspiration) and by using materials with a low sodium content during manufacture (such as quartz parts, targets, photoresist and photoresist developer). The second approach is to trap existing sodium ions in a gettering layer (usually phosphorus glass or boron phosphorus glass) and to prevent sodium penetration using passivation films (e.g. nitrides). In addition, introducing HCl into the thermal oxidation process (see Sect. 3.1.2) neutralizes the harmful effect of sodium diffusing out of the heating cassettes and the ceramic tubes in furnaces.

The diffusion of oxygen in SiO_2 has been discussed in the context of thermal oxidation in Sect. 3.1.2, whilst diffusion of oxygen in silicon was dealt with in Sect. 3.2.3 with regard to producing oxygen-depleted zones in melt grown silicon.

The diffusion of hydrogen in SiO_2, silicon and aluminium becomes a significant factor during annealing. At the SiO_2-Si interface the hydrogen can saturate free valencies and thus reduce the surface state density. Nitride films produced by LPCVD (Sect. 3.1.1) at 750 °C act as diffusion barriers to hydrogen. Plasma nitride films (Sect. 3.7.4) behave differently, however, because they contain a high level of hydrogen and thus act as a source of hydrogen.

7
Cleaning technology

In the earlier chapters, frequent mention was made of the damaging effect of contaminants in the silicon substrate, in the various layers and on the surfaces of the integrated circuit. This chapter summarizes the various types of contamination, and describes the measures employed to prevent or remove contaminants.

7.1
Contaminants and their effect

Table 7.1 shows the four major sources of contamination and their possible way to reach the wafer surface.

The damaging effect of sodium and heavy metal contaminants on oxide stability has already been discussed in Sect. 3.4.14. Heavy metals are also a problem in monocrystalline silicon, where they act as generation-recombination centres for electron-hole pairs. If present in the base region of a bipolar transistor, these centres reduce the current gain. In depletion zones they cause increased leakage

Table 7.1. Contamination of the wafer surface

Type of contamination	Cause, origin
Particles	Incorporation of particles from gases (surrounding air, process gases), liquids (water, etching solutions, developers) and solids (abrasion, flaking of films, detachment of loosely attached particles)
Heavy metals	Incorporation of heavy metals from gases (surrounding air, process gases, plasmas), liquids (water, etching solutions, developers) and solids (abrasion, sputtering, etching and diffusion from solids)
Organic residues	Incorporation of organic contaminants on the wafer surface during plasma etching (polymerization), during sputtering and electron beam processes (oil crack products), when rinsing with water (bacteria, algae) and from the clean room air. Residues from partially removed resist.
Sodium ions	Incorporation of alkali compounds from gases (surrounding air, process gases), liquids (developers, hand perspiration) and solids (abrasion, sputtering, etching and diffusion from solids)

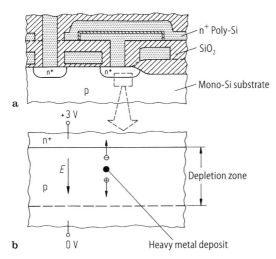

Fig. 7.1.1 a, b. Illustration of the accelerated charge loss from a dynamic memory cell (**a**) as a result of a heavy metal deposit in the depletion zone of the "storage node" (**b**). Electron-hole pairs are generated at the location of the heavy metal deposit. These pairs are separated in the electric field E of the depletion zone, and thus create a leakage current from the storage node to the Si substrate

currents. This means a larger subthreshold current in the channel region of a MOS transistor, and an increased cut-off current in a pn junction. In the "storage node" of a dynamic memory cell (see Sect. 8.4.2), the increased leakage current means that the stored charge is lost more rapidly (Fig. 7.1.1). Charge loss is critical in a DRAM cell, since it can only store about 10^6 elementary charges, which should be retained for about 1 second. A DRAM process must therefore meet very tight specifications for avoiding or removing heavy metal contaminants. For instance, no metal surfaces should be exposed in a plasma etching chamber where oxide layers are being etched down to the silicon substrate.

The major source of organic residues is the incomplete removal of resists. These resist deposits are particularly critical after etching a via hole over aluminium, because the presence of aluminium means that aggressive cleaning agents cannot be used (see Table 7.4). Figure 7.1.2 shows how polymer residues, known as "via crowns", are formed around the top of vias. If these polymer residues are not completely removed, then the second aluminium film is thinned or even interrupted at these points.

Whilst heavy metal contamination and organic residues can be tackled in state of the art processing by process control and especially by effective cleaning procedures (see Sect. 7.1.2), particulate contamination presents a continuous challenge to every advanced process line. Figure 7.1.3 shows how even particles smaller than the minimum feature dimension b_{min} can cause total failure of the

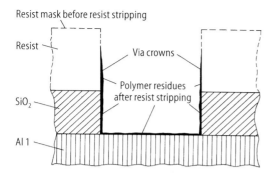

Fig. 7.1.2. Formation of polymer residues at resist edges ("via crowns") during reactive ion etching (RIE) of vias. The reaction gases and the gaseous reaction products of SiO_2 etching create a polymer film which is left behind at vertical edges, since vertical ion bombardment cannot act at these angles (see Fig. 5.2.1)

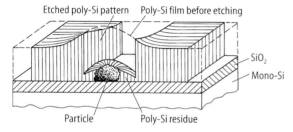

Fig. 7.1.3. Illustration of how a short-circuit can occur between two poly-Si tracks because of a particle measuring about half of the minimum feature dimension. Conformal deposition of the poly-Si means that the vertical thickness of the poly-Si is larger around the particle. This extra thickness may not be completely removed when the poly-Si is etched anisotropically, so that some poly-Si is left behind. If this poly-Si residue extends between adjacent poly-Si tracks, the tracks are shorted, and the chip concerned exhibits total failure

chip concerned by creating a local short-circuit. Fig. 7.1.3 illustrates that particles with a size $d_p > \frac{1}{3} b_{min}$ may be fatal[1].

In CMOS technology, however, only some 4 to 5 of the total 12 to 25 pattern planes are critical concerning particulate sensitivity. These are the "active area" plane, the "poly-Si gate" plane and the interconnection planes (cf. Sect. 8.2.1). All other pattern planes in a CMOS process are far less sensitive to particulate contamination.

---

1 For a chip of size 1 cm² with minimum feature size 0.5 µm, this means that a particle just 0.17 µm in size can cause the whole integrated circuit to fail. To put this in perspective, if the chip were magnified to the size of a soccer field, then the "fatal" particle would be the size of a pin head.

If the fatal defects in each of the n critical planes are distributed over the wafer surface with an average density D_o, which takes into account a certain degree of defect clustering, then for a chip size A_{chip}, the chip yield Y_{chip} affected by defects is given by the Price formula [7.1] as:

$$Y_{chip} = (1 + A_{chip} \cdot D_o)^{-n}.$$

For high yields (Y_{chip} > about 70%), the Price formula can be simplified as follows:

$$Y_{chip} \approx 1 - n \cdot A_{chip} \cdot D_o, \text{ for } Y_{chip} > 70\%.$$

The fatal defect density D_o can be found from the actual defect density D ($d_p > b_{min}/3$) for a particle size $d_p > b_{min}/3$, by multiplying D ($d_p > b_{min}/3$) by the probability w_{fatal} of the particle coming to rest in a position which will cause failure. For densely packed memory chips one can set

$$w_{fatal} \approx 0.4,$$

whilst for the less densely packed logic chips one can use $w_{fatal} \approx 0.2$. Practical experience has shown that for a given defect density level, the defect density for defect size D_p, denoted by $D(d_p)$, is proportional to d_p^{-3} [7.1]. Using this fact, one obtains the following relationships[2]:

$$D(d_p > b_{min}/3) = D(d_p > 0.33 \mu m) \cdot \left(\frac{\mu m}{b_{min}}\right)^2$$

$$D_o = w_{fatal} \cdot D(d_p > 0.33 \mu m) \cdot \left(\frac{\mu m}{b_{min}}\right)^2$$

$$Y_{chip} \approx 1 - n \cdot A_{chip} \cdot w_{fatal} \cdot D(d_p > 0.33 \mu m) \cdot \left(\frac{\mu m}{b_{min}}\right)^2$$

Thus, if one wants to achieve an 80 % particle-related yield, say, for a 16M DRAM chip with a 75 mm² chip area and 0.5 µm feature size with $n = 4$ critical planes, then only 0.04 particles per cm² are permitted with a size > 0.33 µm.

From the last equation one can also deduce that the particle-dependent chip yield Y_{chip} remains the same if for an existing integrated circuit and a given defect density level D ($d_p > 0.33$ µm) one reduces the feature size by the factor K (shrink factor K, see Table 8.6). This can be proven as follows:

---

2 If $D(d_p) \sim d_p^{-3}$, then $D(>d_p) = \int_{d_p}^{\infty} D(d_p) d(d_p) \sim d_p^{-2}$

$$A_{\text{chip}}(K) \cdot D_o(K) = \frac{A_{\text{chip}}}{K^2} \cdot w_{\text{fatal}} \cdot D(> 0.33 \mu m) \cdot \left(\frac{\mu m}{b_{\min} / K}\right)^2$$

$$= \frac{A_{\text{chip}}}{K^2} \cdot w_{\text{fatal}} \cdot D(> 0.33 \mu m) \cdot \left(\frac{\mu m}{b_{\min}}\right)^2 \cdot K^2 = A_{\text{chip}} \cdot D_o$$

Another conclusion from this relation is that in an advanced process line, the defect density level, which can be characterized by the defect density $D(d_p > 0.33\ \mu m)$, must be reduced by 30 % every year if one intends to keep pace with the progress in CMOS technology, i.e. if one is to produce ever larger chips (double the chip area every six years, cf. Table 8.9) with ever finer features (halving every six years, cf. Fig. 1.1) with the same chip yield. This figure does not take into account the increasing technology complexity, using e.g. additional interconnection levels.

Modern wafer processing involves two ways to reduce the contamination on wafers. On the one hand, the causes of contamination and the transfer of contaminants to the wafer surface are tackled (see Table 7.1) by ensuring clean rooms are actually clean, and that materials and processes are contaminant-free (see Sect. 7.2). On the other hand, cleaning steps are integrated into the chip manufacturing process, so that any contaminants reaching the silicon wafers are removed again (see Sect. 7.3).

7.2
Clean rooms, clean materials and clean processes

As the summary in Table 7.1 shows, contaminants reaching the silicon wafer originate from gases, liquids and solid materials coming into contact with the wafers. The air in the process line (clean room), the process gases and liquids, and both handling and processing of the silicon wafer thus have to satisfy the very highest standards of cleanliness.

7.2.1
Clean rooms

The air in clean rooms must be temperature controlled (23±0.5 °C), humidity controlled (42 %±3 % relative humidity, especially in the lithography and dry etch areas) and contain extremely low levels of particulates.

The particulate contamination of the air in a clean room is defined by the particle density and the particle size. Thus for instance a 10/0.1 μm classification means that less than 10 particles having a diameter greater than 0.1 μm are contained in 1 ft³ of air (about 28 l) [3]. A clean room of about this classifi-

---

[3] The early clean room classification based on US Federal Standard 209b which only covered particles bigger that 0.5 μm, is no longer appropriate for specifying modern clean rooms.

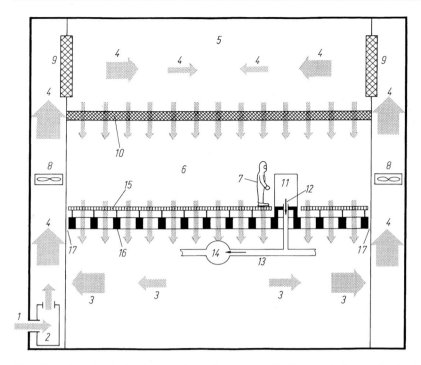

Fig. 7.2.1. Air circulation system in a clean room (ballroom design). *1* fresh air (same amount as outgoing air); *2* fresh air treatment (intake, filtering, dehumidifying, humidifying, temperature setting); *3* circulating air; *4* incoming air to clean room (= circulating air + fresh air); *5* plenum; *6* clean room with vertical laminar air flow (0.4 m/s); *7* clean room clothing; *8* fan; *9* coarse filter; *10* fine filter; *11* equipment containing exhaustion device (e.g. cleaning bank); *12* exhausted air; *13* collecting pipe for outgoing air; *14* outgoing air duct to air washer, fans and chimney for outgoing air; *15* elevated grid floor; *16* solid concrete honeycomb floor (for low floor vibrations); *17* separating gap (to decouple vibrations between building and honeycomb floor)

cation is required for the manufacture of VLSI circuits with 1 µm features. In the 0.25 µm region clean rooms with a classification of 1/0.03 µm may be required.

Figure 7.2.1 shows the design of a typical air circulation system in a 1/0.03 µm grade clean room. The entire ceiling area of the clean room is made up of ULPA filters (*Ultra Low Penetration Filters*), of the design shown in Fig. 7.2.2. A sufficiently high vertical air speed must be maintained (typically 40 cm/s) to prevent particles e.g. from the floor, reaching the silicon wafers against the direction of air flow. In order to avoid crossflow and turbulence, the filter elements in the ceiling are replaced by blanket shields at those locations, where a piece of equipment prevents air from penetrating the grid floor. In addition, people and mov-

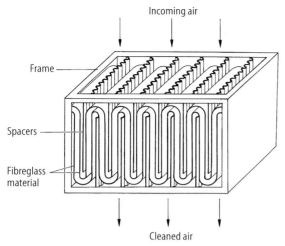

Fig. 7.2.2. Design of a ULPA filter for fine filtering of the surrounding air in a clean room (e.g. for 1/0.03 µm classification). The filtration efficiency for 0.1 µm particles is > 99.9999 %

ing parts must not perform rapid movements in the clean room. A grid floor with sufficient flow resistance also prevents crossflow, and ensures there is a higher pressure in the clean room than outside.

The electrostatic charging of insulators and non-grounded parts in the clean room (slight friction can easily generate several kV for instance), can induce particles to move against the air flow due to the strong Coulomb forces. In order to avoid such charges[4], electrically conductive materials are used wherever possible in clean rooms. Another measure is to install tip discharge devices directly below the filter ceiling. These ionize the air in the room so that charged surfaces are discharged by the passing air [7.2].

Numerous safety precautions have to be taken in clean rooms. For instance, sprinkler jets are arranged over the whole clean room ceiling, which open if the relevant fire alarm indicates a raised temperature. Emergency showers are positioned at regular intervals. Toxic process gases are prevented from entering the circulating air by enclosing all gas fittings in exhausted boxes, i.e. enclosing all screw fittings, pressure reducers, valves and flow meters (MFC = mass flow control). Gas sensors in the extracted air then indicate whether a leak in the process gas network has occurred.

An alternative clean room concept to the "ballroom" design shown in Fig. 7.2.1, is the "mini-environment" design [7.10]. In this approach, the pure air

---

4 Electrostatic charges in the clean room are also unwanted because they can cause electrical breakdown of the thin insulating layers on silicon wafers.

is fed through ducts to the loading area of the processing machines, whilst relaxed cleanliness requirements are permitted in the room itself. The wafers are transported in air-tight boxes. In order to load the wafers into a process chamber, a SMIF (standard mechanical interface) robot is used.

7.2.2
Clean materials

All process gases (e.g. oxidation, annealing, doping and etching gases), process liquids (e.g. water, cleaning fluids, photoresists, developers) and solid materials (e.g. sputter targets) that come into direct or indirect contact with the silicon wafers, must have the highest level of cleanliness. The waste gases and water that are returned to the environment must also be treated to make them environmentally safe.

Figure 7.2.3 shows as an example the gas and cool water supply of a dry etch facility, and the disposal system for the process gases.

The permitted degree of contamination for sub-0.5 µm processes is 1 ppb (ppb = parts per billion) for base gases (N_2, Ar, H_2) and 100 ppb for process gases.

In order to filter out particulates, filters with a pore size of 0.03 µm are fitted in the gas lines as close as possible to the point where the relevant gas reaches the silicon wafers (point-of-use filters). To prevent contaminants, the gas supply system should be made of vacuum-melted special steel with inert gas shielded welding and internal polishing.

The cleanliness of fluids coming into contact with the silicon wafers is particularly important, because most contaminants prefer to remain on the silicon surface. Of all the process liquids, it is the water that must meet the highest standards, as this is used at the end of each and every wafer cleaning process, and is by far the most frequent material to come into contact with the wafers. The ultra-pure water that is used (deionized water) is normally produced by treating water from public supply in a purifying plant close to the process line. It is then fed into a ring pipeline and continuously circulated around the process line in order to inhibit bacterial growth. The ring pipeline is ideally made of PVDF, with minimum length pipe branches leading to the points of use via point-of-use membrane filters (0.1-0.2 µm). The key elements in an ultra-pure water plant are the ion exchangers (cations and anions), the reverse osmosis process (removes all particles of molecular weight > 56), vacuum degassing (to remove CO and CO_2), a second ion exchange stage, ultraviolet irradiation (to kill bacteria) and ultra-filtration. Table 7.2 shows an example of the water contamination specification for the ring pipeline in a modern facility producing ultra-pure water.

By far the largest amounts of water are used in the rinsing procedures of wafer cleaning. In order to recycle some of the precious water, it is diverted to a recycling system once it reaches a certain level of conductivity during rinsing, and

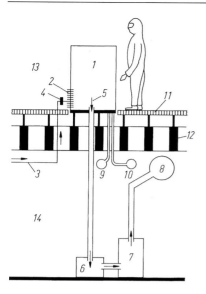

Fig. 7.2.3. Gas and cool water supply of a dry etch facility, and the disposal system for the process waste gases. *1* vacuum reaction chamber of the dry etch facility; *2* gas connections; *3* gas line to gas bottle (e.g. CF_4) or gas tank (e.g. N_2); *4* point-of-use filter (e.g. 0.03 µm); *5* exhaust gases; *6* pump; *7* waste gas cleaning unit, in which combustion of waste gases occurs, and combusted gases are washed out [7.3]; *8* outgoing gas duct; *9, 10* cool water collection pipes for inward and return flow; *11* grid floor; *12* solid honeycomb floor; *13* clean room; *14* supply and disposal floor

Table 7.2. Specified water contamination figures for a modern plant providing ultra-pure water. ppb: parts per billion (10^{-9}); ppm: parts per million (10^{-6})

resistivity	> 18 MΩ cm (at 25 °C)
density of 0.05 µm particulates	< 10/ml
bacterial cells	< 50/l
TOC (total organic carbon)	< 10 ppb
Metal content	< 0.05 ppb
CO_2 content	0
dissolved oxygen	< 5 ppb

then fed back into the ultra-pure water network. The waste water is returned to the public sewage system after passing through a continuously monitored neutralization plant.

In comparison with the ultra-pure water, the purity requirements for the other process liquids (e.g. HF, HCl, H_2O_2, H_2SO_4, NH_4OH, choline) are slightly more relaxed. For instance the residual metal contamination level in these liquids is specified as 1 ppb. Like the gas supply to process facilities (see Fig. 7.2.3), the liquid tanks in modern process lines are also located on the floor beneath the clean room, together with the associated pumping equipment plus filtration, control and measuring devices.

7.2.3
Clean processing

Even if the clean room air and process materials are perfectly clean, the silicon wafers can still be contaminated during processing and wafer transport.

Contamination from particulates loosely attached to personnel or objects in the clean room is the main problem. Table 7.3 shows the various means by which these particulates can reach the silicon wafers. Table 7.3 shows that appropriate clean room dress, suitable clean room working practices, and adequate processing and wafer handling techniques are crucial for low particulate contamination. In the future, automated processing will play an increasingly important role, which is expected to provide better control over the particulate contamination problem.

Apart from contamination in particulate form, metal contaminants can reach the wafers by diffusion during high-temperature processes, from contact with vacuum pick-up tools, chucks and loading devices, and from hand perspiration. As has been mentioned already, countermeasures include making pick-up tools from materials that do not contain heavy metals, and wearing gloves. Section 3.4.4 also described the gettering techniques which can neutralize the harmful effects of metal contamination in silicon.

7.3
Wafer cleaning

The measures described in Sect. 7.2 for providing rooms, materials and processes of optimum cleanliness, are still not sufficient to completely avoid contamination. This is why several cleaning steps are incorporated in the manufacturing sequence for integrated circuits (see Sect. 8.6), aimed at removing any contamination from the wafer surface.

Table 7.4 shows, as an example, the wafer cleaning sequence which is widely used in CMOS processes immediately before thermal oxidation, e.g. before gate oxidation (cf. Fig. 3.4.3). This cleaning procedure is referred to as "RCA cleaning" or "Huang cleaning", and is attributed to W.Kern [7.4]. The alkaline cleaning stage is primarily intended to remove particulate and organic residues, whilst the acidic oxidizing agent in the second cleaning stage dissolves the heavy metals. After rinsing in water, and drying, the Si surface is in a defined state, which ensures that a reproducible thermal oxide layer can be produced having a defined interface with the silicon (see Sect. 3.4.3).

The acid cleaning step of RCA cleaning may be omitted if a suitable chelating agent that can bring the heavy metals into solution even in an alkaline environment, is added to the alkaline cleaning solution[5] [7.6].

---

5 In this case choline is preferred as the alkaline substance instead of NH_4OH [7.8].

Table 7.3. Particulate contamination during processing

Particulate source	Route onto wafer?	Countermeasures
loosely attached particles on clean room garments	in the air stream and by electrostatic forces	• suitable material for clean room clothes (e.g. polyester fabric interwoven with conducting fibres) • suitable design of clean room clothes (tight-fitting openings, providing as complete cover as possible) • training of personnel how to behave in a clean room • adequate interlock design (e.g. air showers) • replacement of staff by automation equipment
loosely attached particles on cassettes, boxes, chucks and objects	during wafer transport in air stream and by mechanical impact as well as friction and electrostatic forces	• avoiding particle-generating surfaces • avoiding mechanical impact (e.g. when loading wafer into cassettes) • consistent cleaning approach for cassettes, boxes etc. • keeping wafer environment free of any unnecessary objects
loosely attached particles on the interior walls of processing equipment	particles falling directly onto the wafers, or transported to the wafers in the gas flow	• frequent cleaning of equipment if films are deposited on the interior walls of the machines (during CVD, sputtering and dry etching procedures) • avoiding vibrations • frictionless loading devices for furnaces (cantilevers) • low pumping speed to avoid turbulence • vertical arrangement of wafers
chipping of silicon wafers	particles can fall directly onto wafers	• careful wafer handling • automated wafer transport with smooth end stops
chipping of photoresist films	particles can fall directly onto wafers	• keep photoresist off edge of wafers (see Sect. 4.2.1) • careful wafer handling • automated wafer transport

Table 7.4. Sequence and effect of RCA cleaning. The main purpose of RCA cleaning is to produce a defined silicon surface, free of particulate, organic or heavy metal contamination, ready for subsequent thermal oxidation of the silicon

Processing sequence	Cleaning agent	Apparatus	Cleaning effect
Alkaline cleaning	$NH_4OH : H_2O_2 : H_2O = 1:1:1.5$ 70 °C Megasonic	Si water / Process tank / Cold water tank (< 50 °C) / Ultrasonic waves (Megasonic) / Transducer (piezo material) / 6. MHz generator	Particulates are removed by slight etching of Si surface, and lifted off the surface by mechanical effect of ultrasonic waves [7.5]. Organic residues are dissolved
Rinsing	H_2O 23 °C Megasonic		The water displaces the alkaline solution from below (overflow)
Acid cleaning	$HCl : H_2O_2 : H_2O = 1:1:5$ wetting agent 70 °C		Heavy metals are dissolved by chelation. The wetting agent is required in order to prevent the adsorption of particles on the Si surface.
Rinsing	H_2O; 23 °C Megasonic		Generation of a defined Si surface
Drying	N_2 (possibly hot)	spin dryer	Residue-free drying of Si and SiO_2 surfaces

Aimed at cutting the cost of RCA cleaning, T. Ohmi has proposed a cleaning procedure which is performed at room temperature and which reduces the consumption of chemicals [7.7]. The essential features of the Ohmi cleaning procedure are the treatment of wafers in ozone-containing deionized water in order to remove the organic carbon, and also the use of HF instead of HCl during the acid cleaning step.

RCA cleaning can also be performed in a spray cleaner instead of the tank shown in Table 7.4. The benefits of this system are that the spin dryer can be incorporated in the cleaner, and smaller quantities of chemicals may be used. The disadvantages are the increased process complexity and the fact that Megasonic

Table 7.5. Sequence and effect of cleaning after reactive ion etching of contact holes. The rinsing stages are not shown

Processing sequence	Cleaning agent	Apparatus	Cleaning effect
Stripping of resist	O_2 plasma	Barrel reactor (see Fig. 5.2.2)	"etching" of resist in O_2 plasma
Removal of polymer layer	$H_2SO_4 : H_2O_2$ = 6:1 (Caro´s acid)	Tank (see Table 7.4)	Polymers are removed by oxidation
Removal of damaged layer	Choline: H_2O = 1:1500 wetting agent; 70°C	Tank (see Table 7.4)	Si and sub-oxides are etched at a rate of about 1 nm/min
RCA cleaning	see Table 7.4	see Table 7.4	see table 7.4
Removal of naturally occurring oxide layer on Si	HF: H_2O = 1:200	Tank (see Table 7.4)	Creation of an oxide-free Si surface

treatment cannot be used here. If one installs a Megasonic transducer in the spray head of the fluid inlet, however, the drops of fluid can be made to oscillate, which they continue to do as they hit the wafer surface (Finesonic). This does improve removal of the particles, but is still distinctly less effective[6] than the Megasonic cleaning method shown in Table 7.4.

"Marangoni" drying [7.9] has been proposed as an alternative to spin-drying. In this technique the wafers are drawn out of a mixture of isopropanol and water and dried in hot nitrogen.

Another important cleaning step in wafer processing is the total removal of photoresists. The resist removal procedures after contact hole etching and after via etching are described below.

As was mentioned in Sect. 5.3.5, the contact hole etching process is usually designed to leave a polymer film on the exposed Si surface at the bottom of the contact hole. This is done to achieve a high selectivity, i.e. a large SiO_2:Si etch rate ratio. In order to provide a low resistance contact, this polymer film must be removed during cleaning. Furthermore, the cleaning process must also remove the layer of monocrystalline silicon damaged by ion bombardment. This layer extends to a depth of several nm for reactive ion etching, where ions hit the wafer surface at energies of several 100 eV (see Sect. 5.2.3). Table 7.5 shows one possible cleaning sequence following contact hole etching [7.6].

This cleaning process cannot be used after via etching, however, because Al instead of Si is in place at the bottom of the via holes. Table 7.6 shows a possible option for cleaning vias after reactive ion etching [7.6].

---

6 The reduced ultrasonic effect can actually be an advantage, e.g. if aluminium tracks are on the wafer surface.

Table 7.6. Sequence and effect of cleaning after reactive ion etching of vias

Processing sequence	Cleaning agent	Apparatus	Cleaning effect
Stripping of resist	O_2 plasma	Barrel reactor (see Fig. 5.2.2)	"Etching" of resist in O_2 plasma
Removal of polymer residues	30 % dimethyl-sulphoxide + 70 % monoethanolamine; 90 °C	Spray cleaner	"Etching" of polymer residues (see Fig. 7.1.2)
Intermediate rinsing	IPA	Spray cleaner	Intermediate rinsing is necessary, otherwise the Al would be attacked if immediately rinsed with water
Rinsing	H_2O	Spray cleaner	Creation of a defined Al surface
Drying	N_2	Spin dryer incorporated in spray cleaner	Residue-free drying of wafer surface

A final reference should be made to scrubber cleaning, which is a mechanical cleaning technique used for removing particulates after CMP planarization for instance (see Sect. 5.1.2). The particulates are removed from the wafer surface using a brush (brush cleaning) or a focused jet of water (jet scrubbing), or using a "particle jet" containing CO_2 particles (below −80 °C).

8
Process integration

The basic elements of an entire manufacturing process for an integrated circuit have already been described in Chap. 2. This chapter provides a more detailed description of the architecture of the key technologies currently in world-wide use, or to be introduced in the near future.

8.1
The various MOS and bipolar technologies

The different technologies are classified according to the active devices within the integrated circuits.

8.1.1
Active components in integrated circuits

Table 8.1 shows the most important active components in integrated circuits. A sensor has been included in addition to the 3 basic types of MOS transistor (standard, floating-gate and DMOS) and the bipolar transistor.
Integrated sensors, i.e. the integration of sensor, signal processing circuits and power switches on one silicon chip, are not widely used as yet, but may be important in the future. Apart from the magnetic field sensor shown in Table 8.1, acceleration sensors and temperature, pressure and radiation sensors (e.g. infrared) are being considered as suitable for silicon integration (e.g. [8.21]).

8.1.2
Comparison of MOS and bipolar technologies

The complete manufacturing process is referred to as a CMOS, E^2PROM, smart-power, bipolar, BICMOS or smart-sensor technology, depending on which of the active components listed in Table 8.1 are integrated into the integrated circuit. Table 8.2 summarizes the technologies with their basic properties and areas of use.
CMOS is by far the most important technology. The main reasons for its dominance are the small space occupied by the actual MOS transistors, and the extremely dense packing density in the integrated circuit that can be obtained with

8 Process integration

Table 8.1. The active devices used in integrated circuits

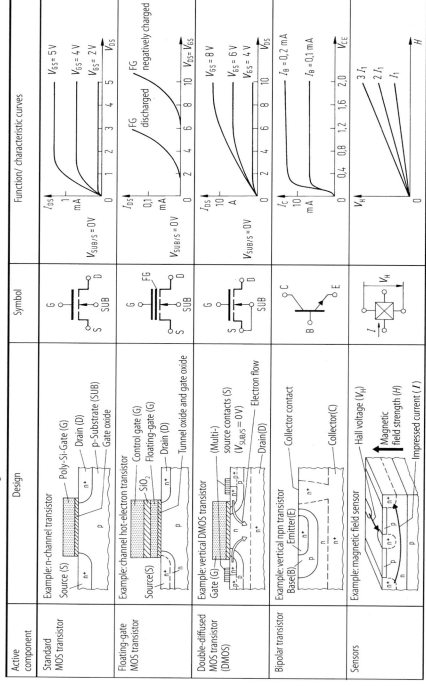

Table 8.2. Integrated circuit technologies, their properties and areas of use

Active devices in the integrated circuit	Name of technology	Outstanding properties of the technology	Technology applications
n-channel MOS transistors + p-channel MOS transistors	• CMOS (= complementary metal oxide semiconductor)	• very high packing density • low power consumption • scaleable	• complex logic circuits (microprocessors, microcontrollers, digital signal processors, ASICs) • static memories (SRAMs) • dynamic memories (DRAMs)
Floating-gate MOS transistors + CMOS transistors	• NVM (non-volatile memory) • E²PROM (electrically erasable programmable read-only memory) • Flash E²PROM	• non battery-backed data storage • erasable and reprogrammable memory	• chip cards • replacement for magnetic data storage
Bipolar transistors	• Bipolar	• very good analogue properties • high current drive capability • high speed	• operational amplifiers • power output stages • tuners, mixers • fast gate arrays • multiplexers/demultiplexers
DMOS transistors + CMOS/bipolar transistors	• SPT (= smart power technology) • BCD (bipolar/CMOS/DMOS)	• combination of controller logic and high switching power • high voltage capability (> 100 V)	• motor control • coupling devices in communications systems • 220 V devices (e.g. lamp control)
MOS transistors + Bipolar transistors	• BICMOS	• combination of CMOS and bipolar properties possible	• fast filters • analogue/digital converters
Sensors + CMOS/DMOS/bipolar transistors	• Smart sensor technology	• combination of sensor, signal processing and power output	• speed control • antilocking braking system (ABS) • airbag control

them (see Sect. 8.3.1). Other benefits are the low power consumption, and also the high switching speed that improves with progressive miniaturization.

The other technologies are used when the particular features they offer cannot be met by CMOS, for instance non-volatile memory, high voltage and high current applications, current drive capability, ultra-high speed or excellent analogue performance.

8.1.3
Passive components in integrated circuits

Apart from the active devices, integrated circuits also contain passive components, in particular diodes, resistors and capacitors. Table 8.3 provides a summary of the various designs and properties of the different passive components.

Except for the high-value resistor and the poly-Si1/poly-Si2 capacitor (cf. Table 8.12), the other passive components shown in Table 8.3 normally do not require any additional processing steps, as they can be regarded as sub-elements of a MOS or bipolar transistor.

8.2
Technology architecture

As was already highlighted in Table 2.1, a complete manufacturing process is made up of numerous individual processing steps (200-500). These steps are themselves combined into processing blocks or process modules, which incorporate a distinct functional part of the integrated circuit to be produced.

8.2.1
Architecture of MOS technology

In a basic CMOS process (Tables 2.1 and 8.4), the p- and n-type wells are created first to provide the substrate regions of the n-channel and p-channel MOS transistors (well process module). The next step in the process is the isolation of neighbouring transistors by creating a so-called field oxide regions between the transistors. The MOS transistors are then formed in the areas not covered by field oxide, which are called the active regions. This completes the first stage of the complete process, referred to as the FEOL stage (*Front End Of Line*) where the transistors and their mutual isolation are produced. In the BEOL stage (*Back End Of Line*), the individual regions of monocrystalline and polycrystalline silicon produced in the FEOL stage are interconnected and contacts produced according to the integrated circuit design. With two or more metal layers being the norm in today's integrated circuits, this is usually referred to as multi-layer metallization. The final step in the complete process is passivation, designed to protect the integrated circuit against mechanical damage and contamination by foreign matter.

8.2 Technology architecture

Table 8.3. The passive components used in integrated circuits

		Types of design	Properties	More details in section
Passive component	Diode	pn junction	Forward voltage 0.7 V; Breakdown voltage 5 to > 100 V	
		Emitter-base diode formed by short-circuiting C-B in bipolar transistor	Low series resistance; Forward voltage 0.7 V; Defined breakdown voltage 5-8 V (Z-diode)	8.3.3
		Gated diode	Diode characteristic dependent on gate voltage and oxide thickness	
		Schottky diode	Low forward voltage (approx. 0.4 V)	3.11.4
	Resistor	pn-isolated diffusion regions	Sheet resistance of $10-10^3$ $\Omega/\square$ achievable; pn-isolation takes up space	6.3.7
		Doped poly-Si tracks	Sheet resistance of $10-10^9$ $\Omega/\square$ achievable; limited stability of high-ohmic resistors	3.8.3
		Polycide tracks	Sheet resistance of 1-10 $\Omega/\square$ achievable;	3.9.2
		Silicided diffusion regions	Sheet resistance of 2-10 $\Omega/\square$ achievable;	3.9.3
	Capacitor	Reverse-biased pn junction	Non-linear capacitance (voltage dependent); $1-10^{-4}$ fF per μm^2 of pn surface area depending on level of p-type doping	8.4.2
		Poly-Si/diffusion capacitance	Linear capacitance (not voltage dependent); for SiO_2 dielectric 3.5 fF per μm^2 surface area for 10nm oxide thickness	
		Poly-Si1/poly-Si2 capacitance	Linear capacitance; SiO_2 or ONO as dielectric; 35 fF per μm^2 of surface area for 10 nm oxide thickness	3.7.3 8.4.2
		Metal1/metal2 capacitance	Linear capacitance; planarization of dielectric makes reproducibility more difficult	8.5.1

Table 8.4. Process modules making up complete CMOS processes. Analogue, E²PROM or DRAM technologies can be developed from the CMOS [baseline] process by modifying or adding process modules (shown dotted or cross-hatched respectively). The area of the rectangle assigned to a process module corresponds approximately to the number of processing steps that it involves

CMOS baseline process	CMOS analogue	EEPROM	DRAM
Passivation	Passivation	Passivation	Passivation
Metal 2	Metal 2	Metal 2	Metal 2
Me1/Me2 contacts	Me1/Me2 contacts	Me1/Me2 contacts	Me1/Me2 contacts
Metal 1	Metal 1	Metal 1	Metal 1
Si/Me1 contacts	Si/Me1 contacts	Si/Me1 contacts	Si/Me1 contacts
	Poly-Si2 for R and C		/// Bit line (metal 0) ///
			/// Bit line contacts ///
CMOS transistors	CMOS transistors	CMOS transistors	CMOS transistors
		/// Floating gate ///	/// Trench capacitor ///
Insulation	Insulation	Insulation	Insulation
Wells	Wells	Wells	Wells

Extra circuit functions can be created by adding specific process modules to the CMOS baseline process. In some cases the process modules in the baseline process have to be modified. Table 8.4 shows key examples of CMOS process architectures providing analogue functions (using linear resistors and capacitors), non-volatile memory cells (E²PROM) or dynamic memory cells (DRAM = *Dynamic Random Access Memory*). Tables 8.12 and 8.14 show the detailed processing sequence for an analogue-digital technology and a DRAM technology respectively.

8.2.2
Architecture of bipolar and BICMOS technologies

Table 8.5 shows the bipolar, BICMOS and smart-power processes, again as derivatives of a CMOS baseline process. These spin-off processes are by no means intended as purely illustrative. Rather they indicate the opportunity, often exploited in a real CMOS process line, of adding specific process modules to produce integrated bipolar, BICMOS and power circuits. Apart from the epitaxy step, the added process modules do not contain any individual processing steps that are not in principle part of the CMOS process.

Table 8.5. Process architecture of the bipolar, BICMOS and smart-power technologies, shown as derivatives of a CMOS baseline process. Those modules that have been modified compared to the CMOS baseline process are shown dotted, whilst additional process modules are cross-hatched. The area of the rectangle assigned to a process module corresponds approximately to the number of processing steps that it involves

CMOS baseline process	Standard bipolar process	High speed bipolar	Analogue/digital BICMOS	High speed BICMOS	Smart power (DMOS + CMOS + bipolar)
Passivation	Passivation	Passivation	Passivation	Passivation	Passivation
Metal 2	Metal 2	Metal 3	Metal 2	Metal 3	Metal 2
Me1/Me2 contacts	Me1/Me2 contacts	Me2/Me3 contacts	Me1/Me2 contacts	Me2/Me3 contacts	Me1/Me2 contacts
Metal 1	Metal 1	Metal 2	Metal 1	Metal 2	Metal 1
Si/Me1 contacts	Si/Me1 contacts	Me1/Me2 contacts	Si/Me1 contacts	Me1/Me2 contacts	Si/Me1 contacts
CMOS transistors	Base/Emitter	Metal 1	Poly-Si for R and C	Metal 1	CMOS-, DMOS transistors and base/emitter
		Si/Me1 contacts	CMOS transistors and base/emitter	Si/Me1 contacts	
		Self-aligned poly-Si base/emitter		Self-aligned poly-Si base/emitter	
				CMOS transistors	
	Collector lead	Collector lead	Collector lead	Collector lead	Collector lead
Insulation	Insulation	Insulation	Insulation	Insulation	Insulation
Wells			Wells	Wells	Wells
	Epitaxy	Epitaxy	Epitaxy	Epitaxy	Epitaxy
	Buried Layer	Buried Layer	Buried Layer	Buried Layer	Buried Layer

8.3
Transistors in integrated circuits

8.3.1
Design of MOS transistors and their isolation

Standard MOS transistors have a very simple and space-saving design compared with bipolar devices, and offer easy and compact integration. They are therefore the ideal active components for VLSI circuits.

Figure 8.3.1 shows two n-channel MOS transistors and one p-channel MOS transistor forming part of an integrated circuit before metallization. In the layout view, one can see the two main pattern planes (masks) used in the front end of the CMOS process, namely the isolation mask and the gate mask (cf. the detailed processing sequence in Tables 2.1 and 8.11, in Sect. 2 and 8.6 respect.).

The isolation mask divides the chip surface into the active regions, which contain the transistors and diffused regions, and into the intervening field oxide regions. The field oxide, which is typically 0.5 μm thick, is effectively the gate oxide

Fig. 8.3.1. Example showing the layout of two n-channel MOS transistors and one p-channel MOS transistor in a CMOS circuit. S = source region; G = gate region; D = drain region; L_G = gate length, geometric channel length, L_{eff} = effective channel length; W = channel width

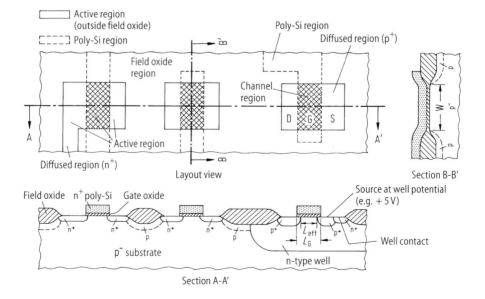

of the parasitic MOS transistors in the integrated circuit. These parasitic transistors must be permanently cut-off otherwise electrical isolation between neighbouring diffused regions and active channel regions cannot be guaranteed. Cut-off must be guaranteed for all possible voltages carried by the poly-Si or metal tracks crossing the field oxide, which are effectively the gates of the parasitic transistors. A LOCOS technique[1] is still the most common method for producing the field oxide (see Sect. 3.4.2). Trench isolation (see Sect. 3.5.4) may be needed for field oxide ribs less than 0.3 μm wide.

Apart from the isolation mask and the gate mask, three further CMOS-specific implantation masks are required for the layout shown in Fig. 8.3.1, although these are not critical from the aspect of pattern size. They either mask the n-channel transistor regions (during implantation of the n-type well and p$^+$ diffusion regions) or the p-channel transistor regions (during implantation of the n$^+$ diffusion regions). It is usual for poly-Si gates not to be covered by a masking layer when source/drain implantation occurs, which means they receive the same source/drain implantation dose. Since the poly-Si is usually already heavily doped with phosphorus, however (cf. Fig. 3.6.1, right-hand column), the p$^+$ implantation does not affect the n$^+$ doping level of the poly-Si.

Whilst the isolation between active MOS transistors is provided at the silicon surface by cut-off parasitic MOS transistors, it is reverse-biased pn diodes that must ensure isolation within the bulk monosilicon. This is why the p-type substrate in Fig. 8.3.1 is held at 0 V, or is given a negative bias voltage, whilst the n-type well is held at the largest positive potential, which is normally the supply voltage, e.g. +5 V (see Fig. 8.3.1). Since this means that the voltages U_{DS} and U_{GS} are always negative, the p-channel transistors are permanently in their active state.

Fixing the potential of the p-type substrate and n-type well at 0 V and +5 V respectively, is still not sufficient to guarantee that the n- and p-channel MOS transistors will always be isolated from each other. As can be seen from Fig. 8.3.2, between an n-channel transistor and a neighbouring p-channel transistor in a CMOS circuit there exists a parasitic thyristor, formed from one parasitic npn transistor and one parasitic pnp transistor.

As soon as either of the two emitter-base pn junctions is forward biased at any point[2], the thyristor is "triggered" (latch-up effect). The high current can cause localized damage (e.g. melting of metallization) in the integrated circuit. The most important measures for preventing the latch-up effect are:
- maintaining a minimum distance between neighbouring n$^+$ and p$^+$ regions, the distance being dependent on the technology (this is an important design rule in CMOS circuits),

---

1 In rare cases, an n$^+$ poly-Si field plate at a voltage of e.g. 0 V is used above a thin gate oxide instead of the field oxide.
2 The pn junction can be forward biased by e.g. a parasitic current generated by the ionizing impact of hot electrons, setting up a voltage drop in the base region of the bipolar transistor.

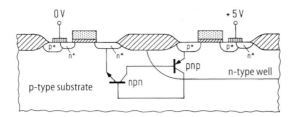

Fig. 8.3.2. Parasitic thyristor that exists between an n-channel MOS transistor and an adjacent p-channel MOS transistor in a CMOS circuit. If either of the two emitter-base pn junctions of the two parasitic bipolar transistors is forward biased locally, the thyristor is triggered (latch-up). The high current that then flows can cause damage

- applying a negative bias to the p-type substrate (for an n-type well)
- using a p^+ substrate with a p-type epitaxial layer (for an n-type well)
- using "retrograde" wells (see Sect. 6.2.2)
- using as many well contacts as possible in order to maintain a constant potential throughout the well, and thus avoiding voltage drops which could otherwise trigger latch-up.

When integrated MOS circuits are designed, the diffused areas and poly-Si regions, which are isolated with respect to their surrounding areas, are extended beyond the actual transistor region to form conducting links to a metal contact or a neighbouring transistor for instance. Such links can be seen in the lower left (diffused region) and the upper right (poly-Si region) of the layout view of Fig. 8.3.1. A fundamental reason for the excellent integration capability of CMOS technology is this ability to implement some of the CMOS circuit interconnections directly without resorting to metallization, even with the limitation that diffused regions and poly-Si tracks are not allowed to cross over each other[3]. Taking the static memory cells shown in Fig. 8.4.1 as an example, only 6 metal contacts are required to connect together the 6 MOS transistors within the memory cell.

The real driving force behind the rapid development of integrated CMOS circuits is miniaturization, or down-scaling as it is called. As Table 8.6 shows, scaling down the feature size by the factor K not only means that one can integrate K^2 more transistors per unit area, but the speed of the integrated circuits themselves also increases, indicated by the shorter delay time for a logic gate. If one then reduces the supply voltage by the factor K as well, the power dissipation also falls by the factor K^2. The energy per logic operation, referred to as the power-delay product, is then actually reduced by the factor K^3.

---

3 An active MOS transistor would be produced by the crossover.

Table 8.6. The scaling principle applied to the two cases "supply voltage V_{DD} is also scaled" and "supply voltage V_{DD} remains constant". The scaling factor is K (K>1)

Parameter to be scaled	Multiplication factor	
All lateral and vertical dimensions	1/K	
All dopant atom concentrations	K	

Based on the well-known MOS equations, when this scaling is performed the parameters listed below are transformed approximately as follows:

Transformed parameter	Multiplication factor	
	$V_{DD} \rightarrow V_{DD}/K$	V_{DD} constant
packing density	K^2	K^2
drain current per channel width	1	K^2
current densities	K	K^3
field strengths	1	K
oxide capacitance per unit area	K	K
pn capacitance per unit area	K	$\sqrt{K}$
power dissipation density	1	K^3
power dissipation per gate	$1/K^2$	K
delay time per gate	1/K	$1/K^2$
power-delay product	$1/K^3$	1/K

The critical significance of the feature size means that CMOS technologies are usually identified by a dimension, such as 0.7 µm. Unfortunately there is no standardized definition of this characteristic dimension, which is a permanent source of confusion.

The following definitions can be found for a "0.7 µm technology":
– the minimum effective channel length of the MOS transistors is 0.7 µm;
– the minimum feature size in the integrated circuit that can be produced lithographically is 0.7 µm;
– the mean value of half the minimum grid pitch and the contact hole dimensions in the "critical" pattern planes equals 0.7 µm.

In this book we intend to use the third definition, because this represents the best measure of the complexity of the production equipment and processes.

As Fig. 8.3.3 shows, the supply voltage of CMOS circuits has remained constant down to the 0.5 µm technology generation. This means that feature sizes have reduced by approximately a factor of 5 over the last 10 years (K = 5) without a reduction in voltage. If one had applied the scaling principle consistently (see Table 8.6, right-hand column), then at critical points in the integrated circuit, the current density (increased by the factor K^3) and the electric field strength (increased by the factor K), would have become so large that circuit reliability could no longer have been guaranteed. Two approaches have been adopted in order to minimize the failure risk. Either the scaling principle has not been strictly observed, e.g. the gate oxide thickness has been reduced less sharply than the lateral feature dimensions (see Fig. 8.3.3), or new materials for metallizations or a new doping profile design

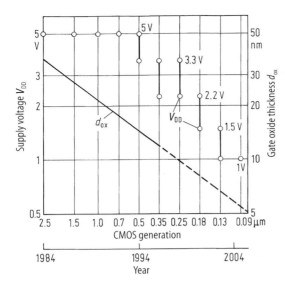

Fig. 8.3.3. Change in supply voltage and gate oxide thickness of CMOS circuits as miniaturization progresses

Table 8.7. Technological measures for ensuring reliability of CMOS circuits when miniaturization occurs without a reduction in the supply voltage

Reliability risk	Cause	Technological countermeasures
Gate oxide short circuit	Local degradation of the gate oxide caused by Q_{bd} at weak spots, see Sect. 3.4.3.	Avoid weak spots in gate oxide, particularly by suitable cleaning prior to gate oxidation, see Sect. 7.3
Drain current drift	The peaks in field strength at the drain end of the channel create hot electrons (hot electron effect). They tunnel into the gate oxide and are "trapped" there.	A lightly doped drain (LDD) reduces the field strength at the drain end, see Fig. 3.5.2 a
Leaking pn junctions	"Spiking" at flat pn junctions, see Sect. 3.11.4	1. Silicon added to aluminium, see Sect. 3.11.4 2. TiN barrier layer under the Al film, see Sect. 3.11.4.
Breaks in aluminium tracks	Electromigration if high current densities in the Al, see Sect. 3.11.3.	1. Al film thickness is not down-scaled and remains at about 1 μm 2. Copper added to aluminium, see Sect. 3.11.3 3. TiN barrier layer under Al film, see Sect. 3.10.

Table 8.8. Possible technological measures to enable further miniaturization of CMOS circuits from 0.5 μm down to 0.1 μm

Objective	Possible technological measures	Effects of measures
Shallow channel doping profiles (approx. 0.1 μm)	1. Implantation of heavy ions (BF_2^+, In^+, As^+) 2. Selective epitaxy, see Sect. 3.3.1 3. SOI substrate, see Sects. 3.1.6, 3.1.7	
Flat source/drain doping profiles with low sheet resistance	1. Silicided S/D regions (salicide), see Sect. 3.9.1 2. Selective epitaxy, see Sect. 3.3.1 3. Selective $TiSi_2$ CVD deposition, see Sect. 3.9.1	
Reliability against latch-up for small n^+/p^+ separation	1. Retrograde well doping profiles, see Sect. 6.2.2 2. SOI substrate	
Better short-channel properties of the p-channel transistor by using surface channel instead of buried channel	1. p^+ poly gate for p-channel transistors 2. SOI substrate	Polycide gate necessary, see Sect. 6.3.6
Short LDD doping profiles	1. Oblique implantation (approx. 45°) through the spacer	RTP technique required, see Sect. 6.3.1
Short field oxide ribs	1. Advanced LOCOS technique, see Sect. 3.4.2 2. Trench isolation, see Sect. 3.5.4	More processing steps
Relaxed tolerances for overlay accuracy	1. Overlapping contacts, see Sect. 8.5.2 2. Non-capped contacts, see Sect. 8.5.2	More processing steps
Relaxed tolerances for linewidth control	1. Global planarization, see Sect. 8.5.1	More processing steps
Reliable fine-feature metallization	1. Planarization of insulating films, see Sect. 8.5.1 2. Introduction of local interconnects, see Sect. 8.5.3 3. Introduction of additional metal pattern planes	

(LDD) have been adopted. The most important of these measures are listed in Table 8.7. Thanks to the tremendous opportunities offered by miniaturization, further down-scaling of CMOS circuits will continue in the future. A 0.1 μm technology looks perfectly feasible [8.1]. The main challenge lies in the production control of commercial manufacturing processes for integrated circuits.

In order to guarantee the reliability of integrated circuits for feature sizes as low as 0.5 μm and below, several processing steps must be modified, and/or new processing steps and materials added. Table 8.8 summarizes the innovative measures that may be used.

The opportunities offered by SOI substrates (see Sect. 3.2.4) also deserve special mention. Thanks to the lateral and vertical isolation provided by the oxide (and not by pn junctions or cut-off field oxide transistors) there is no latch-up effect. One can therefore place n-channel and p-channel MOS transistors as close together as lithographic restrictions allow. Another advantage of SOI substrates is that there are almost no parasitic pn capacitances. This means that SOI substrates can also provide faster digital circuits.

One final option, which increases the transistor packing density for a given minimum feature size, is to use a vertical transistor design [8.2], in a similar way to trench capacitors in memory cells (cf. Fig. 8.4.2).

8.3.2
Design of DMOS transistors

Unlike the standard MOS transistor, the channel length in the DMOS transistor (DMOS = Double Diffused MOS) is not determined by the length of the poly-Si gate, but by the difference in penetration depths of a p- and n-type diffusion (cf. Table 8.1). Another feature of the DMOS transistor is its long and lightly doped drain drift region, which provides high voltage stability. This is why DMOS transistors are integrated in those circuits requiring a high-voltage power switch at their output (see Table 8.2).

To achieve the 1 A current range, numerous DMOS transistors are connected together in parallel [8.3]. Ideally one would obviously like the low-resistance power switch to occupy the minimum possible chip area[4]. The arrangement shown in Fig. 8.3.4 is particularly economical regarding chip space. Since the drain region is designed as a blanket buried layer, this does not take up any chip area. Space-saving self-aligned contacts (see Fig. 3.5.2 d) are used to connect each of the source regions and each of the substrate (body) regions. An Al film covering the whole of the DMOS structure connects the individual source and substrate regions together.

The vertical DMOS transistor in Fig. 8.3.4 is not suitable once cut-off voltages of over 300 V are required. This is because the epitaxial layer which must carry the high voltage drop would have to be made too thick (> 30 μm) or the epitaxial doping would have to be uncontrollably small (< 10^{14} cm^{-3}). This is why a differ-

---

4 A measure for the space occupied by the power switch is the resistivity $R_{ON} = R \cdot A_{Chip}$. R is the on resistance and A_{Chip} is the chip area required. For the design shown in Fig. 8.3.4, R_{ON} values of 0.1 Ω mm^2 are obtained for 0.7 μm technology.

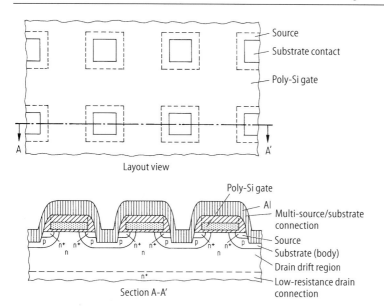

Fig. 8.3.4. Vertical multi-source DMOS transistor with space-saving self-aligned source/substrate contacts [8.3]

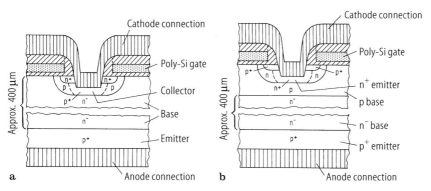

Fig. 8.3.5 a, b. DMOS controlled power switches. **a** IGBT (Insulated Gate Bipolar Transistor) with n-channel DMOS transistor; **b** MCT (MOS Controlled Thyristor) with p-channel DMOS transistor

ent device design is used in power electronics, which is based on a lightly n-doped silicon wafer[5] with the anode connection on the back of the wafer. Figure 8.3.5 shows two DMOS controlled designs: the DMOS controlled bipolar transistor (IGBT = *I*nsulated *G*ate *B*ipolar *T*ransistor) and the DMOS controlled thyristor (MCT = *M*OS *C*ontrolled *T*hyristor). Both power devices are currently

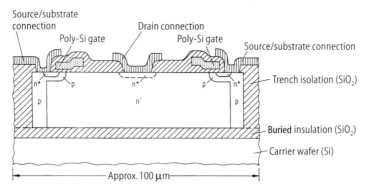

Fig. 8.3.6. Lateral high voltage DMOS transistor with surrounding dielectric isolation

available only as stand-alone power switches [8.4], although in principle they could be integrated with standard MOS transistors and bipolar transistors on a single chip.

Figure 8.3.6 shows a lateral DMOS transistor without backside connection, which operates in the 500-1000 V region [8.5]. The SOI wafer (produced by wafer bonding, see Sect. 3.1.7) and lateral trench isolation (see Sect. 3.5.4), ensures that the DMOS transistor is entirely isolated by a thick SiO_2 layer.

8.3.3
Design of bipolar transistors and their isolation

As one can see from the feature summary in Table 8.2, bipolar transistors are less suited to very large scale integration than MOS transistors because they take up a large amount of space. They are still used in integrated circuits, however, where there is a need for those characteristics that CMOS cannot provide but which bipolar technology can. These advantages are a high switching speed, excellent analogue performance and the option to integrate power output stages or Hall sensors without increasing the process complexity.

Figure 8.3.7 shows the design of a standard npn bipolar transistor, assumed in this example to be part of an integrated BICMOS circuit. This basic type of bipolar transistor is found in all high-speed bipolar, BICMOS or SPT circuits (see overview in Table 8.5).

One contact hole is usually required for each emitter, base and collector connection. These transistor areas are connected to other bipolar or MOS transistors in the integrated circuit using aluminium tracks, which means that "wiring" is much more expensive than in MOS circuits (see Sect. 8.3.1). It also means that

---

5   Using a technique called neutron transmutation (conversion of Si into P), low phosphorus doping levels ($< 10^{14}$ cm^{-3}) can be achieved with extremely good uniformity. There is no equivalent technique available for p-type doping.

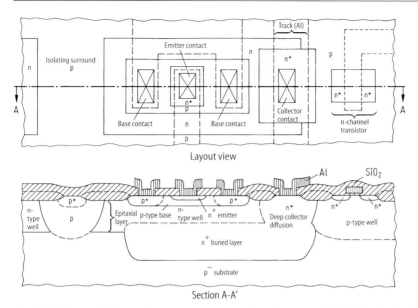

Fig. 8.3.7. Example layout for an npn transistor in an integrated BICMOS circuit. An n-type well is shown to the left of the bipolar transistor, which can take an npn bipolar transistor or a p-channel MOS transistor. To the right is an n-channel MOS transistor. In the layout view, the continuous lines indicate the doping boundaries, the dashed lines the metallization or poly-Si tracks

for the same feature size, the maximum packing density that can be achieved with bipolar transistors is only one tenth of that for CMOS technology. Moreover, only a marginal increase in packing density is achieved when the feature size is reduced in bipolar circuits. This is mainly because of the space occupied by the isolating surround and the collector diffusion, since these diffused regions extend at least as far laterally as the thickness of the epitaxial layer (usually a few µm). Since miniaturization is not a major factor, however, bipolar circuits can normally be produced using conventional process technology.

The vertical npn transistor shown in Fig. 8.3.7 is by far the most commonly used bipolar type. Occasionally, lateral or vertical pnp transistors are also integrated into the circuit. The lateral pnp transistors can be produced without extra complexity using the process architectures for vertical npn transistors[6]. In BICMOS processes one obtains vertical pnp transistors by adding a p^+ buried layer.

Figure 8.3.8 shows a transistor designed for maximum speed (transit frequency of 30 GHz achievable) [8.6]. In this design the emitter and base connec-

---

6 A lateral pnp transistor is created by using the p-type base region of an npn transistor as the emitter or collector of the pnp transistor, and letting the n-type well act as the base region of the pnp transistor.

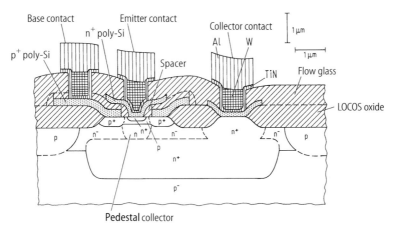

Fig. 8.3.8. Bipolar transistor designed for maximum speed, with self-aligned poly-Si base contacts and poly-Si emitter contacts, and a pedestal collector

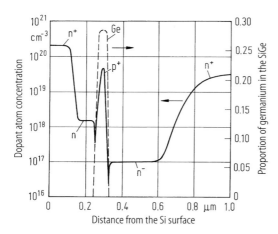

Fig. 8.3.9. Doping profile of an Si/SiGe HBT with a possible transit frequency of over 100 GHz [8.8]. HBT = Heterojunction Bipolar Transistor

tions are produced by n^+ and p^+ doped poly-Si films, which are aligned with each other using the spacer technique (see Fig. 3.5.2 c). The far smaller surface area of the base-collector junction means that the associated depletion layer capacitance is drastically reduced. The short distance between the outer base and the edge of the emitter reduces the bulk resistance of the base. The bulk resistance of the inner base (below the emitter) is also reduced because the emitter is now narrower than the minimum lithographic dimension by an amount equal to two spacer widths. Furthermore, using a poly-Si emitter reduces the emitter junction depth in the monocrystalline silicon, and thus increases the transit frequency. Finally, implantation of phosphorus through the emitter window into

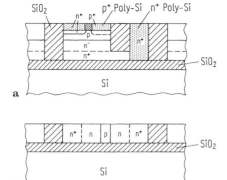

Fig. 8.5.10 a, b. Cross-sectional diagram of two possible bipolar transistor designs on an SOI substrate. **a** Vertical npn transistor, **b** lateral npn transistor [8.9]

the lightly n-doped collector region below the emitter (pedestal collector or SIC = Selective Implanted Collector) increases the permitted current density in the transistor.

There are other measures for increasing the speed of this type of bipolar transistor still further. These include:
– reducing the base lead resistance by using polycide films (see Sect. 3.9.2)
– using selective epitaxy (see Sect. 3.3.1) for the base and collector [8.7]
– using SiGe epitaxy for the base (HBT = Heterojunction Bipolar Transistor) [8.8], Fig. 8.3.9.

Just as for CMOS technology, introducing SOI substrates for bipolar technology opens up completely new approaches. In particular, drastic space savings can be realized in the isolation of bipolar transistors. Figure 8.3.10 shows two possible versions of a bipolar transistor on an SOI substrate [8.9].

8.4
Memory cells

Semiconductor memories are classified into three basic types: static, dynamic and non-volatile memory. Each type differs fundamentally in its physical operation. The following sections describe their various designs.

8.4.1
Design of static memory cells

A static semiconductor memory cell is created from a bistable flip-flop with 2 selection transistors (Fig. 8.4.1). The flip-flop has only two stable states. In one state the "memory nodes" **6** and **10** are at 0 V and the V_{DD} potential respectively, whilst in the other stable state they are at V_{DD} and 0 V respectively. As long as the supply voltage V_{DD} is present, the memory state is stable over time, hence the name "static" memory.

8 Process integration

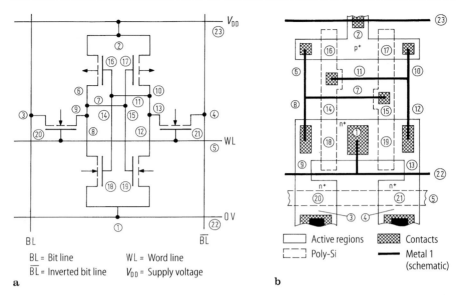

Fig. 8.4.1 a, b. Circuit diagram (**a**) and layout (**b**) of a static memory cell comprising 6 transistors in CMOS technology. The circled numbers indicate the respective position of these circuit elements in each diagram. For reasons of clarity, the layout view (**b**) only shows the metal *1* tracks schematically as thick lines. The bit lines are implemented as metal *2* and would run from top to bottom in the layout view (**b**), but are not shown, again for reasons of clarity

To provide random access, each static memory cell not only requires the 0 V and V_{DD} connections, but also two bit lines and one word line (SRAM = *Static Random Access Memory*).

The static memory cell shown in Fig. 8.4.1 can be produced with a CMOS [baseline] process (see Table 8.4) with no extra processing steps. This version is thus particularly common in "embedded SRAM" logic circuits (e.g. microprocessors), where the SRAM is integrated in the circuit. In "stand-alone SRAMs", however, the key requirement for the SRAM cell is that it occupies a minimum of space. Thus although the 6-transistor CMOS cell dominates, there are space-saving versions available, where the p-channel transistors (see Fig. 8.4.1) are replaced by depletion transistors (n-channel transistors with a negative threshold voltage), or by "poly-loads" (high-resistance poly-Si resistors, which can be packed above the transistors in a second poly-Si plane), or by TFTs (TFT = *Thin Film Transistor*: this is a MOS transistor having a poly-Si channel region, which can also be placed above the n-channel transistors, cf. Fig. 3.3.2).

For the same process technology, static memory cells occupy almost three times the chip area of dynamic memory cells (see Sect. 8.4.2). Their advantage lies in the fact that they can be made using CMOS [baseline] technology, and that they have a shorter access time (a few ns for SRAMs compared with 40-70 ns for DRAMs).

8.4.2
Design of dynamic memory cells

Compared with static memory cells, dynamic memory cells have a very simple design. They consist solely of a selection transistor and a storage capacitor (Fig. 8.4.2). The memory states "0" and "1" correspond to the capacitor being positively or negatively charged respectively. In modern memory cells the capacitor charge can be lost within a period of about 1 second, which means that the charge must be repeatedly refreshed. The data also need to be re-written after a read operation. The refresh process is performed automatically using a special circuit integrated onto the chip. Data refresh is typical to this type of memory, which is consequently called "dynamic memory" (DRAM = Dynamic Random Access Memory).

The main complexity in the development of DRAM technology lies in the storage capacitor. The memory capacitance needs to equal about 35 fF in order to provide a sufficiently large read signal, and to remain insensitive to alpha particles[7]. According to the capacitance formula

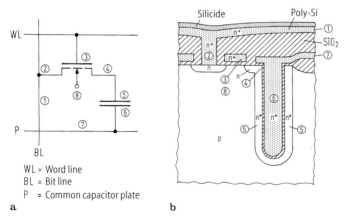

Fig. 8.4.2 a, b. Circuit diagram (a) and cross-section (b) of a dynamic memory cell with trench capacitor. The circled numbers indicate the corresponding position of circuit elements in each diagram

---

7 Alpha particles are helium nuclei with an energy of 5.5 MeV. They are present in space, and occur as trace elements, particularly in metals. An alpha particle reaching a silicon surface can penetrate up to 35 μm deep into the silicon, creating a large number of electron/hole pairs in the process. These charge carriers can recombine with the stored charge and thus destroy it (soft error).

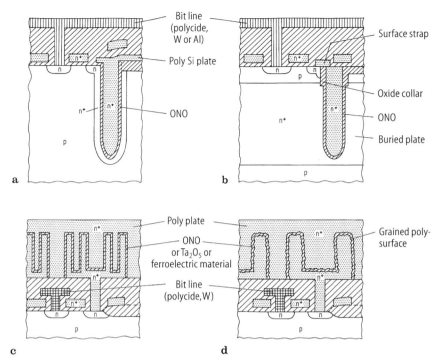

Fig. 8.4.3 a–d. Four important designs for the capacitor in DRAM memory cells. **a** Trench capacitor with poly-Si plate (poly plate trench capacitor); **b** buried plate trench capacitor; **c** crown stacked capacitor; **d** hemi-spherical grained silicon stacked capacitor

$$\left(\frac{C}{\text{fF}}\right) \approx 9\varepsilon_\text{r} \left(\frac{\text{nm}}{d}\right)\left(\frac{A}{\mu\text{m}^2}\right)$$

one would need a capacitor area $A \approx 10\ \mu\text{m}^2$, for an SiO$_2$ dielectric ($\varepsilon_\text{r} = 4$) of thickness $d = 10$ nm. Less than half of the memory cell surface area can be used to provide this capacitance. For a 4 M DRAM, however, the total surface area of a memory cell is already smaller than 10 µm². Thus a planar layout for the capacitor can no longer be an option. This is why storage capacitor designs have been developed that use the third dimension. Figure 8.4.3 shows two versions each of a trench capacitor design and of a stacked capacitor design.

The total surface area of the 4 million capacitors in a 4 M DRAM equals about 0.4 cm², and should contain practically no defects. This is almost impossible to achieve with a thermal or deposited SiO$_2$ film, which is why the ONO dielectric film (oxide/nitride/oxide) is now the established alternative for both trench and stacked capacitors, since it can be manufactured practically without defects (see Fig. 3.7.2).

As the area of memory cells continues to shrink, the trend is likely to be towards using a higher dielectric constant ε_r, instead of ever deeper trenches or

Table 8.9. Previous and predicted developments in large-scale production of dynamic memory. The "feature density" specifies the total feature edge length, summed over all pattern planes, per unit area of layout of the memory cell field. It is a measure of the process complexity (shrink = miniaturized version)

Chip complexity (DRAM bits)	Year of introduction	Minimum feature size (μm)	"Feature density" ($\mu m/\mu m^2$)	Cell area (μm^2)	Chip area (mm^2)	Wafer diameter (mm)	Cost of manufacture (US $/10^6$ bits)
1 M start product	1987	1.20	4.5	25	60	150	4
1 M last shrink	1994	0.80	6.0	10	25	150	1
4 M start product	1990	0.85	6.0	10	90	150	2
4 M last shrink	1997	0.50	8.5	4	35	150	0.5
16 M start product	1993	0.60	8.5	4	135	200	1
16 M last shrink	2000	0.35	12	1.5	50	200	0.25
64 M start product	1996	0.40	12	1.6	200	200	0.5
64 M last shrink	2003	0.25	17	0.6	80	200	0.12
256 M start product	1999	0.30	17	0.65	300	300	0.25
256 M last shrink	2006	0.18	24	0.25	120	300	0.06
1 G start product	2002	0.20	24	0.25	430	300	0.12
1 G last shrink	2009	0.13	33	0.10	170	300	0.03
4 G start product	2005	0.15	33	0.10	600	500	0.06
4 G last shrink	2012	0.09	50	0.04	230	500	0.015
16 G start product	2008	0.11	50	0.04	850	500	0.03
16 G last shrink	2015	0.07	70	0.015	330	500	0.008

ever higher stacks. Materials such as Ta_2O_5 or barium strontium titanate (BST) may well be used [8.10].

Thanks to its simple design, the dynamic memory cell is by far the cheapest memory cell available. The mass memory in today's computers, from the PC to the mainframe, is therefore composed of DRAMs. The increasing demand for low-cost memory has been the driving force behind advances in semiconductor technology since the middle of the 1970s, and continues to push the pace of progress. Table 8.9 shows past developments in DRAMs and predicts future trends. These predictions assume that it will still be possible to halve the manufacturing cost per bit every 3 years or so. Achieving this target will depend on the same key factors as before: increased chip complexity, feature miniaturization, developments in the art of process architecture, better process control, larger wafer diameters and a greater level of redundancy[8].

---

8 Redundancy in this situation means providing extra memory cells on the DRAM chip in order to replace faulty memory cells with working cells.

The rapid development in DRAM technology shown in Table 8.9, has led to DRAM assuming the global role of a "technology driver". No other semiconductor product needs, for instance, down-scaling as early as the DRAM. One reason for this is that for the same minimum feature size, the "feature density" (see Table 8.9) is at least twice as large in a DRAM cell as it is in a logic circuit, say. Moreover, its dynamic charge storage principle makes the DRAM far more sensitive to contamination introduced during manufacture (see Fig. 7.1.1) than any other type of integrated circuit.

The technology driver role of the DRAM is only relevant concerning the silicon base material and the unit processes (lithography, etching technique, film deposition), but is not relevant for CMOS process integration. Although the DRAM technology involves a CMOS [baseline] process (see Table 8.4), MOS transistors used in a microprocessor circuit, for instance, must be optimized for this purpose, and will thus have a design that differs from the DRAM (doping, geometry). Another area of development not driven by advances in DRAM has been multilayer metallization. While today's DRAMs use two metallization levels complex logic products have five or more metal layers.

8.4.3
Design of non-volatile memory cells

The main feature of non-volatile memory (NVM) is that the data held in the memory cell remains for a long period of time (> 10 years), even when the power supply has been switched off.

The simplest type of non-volatile semiconductor memory is a ROM (Read-Only Memory). The memory cell field can be produced by a matrix of n-channel MOS transistors, for instance. Data is written into a memory cell right at the time of manufacture. A "1" is produced by implanting phosphorus in the relevant transistor so that the threshold voltage of the transistor becomes negative (depletion transistor). Since those memory cells that are to retain a "0" are covered by a resist mask during phosphorus implantation, this process is also called mask programming. As the name implies, one can only read from a ROM, but not erase or reprogram it.

In a PROM (*P*rogrammable *R*ead-*O*nly *M*emory) the user can program the memory just once, e.g. by blowing fuses on a chip with a current surge.

The non-volatile semiconductor memory in universal use is the E^2PROM (Electrically Erasable Programmable Read-Only Memory). This is designed to allow the user to repeatedly read, electrically erase and reprogram the memory.

Three different physical properties have been employed in the three non-volatile memory designs available in silicon technology.
– In the MNOS memory (MNOS = *M*etal *N*itride *O*xide *S*emiconductor) the charge is stored in traps at the nitride/oxide interface. Electrons tunnelling through the 2 nm thick oxide are used to transfer the charge.
– In the floating-gate memory (see Table 8.1) the charge is stored in a totally isolated poly-Si structure (floating gate). Once again, the charge is transferred

by electrons tunnelling through the thin oxide film between semiconductor and floating gate.
– In the ferroelectric memory, the hysteresis of a ferroelectric material lying between two capacitor electrodes provides the non-volatile storage effect, in a similar way to ferromagnetic storage.

The MNOS memory was actually developed first, but has hardly been used in practice. Modern non-volatile memories are almost entirely floating gate devices, although the ferroelectric memory appears to have great potential. The remainder of this section therefore concentrates on the last two designs of non-volatile memory.

Figure 8.4.4 shows the design of a floating-gate memory cell (FLOTOX = *FLO*ating-gate *T*unneling *OX*ide). In this case the DRAM storage capacitor is replaced by the floating-gate storage transistor. When a cell is being programmed, a large positive voltage (e.g. +15 V) is applied to the relevant word and bit lines, whilst the control gate is held at 0 V and the source line is left floating. Since these voltage levels switch on the selection transistor, the n$^+$ region below the tunnel oxide is also taken to the high positive voltage, so that the electric field

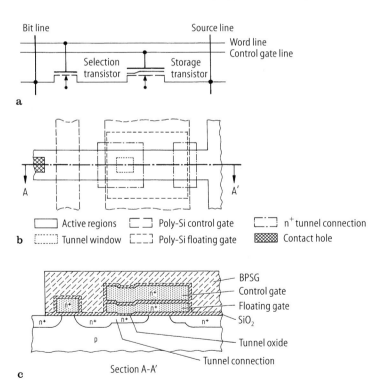

Fig. 8.4.4 a–c. Circuit diagram (**a**), layout (**b**) and cross-section (**c**) through AA′ of a non-volatile floating-gate memory cell (FLOTOX)

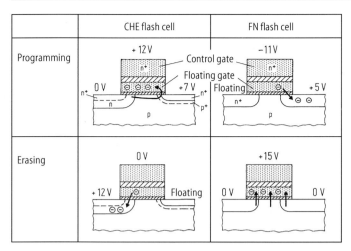

Fig. 8.4.5. Design and operation of the CHE flash cell (*Channel Hot Electron*) and the FN flash cell (*Fowler-Nordheim tunnelling*). In the CHE cell the peak field strength at the drain edge is increased by producing as abrupt a pn junction as possible. This produces more hot electrons for tunnelling through the gate oxide (opposite of the lightly doped drain, see Sect. 8.3.1)

strength in the tunnel oxide – but nowhere else – approaches the breakdown field strength (about 10^7 V/cm).

As a result, electrons tunnel out of the floating gate into the underlying n^+ region, and can then flow to the bit line via the conducting selection transistor. The positive charge in the floating gate cannot flow away, however. This charge can remain for very long periods (> 10 years), even when the applied voltage has been removed. Since the floating gate forms part of the storage transistor, the threshold voltage for the control gate is permanently reduced. Thus when a read operation is performed, the storage transistor is conductive and a current flows in the bit line.

A particularly space-saving version of the non-volatile memory cell is the flash-E²PROM [8.11]. It gets its name from an earlier design, the UV EPROM, where all the floating gates in the memory cell field were discharged by exposing them to a flash of UV light[9]. Figure 8.4.5 shows two different versions of a flash E²PROM memory cell. The difference lies in their programming mechanism. In the CHE flash cell (CHE = *Channel Hot Electron*) hot electrons tunnel through the gate oxide near the drain to the floating gate. In the FN flash cell (FN = *Fowler- ordheim*) a high electric field is required for the electrons to tunnel into the gate oxide (Fowler-Nordheim tunnelling), as is used in the FLOTOX cell. In both flash E²PROM designs the data is erased in blocks.

---

9 The UV light makes the SiO_2 conductive.

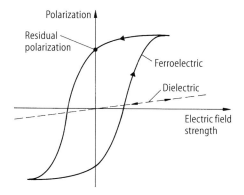

Fig. 8.4.6. Characteristic of a ferroelectric material compared with a dielectric. After removing a positive electric field, a positive polarization remains in the ferroelectric (remanence), whereas the polarization returns to zero in the dielectric. A capacitor containing a ferroelectric material therefore constitutes an E²PROM memory element

CHE memory cells offer the advantage of short programming times, whilst FN cells operate at low power and are suitable for low voltages [8.12].

All floating-gate memory cells are subject to degradation mechanisms arising from current flow through the SiO_2. As shown in Fig. 3.4.13, an SiO_2 film becomes conductive after a charge of a about 1-50 Coulomb/cm² has passed through it. This effect limits, for instance, the maximum number of programming cycles for a FLOTOX cell to 10^5-10^6. Since oxide degradation is more rapid for larger current densities, the programming current must be strictly controlled.

To conclude this section, the ferroelectric memory shall be discussed. The basic design of the cell is identical to the DRAM cell (see Fig. 8.4.2), with the difference that a ferroelectric material (e.g. lead zirconium titanate) is used instead of the dielectric between the capacitor electrodes. As shown in Fig. 8.4.6, the ferroelectric material has a positive or negative residual polarity depending on whether a positive or negative field was applied during programming. Reading can be performed by applying a positive voltage to the bit line, for instance. If the ferroelectric material has a negative polarization, then the polarity is switched to positive causing a packet of charge to flow in the bit line. For a positive residual polarization, the polarization is only changed slightly, so that almost no charge flows to the bit line. As in the DRAM, reading the stored data destroys it. The data must therefore be re-written after every read operation.

Thanks to its simple design[10] and its excellent performance figures (short programming times, $> 10^{12}$ programming cycles, low programming voltages), ferroelectric memories are likely to have a bright future [8.13].

---

10 One problem still needs to be overcome: integrating the ferroelectric into the integrated circuit manufacturing process.

8.5
Multilayer metallization

The last two sections have looked at the design of active and passive devices and their mutual isolation in the integrated circuits. The individual components are produced in the front end of line, and then interconnected in the back end of line to produce the desired integrated circuit.

In this section the most important metallization schemes for integrated circuits are discussed. With advancing miniaturization and increasing number of layers the planarization of steps on the wafer surface is becoming increasingly important. Section 8.5.1 is therefore devoted to planarization techniques.

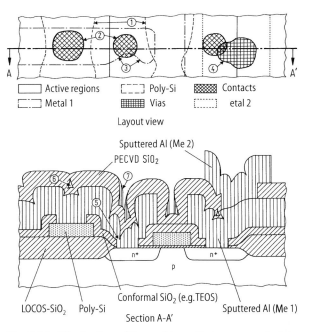

Fig. 8.5.1. Illustration of the problems that can occur in integrated circuits if no planarizing steps are included. *1* Line width variations (here: poly-Si features) caused by Newton's interference effect (see Fig. 4.2.7) or defocusing (see Sect. 4.2.6); *2* different sizes of contact holes caused by Newton's interference effect (see Sect. 4.2.3); *3* narrowing of metal track as a result of light reflection at an edge transition (see Fig. 4.2.8); *4* deformed via as a result of light reflection from the sloping surface of metal 1 at this point (see Fig. 4.2.8); *5* reduced metallization cross-section (risk of electromigration, see Sect. 3.11.3) because of nonconformal sputter coating (see Fig. 3.1.15); *6* cavities (voids) in the intermetal dielectric layer caused by overhanging edges of metal 1 (possible reliability problems); *7* metal 2 residues (stringers) after etching of metal 2, as a result of a localized increase in thickness of vertical metal 2 (cf. Fig. 5.3.1). The stringers can short-circuit adjacent metal 2 tracks

8.5.1
Planarization of surfaces in integrated circuits

Integrated circuits are produced by the precise placement of patterned films one on top of the other (pattern planes). Since features can range in thickness from 0.1-1 μm, the topography would be highly irregular and pitted with steep edges if planarization measures were not taken. Figure 8.5.1 shows an example of what unwelcome side-effects might otherwise result from such a topography. Some of the problems relate to light reflection from the wafer surface during exposure of the photoresist (see Sect. 4.2.3), producing defects in the pattern. The other problems arise from nonconformal edge coverage during sputter coating or from anisotropic etching of these films, leading to circuit shorts or reduced reliability.

Rounding off sharp edges reduces the latter effects but not the lithographic artefacts. Edge rounding techniques are therefore used mostly around contact holes and vias, where the lithographic effects do not come into play. This is because contact holes and vias are completely covered by the overlying metallization, and so no light falls around the contact holes or vias during exposure.

Figure 8.5.2 shows three techniques that are commonly used for rounding off the edges of contact holes and vias. Figure 8.5.2 also shows a bevelled LOCOS edge, which facilitates anisotropic etching of the poly-Si, as well as a bevelled poly-Si edge. The latter technique is inappropriate in the sub-micron region, however, where the width and height of the features are of a similar order of magnitude. Poly-Si bevelling is useful, however, for the poly-Si capacitor plate in the DRAM memory cell shown in Fig. 8.4.2, because this makes it easier to form the gate polysilicon that crosses over it.

Most of the unwanted topographic effects shown in Fig. 8.5.1 arise when aluminium tracks have to be taken over a stepped surface. The only way to solve the problem is to level off the surface to be metallized. Figure 8.5.3 summarizes the planarization techniques used in the manufacture of integrated circuits.
Figure 8.5.2

The flow glass technique is used today almost exclusively for planarizing the surface lying below the metal 1 pattern plane (see Sect. 3.6.2). In fact flow glass (BPSG) has been the key technique enabling feature down-scaling in the 5-0.5 μm region.

As BPSG requires temperatures over 800 °C in order to flow, one must resort to other techniques for planarizing surfaces below the metal 2 plane. The least expensive process is the Dep./Etch PECVD deposition of SiO_2 (see Sect. 3.1.1). This deposition process involves integrated deposition and etching in a plate type reactor, where the wafers lie on the cathode-connected plate (see Fig. 3.1.6). As a result, some of the deposited SiO_2 molecules are removed again by back sputtering. Since PECVD deposition is essentially a conformal process, whilst vertical ion bombardment is more effective on inclined and raised surfaces, steep edges are rounded off and narrow gaps are partially filled (see Fig. 3.1.7).

Flat LOCOS bird's beak (Example: poly buffered LOCOS)	Increased etch rate at surface (Example: poly-Si bevelling)	Isotropic/anisotropic etching (Example: bell-shaped contact holes)	Resist removal etching during SiO$_2$ (Example: sloping vias)	Flow glass reflow after pattern formation (Example: contact holes)
• Thinner nitride and thicker SiO$_2$ layer • Pattern formed in nitride 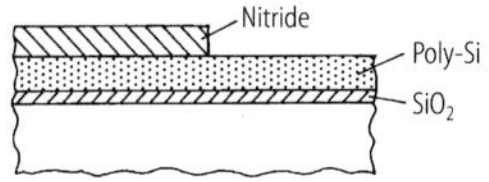 • LOCOS oxidation • Nitride/poly-Si stripping • 100 nm SiO$_2$ stripped 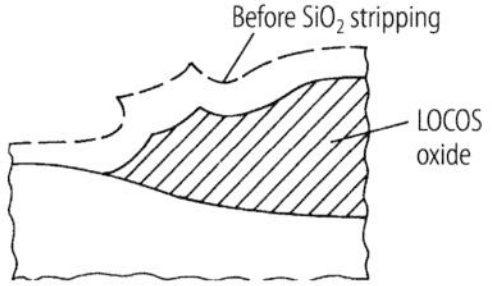	• Poly-Si deposition • Arsenic implantation (damage to surface) • Application of resist mask • Isotropic poly-Si etching 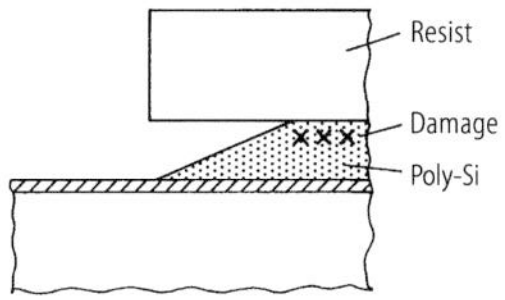	• Production of flow glass (BPSG) • Application of resist mask for contacts • Isotropic etching • Anisotropic etching 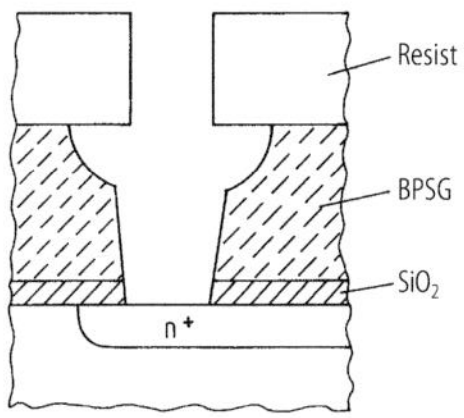	• Production of intermetallic dielectric (SiO$_2$) • Application of resist mask for vias • Anisotropic SiO$_2$, some resist removed in the process 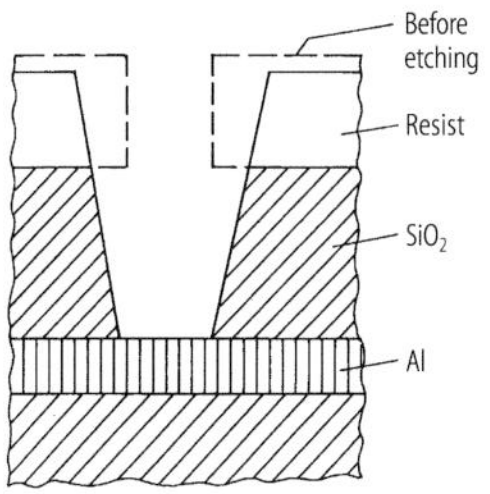	• Production of flow glass • Isotropic + anisotropic etching of contact holes • Slight reflow of BPSG 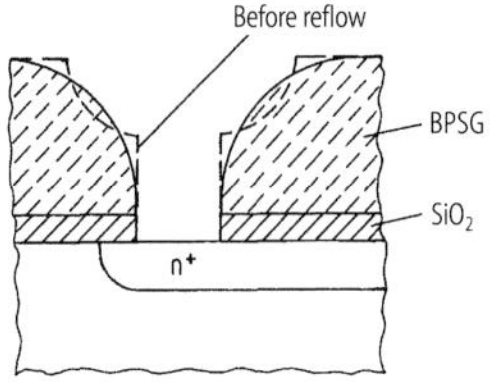
More details in Sect. 3.4.2	More details in Sect. 5.2.2	More details in Sects. 5.2.2 and 5.2.3	More details in Sect. 5.2.3	More details in Sect. 3.6.2

Fig. 8.5.2. Techniques employed for edge bevelling in the manufacture of integrated circuits. The diagram lists the key processing steps

8.5 Multilayer metallization   279

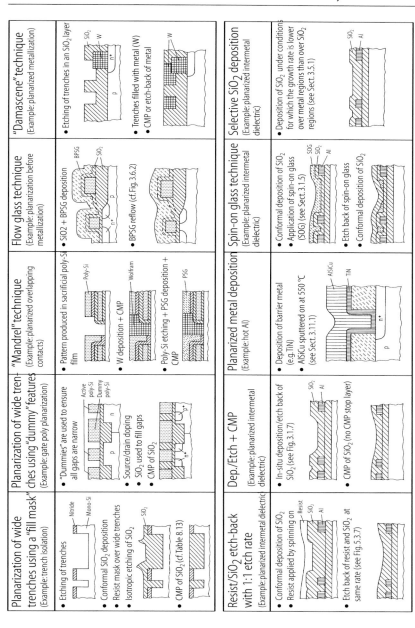

Fig. 8.5.3. Planarization techniques used in the manufacture of integrated circuits. The diagram lists the key processing steps used in planarization. CMP = chemical mechanical polishing

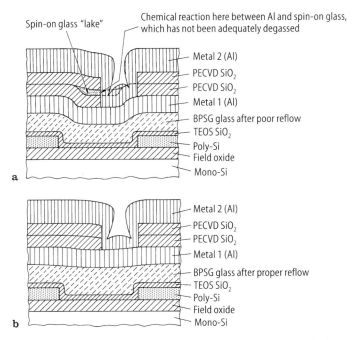

Fig. 8.5.4 a, b. Illustration of how a poisoned via can occur. The flow glass has only flowed slightly (**a**). A spin-on glass "lake" is then formed. The aluminium may then be corroded at the contact surface between the spin-on glass and the aluminium. If the flow-glass has flowed properly (**b**), there is no reason for poisoned vias to occur

If a final CMP step is added (see Sect. 5.1.2), then the planarization result is impressive (Fig. 8.5.3).

Another frequently used technique for planarization of the intermetal dielectric is the spin-on glass process (see Sect. 3.5.5), which is also outlined in Fig. 8.5.3. By spinning on the glass film and then performing partial etch-back, the spin-on glass only remains around steps and in narrow gaps, thereby filling precisely those recesses that are critical (see Fig. 3.5.5). A complication can arise when using the spin-on glass technique if there is inadequate planarization below metal 1, and if the spin-on glass has not been completely degassed. The spin-on glass may then attack the aluminium in the via (poisoned via, Fig. 8.5.4).

Another planarization technique illustrated in Fig. 8.5.3 is the etch-back of both resist and SiO_2 at the same etch rate (see Fig. 3.5.4). In contrast to the spin-on glass technique, the aim is not to leave any resist residues on the finished integrated circuit. The thickness of the SiO_2 film being planarized must therefore be greater than that of metal 1.

The "Mandrel" and "Damascene" processes occupy a special position amongst the planarization techniques shown in Fig. 8.5.3, since the route followed to produce a level surface is fundamentally different from that of the other techniques.

In the "Mandrel" technique [8.14], used to produce planarized overlapping contacts, a sacrificial poly-Si film is employed. This is used because anisotropic etching of poly-Si can be performed selectively with respect to nitride or oxide, and also because it acts as a polish stop in chemical mechanical polishing (CMP, cf. Sect. 5.1.2).

In the "Damascene" technique, used to produce planarized metallizations, the intermetal dielectric (SiO_2) is applied first. Then trenches are etched in the SiO_2 and filled with metal. This is done by initially depositing the metal (e.g. tungsten) over the whole surface and then removing it from the raised areas by CMP or an etch-back process[11]. Unfortunately, there is no practical technique available for achieving conformal deposition of aluminium (cf. Sect. 3.11.1). The Damascene process has therefore only succeeded as yet with tungsten, since conformal CVD deposition of tungsten is possible (see Sect. 3.10). A special application of the Damascene technique is in the filling of contact holes or vias. In this application, selective deposition of tungsten can also be used. Chemical mechanical polishing is then easier, since the tungsten only needs to be removed from above the contact holes and not over the whole surface.

8.5.2
Contacts in integrated circuits

A contact is the intentional connection by a conductor of two pieces of track located in different pattern planes. Contacts are classified as silicon/silicon, metal/silicon and metal/metal contacts depending on the material of the upper and lower conducting pattern planes respectively. Metal/metal contacts are specifically referred to as vias. The conductive silicon regions for connection can be diffusion regions, heavily doped n- and p-type polysilicon, or polycide regions.

Figure 8.5.5 summarizes schematically the various designs of contacts, shown in layout view of the integrated circuits. The contacts are designated as standard, overlapping and uncovered, according to how the position of the contact hole relates to the two tracks being connected.

In the standard contact, the contact hole lies with its whole base area on the lower track (nested contact), and is completely covered by the upper track (capped contact). The contact hole area and the electrical contact area are identical in the standard contact. In order to achieve the same conditions in all contact holes, both during the manufacture of the contact holes and during actual electrical operation later, one usually uses a single contact hole size in one integrated circuit, this being the size of the minimum contact hole. Larger contact areas can be produced by using several standard contact holes.

---

11 The trench is only completely filled if the thickness of the conformally deposited film is at least as great as half the trench width.

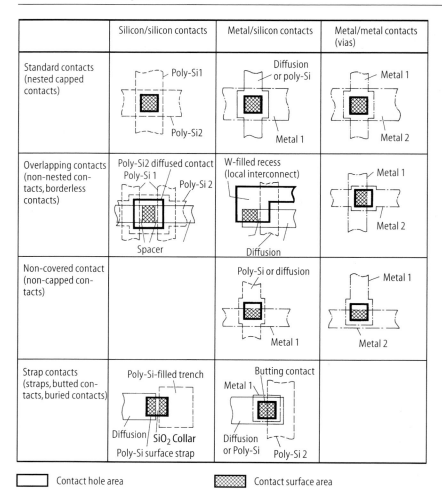

Figure 8.5.5. Various designs of contacts used in integrated circuits

In the overlapping contact [3.34] (non-nested contact or borderless contact) the contact hole area is larger than the electrical contact area. In this case one has to use a suitable processing sequence (e.g. spacer plus etch stop layer, see Fig. 3.5.2 b) to ensure that the isolation layer is not completely etched through in the overlap regions. Otherwise a contact would unintentionally be created between the overlap regions (short-circuit).

The main advantage of overlapping contacts is that almost no additional chip area has to be provided in the pattern planes below the contact, despite having a larger contact hole area. Where overlapping contacts are used, the features in these planes can be as densely packed as the available lithographic process and electrical operation allow. In the poly-Si2/diffusion contact and the local interconnect shown in Fig. 8.5.5, both the field oxide and the poly-Si are overlapped. The main application to date is the overlapping bit line contact in dynamic memory cells (see Fig. 8.4.2).

In a special version of the overlapping contact, the contact and the piece of track leading away from it can be produced as one. In the example in Fig. 8.5.5 (second row, second column), the etched contact hole is filled with tungsten (see Damascene technique in Fig. 8.5.3). If one extends the geometry of the contact hole beyond the actual contact area, then a conductive connection to a neighbouring diffusion region can be created for instance. This is called a local interconnect [8.16], and can cross over poly-Si and field oxide regions, but not over diffusion regions.

Introducing a local interconnect plane in a CMOS technology (see detailed processing sequence in Table 8.11) can save a large amount of chip space. For example, an SRAM cell (see Fig. 8.4.1) with local interconnects requires only about half the cell area of a standard cell under the same design constraints. The deposition, etch-back and polish-back of the tungsten that the process entails, does mean, however, that processing costs are considerably higher than in the standard overlapping contact with no tungsten inserted.

An uncovered contact (non-capped contact) arises when the contact hole area is not completely covered by the track over the contact hole. Under normal circumstances, an uncovered contact is not permitted, since there is a risk that when the metal is etched the etching will advance into the silicon in the exposed part of the contact hole. If, however, the contact hole is filled with tungsten, incomplete coverage is not a problem since the tungsten acts here as an etch stop during aluminium etching. Non-capped contact holes are useful because around the contact holes one can still use the minimum track grid spacing that is possible with the available lithographic technique (grid spacing = feature width + feature separation) (Fig. 8.5.6).

Strap contacts have assumed a special position amongst the various contact types. Whilst the contacts described so far provide a conductive link between two independent conducting planes, a strap contact connects a diffusion region with an adjacent poly-Si region[12]. Figure 8.5.5 (bottom of left column) shows a strap contact where a poly-Si surface strap forms the connection from selection transistor to storage capacitor in a buried-plate trench capacitor DRAM cell (cf. Fig. 8.4.3 b). In order to achieve an ohmic contact, all three areas involved must be heavily p- or n-doped. Instead of using a surface strap, the connection can also be produced by a contact below the mono-Si surface (buried strap). The processing sequence for a buried strap is described in detail in Table 8.14.

The strap contact shown in Fig. 8.5.5 (bottom of centre column) uses an aluminium surface strap to connect a diffusion region with an adjacent poly-Si re-

---

12 The CMOS process architecture does not allow for diffused regions to overlap poly-Si gate regions, since diffused regions occur wherever active regions are not covered by poly-Si (cf. Fig. 8.3.1)

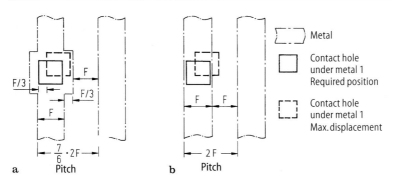

Fig. 8.5.6 a, b. Illustrating the minimum track grid spacing in a covered (**a**) and an uncovered (**b**) contact hole. F is the minimum feature dimension and F/3 is the maximum x- or y-misalignment of the metal 1 plane with respect to the contact hole plane. The minimum track grid pitch is 1/6 larger in covered contacts than in uncovered contacts

gion (butting contact). It is also possible to create the contact without making use of the metal 1 plane, although this does mean paying for an extra contact hole mask. Figure 3.8.6 shows the processing sequence for this type of buried contact.

8.5.3
Metallization in integrated circuits

Although diffused regions as well as tracks made of polysilicon or polycide are used intensively to connect components in integrated circuits (see Sect. 8.3.1), this section is specifically concerned with low resistance metal tracks.

Table 8.10 lists the main metals that are suitable for tracks, together with their properties. Whilst aluminium is the dominant track material, tungsten is indispensable where contact hole or via diameters are less than 0.5 µm, because of its planarizing effect (cf. Fig. 8.5.3). Copper has not yet been introduced industrially, but as miniaturization progresses there is an increasing need for a metal that has a lower resistance and a higher current-carrying capacity than aluminium.

Parasitic capacitances between the tracks present the main problem for multilayer metallizations in the sub-0.5 µm region. With vias having large aspect ratios (via depth: via diameter) it is possible to minimize the capacitance between tracks in different pattern planes by using thick intermetal dielectric layers. The lateral capacitances, however, inevitably increase with increasing miniaturization.

The only option for reducing lateral capacitances using today's proven processes and materials is to shift some of the tracks into an additional (e.g. into the third) metal plane, say. The search for a material with a lower dielectric constant than SiO_2 ($\varepsilon_r<4$, e.g. boron nitride) has not yet yielded any practical results. One possibility that has actually been considered is to completely remove the dielectric between the tracks, and thus produce a free-standing track system resting on support pillars.

Table 8.10. The most important metals used in integrated circuits

	Al (+0.5 % Cu)	Al (+0.5 % Cu +1 % Si)	W	Cu	TiN sputtered	TiN CVD
Resistivity ($\mu\Omega$ cm)	2.5	3	~ 10	1.5	~ 300	~ 10^4
Deposition process	sputtering	sputtering	CVD	elec. chem, sputtering, CVD	sputtering	CVD
Edge coverage	poor	poor	good	a)	poor	good
Barrier layer to Si required?	yes (e.g. TiN)	no	yes (e.g. TiN)	yes (e.g. TiN)	no	no
max. process temperature	450 °C	450 °C	600 °C	a)	600 °C	600 °C
max. current density at 150 °C (10^6 A cm^{-2})	0.5	0.5	3	>3	>3	>3

^{a} depends on deposition process

8.5.4
Passivation of integrated circuits

In order to protect integrated circuits from corrosion and mechanical damage, a passivation layer is applied after patterning the uppermost metal plane. Openings are made in this passivation layer only where the connecting wires (bonding wires) need to be attached (pads). The passivation usually consists of a double layer of plasma oxide (see Sect. 3.5.1) and plasma nitride (see Sect. 3.7.1), each 0.5-1.0 µm thick.

Cracks can arise in the uppermost metallization layer and in the passivation as a result of different mechanical stresses between the layers, inadequate layer bonding or stresses in the moulding material. The following measures can help to cure the problem.
- The passivation film should be as thick as possible (greater than the thickness of the uppermost metal layer)
- Slots need to be cut in wide metal tracks (bus lines)
- There should be no conductors near the outer chip corners
- An extra polyimide film[13] (see Sect. 3.12.2) can also prove effective. It acts as a stress relief buffer and ensures excellent adherence between the moulding material and the chip surface.

---

13 In the past, the polyimide film also served the function of trapping alpha particles from the package moulding material. The film needed to be 35m thick for this purpose. With today's low-alpha moulding materials this function is no longer required, and the thickness of the polyimide film can be reduced to 1 to 5 µm.

8.6
Detailed process sequence of selected technologies

Four selected technologies are described in detail in this section: a digital CMOS process, a analogue/digital BICMOS process, a microwave frequency bipolar process and a DRAM process.

8.6.1
Digital CMOS process

Table 8.11 contains the possible process steps for manufacturing CMOS digital circuits in [8.17]. The diagrams show cross-sections of parts of the Si wafer after the last individual process step that was described. The fine patterning and the use of local interconnects (see Sect. 8.5.2) enable integrated circuits to be produced with very high packing densities using the described process.

8.6.2
BICMOS process

As was explained in Sect. 8.2, BICMOS technology is a combination of bipolar and CMOS technology, enabling a single integrated circuit to benefit from the advantages of both CMOS and bipolar performance (see Sect. 8.1).

Table 8.12 shows a possible process sequence for manufacturing BICMOS circuits for analogue/digital applications [8.18]. The BICMOS circuits contain the following components:
- n-channel MOS transistor
- p-channel MOS transistor
- vertical npn bipolar transistor
- lateral pnp bipolar transistor
- large-value resistor (in poly-Si2)
- capacitor with voltage-independent capacitance (polySi1/poly-Si2).

8.6.3
Microwave bipolar process

As stated in Sect. 8.1, modern integrated bipolar circuits are characterized by a high switching speed/high transit frequencies, good drive performance, high transistor transconductance and good control voltage stability. It is these properties that have ensured the continued importance of bipolar technology in microelectronics alongside MOS technology.

Table 8.13 contains a possible process sequence for manufacturing integrated circuits designed to handle data rates of more than 10 Gbit/s. By using polysilicon emitter and base technology, bipolar transistors with transit frequencies of more than 30 GHz can be produced [8.19].

Table 8.11. Possible process sequence for the manufacture of digital CMOS circuits [8.17]

No.	• Description of each process step • Cross-section through the silicon wafer after the last process step described

1	• Oxidation of the silicon wafers (pad oxides)
	  – Starting material: lightly doped p-type silicon
	• Si_3N_4 deposition
	  – for local oxidation
	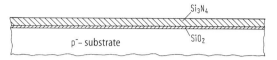

2	• Photolithography with mask 1
	  – to define the N-well
	  – process steps: apply photoresist
	          expose using mask 1
	          develop photoresist
	• Si_3N_4 etching
	  – using photoresist mask
	  – for local oxidation
	• Ion implantation of phosphorus
	  – to dope the N-well
	  – using photoresist mask
	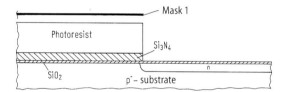

3	• Photoresist etching
	  – to remove the photoresist mask (resist stripping)
	• Local oxidation
	  – using Si_3N_4 mask to mask out N-well
	  – diffusion of phosphorus into the Si substrate during oxidation
	• Si_3N_4 etching
	  – to remove the Si_3N_4 mask
	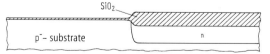
	• Ion implantation of boron
	  – to dope the P-well
	• Diffusion of dopant atoms into the Si substrate (drive in)
	  – to create P- and N-wells
	• SiO_2 etching
	  – to remove the SiO_2 film

Table 8.11. (continued)

No.	• Description of each process step • Cross-section through the silicon wafer after the last process step described

4	• Oxidation
	  – part of the LOCOS technique (Sect. 3.4.2)
	• Poly-Si deposition
	  – part of the LOCOS technique
	• Si_3N_4 deposition
	  – part of the LOCOS technique
	• Photolithography using mask 2
	  – to define the active regions and the field oxide regions
	• Si_3N_4/poly-Si etching
	  – part of the LOCOS technique
	• Photoresist etching
	  – to remove the photoresist mask
	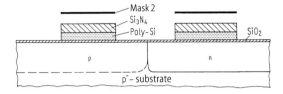

5	• Local oxidation
	  – in those regions not covered by Si_3N_4
	  – to create the field oxide
	• Si_3N_4/poly-Si etching
	  – to remove the Si_3N_4 and poly-Si film
	• Diffusion of the dopant atoms into the Si substrate (drive in)
	  – to set up the required depth for the P- and N-wells
	• SiO_2 etching
	  – to remove the thin oxide
	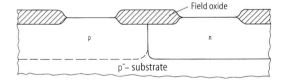

6	• Oxidation
	  – to create the screen oxide for subsequent ion implantation
	• Ion implantation of boron
	  – to set up the required channel doping of the N-channel MOS transistors
	• Photolithography using mask 3
	  – to set up the required channel doping of the P-channel MOS transistors
	• Ion implantation of arsenic and boron
	  – using photoresist mask
	  – to set up the channel doping of the P-channel transistors

8.6 Detailed process sequence of selected technologies

Table 8.11. (continued)

No.	• Description of each process step • Cross-section through the silicon wafer after the last process step described
6	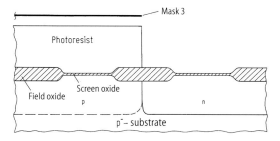

7	• Photoresist etching
	  – to remove the photoresist mask
	• SiO$_2$ etching
	  – to remove the screen oxide
	• Oxidation
	  – to produce the gate oxide
	• Poly-Si deposition and doping
	  – to create the poly-Si gate in the MOS transistors
	• SiO$_2$ deposition
	  – using the TEOS technique (Sect. 3.5)
	• Photolithography using mask 4
	  – to define the poly-Si patterns
	• SiO$_2$/poly-Si etching
	  – to create the poly-Si patterns
	• photoresist etching
	  – to remove the photoresist mask
	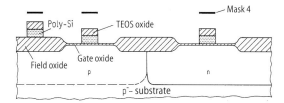

8	• SiO$_2$ deposition
	  – using the TEOS technique to produce spacers
	• SiO$_2$ etching
	  – to create the first SiO$_2$ spacers (Sect. 3.5.3)
	  – using the RIE technique (Sect. 5.2.3)
	• Oxidation
	  – to create the screen oxide for subsequent ion implantation
	• Photolithography using mask 5
	  – to define the NMOS transistors

290   8 Process integration

Table 8.11. (continued)

No.	• Description of each process step • Cross-section through the silicon wafer after the last process step described

8	• Ion implantation of phosphorus
	  – to create the n-doped LDD regions of the NMOS transistors (Sect. 8.3.1)
	  – using photoresist mask
	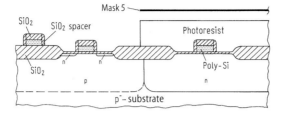

9	• Photoresist etching
	  – to remove the photoresist mask
	• SiO$_2$ deposition
	  – using the TEOS technique (Sect. 3.5)
	• SiO$_2$ etching
	  – to create the second SiO$_2$ spacer (Sect. 3.5.3)
	  – using the RIE technique (Sect. 5.2.3)
	• SiO$_2$ deposition
	  – using the TEOS technique (Sect. 3.5)
	  – to create the screen oxide for subsequent ion implantation
	• Photolithography using mask 5
	  – to define the NMOS transistors
	• Ion implantation of arsenic
	  – to create the heavily n-doped source and drain regions of the NMOS LDD transistors (Sect. 8.3.1)
	  – using photoresist mask
	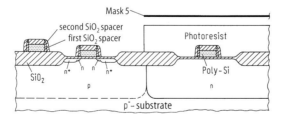

10	• Photoresist etching
	  – to remove the photoresist mask
	• Annealing
	  – to activate the implanted dopant atoms
	• Photolithography using mask 6
	  – to define the PMOS transistors
	• Ion implantation of boron
	  – to dope the source and drain regions of the PMOS transistors
	  – using photoresist mask

8.6 Detailed process sequence of selected technologies   291

Table 8.11. (continued)

No.	• Description of each process step • Cross-section through the silicon wafer after the last process step described

10	------------------------------ [Fig. 10 relates to this last step]
	• Photoresist etching
	  – to remove the photoresist mask
	• Annealing
	  – to activate the implanted boron atoms

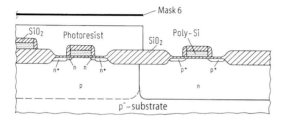

11	• SiO$_2$ etching
	  – wet chemical (Sect. 5.1.1)
	  – to remove the SiO$_2$ film above the source/drain regions of the MOS transistors
	• Ti deposition
	  – using the sputtering technique (Sect. 3.1.4)
	  – to create the TiSi$_2$
	• 1st annealing
	  – to create the "salicide" films (self-aligned silicide, Sect. 3.9.1) over the source/drain regions
	• Ti etching
	  – to remove the Ti film on the SiO$_2$ areas
	• 2nd annealing
	  – to create an optimum barrier layer above the source/drain regions of the MOS transistors

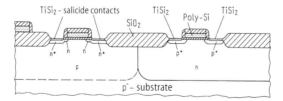

12	• SiO$_2$ deposition
	  – using the TEOS technique (Sect. 3.5.1)
	• Si$_3$N$_4$ deposition
	  – using the LPCVD technique (Sect. 3.7.1)
	• BPSG deposition
	  – using the LPCVD technique (Sect. 3.6.2)
	  – to planarize the wafer surface
	• Reflow of BPSG layer
	  – to planarize the wafer surface

Table 8.11. (continued)

No.	• Description of each process step • Cross-section through the silicon wafer after the last process step described

12	• Photolithography using mask 7 (metal 0)
	  – to define the local interconnects (Sect. 8.5.2)
	• BPSG etching
	  – to create the trenches for the local interconnects
	  – with etch stop at Si_3N_4 layer
	• Si_3N_4/SiO_2 etching
	  – to expose the Si contacts
	• Photoresist etching
	  – to remove the photoresist mask
	• Ti/TiN deposition
	  – using the sputter technique (Sect. 3.1.4)
	  – to create diffusion barriers (Sect. 8.5.2)
	• W deposition
	  – using the CVD technique (Sect. 3.10)
	  – to create local interconnects
	• W/TiN etch-back or CMP (Sect. 5.1.2)
	  – to remove the W layer outside the trenches
	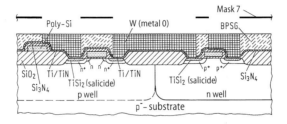

13	• SiO_2 deposition
	  – using the TEOS technique (Sect. 3.5)
	• Photolithography using mask 8 (contacts)
	  – to define the contact holes
	• SiO_2 etching
	  – to produce the contact holes
	• Photoresist etching
	  – to remove the photoresist mask
	• Ti/TiN deposition
	  – using the sputtering technique (Sect. 3.1.4)
	  – to produce the diffusion barrier (Sect. 3.10)
	• W-deposition
	  – using the CVD technique (Sect. 3.10)
	  – to fill the contact holes (W plugs)
	• W etch-back or CMP
	  – to remove the W layer outside the contact holes
	• AlSiCu deposition
	  – to produce the interconnects in the 1st metallization level

8.6 Detailed process sequence of selected technologies   293

Table 8.11. (continued)

No.	• Description of each process step • Cross-section through the silicon wafer after the last process step described

13	• TiN deposition
	  – to produce the diffusion barrier on metal 1
	• Photolithography using mask 9 (metal 1)
	  – to define the 1st metallization level
	• metal 1 etching
	  – to produce the AlSiCu interconnects in the 1st metallization level
	• Photoresist etching
	  – to remove the photoresist mask
	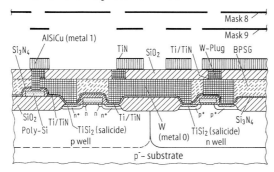

14	• SiO_2 deposition
	  – using the PECVD technique (Sect. 3.1.1)
	• SOG (spin-on glass) deposition
	  – using the spin-on process (Sect. 3.1.5)
	  – to planarize the wafer surface
	• SOG etch-back
	  – so that SOG is only left at steps (Sect. 3.5.5)
	• SiO_2 deposition
	  – using the PECVD technique (Sect. 3.1.1)
	• Annealing
	  – to improve the electrical properties of the transistors
	• Photolithography using mask 10 (via 1)
	  – to define the contact holes between metal 1 and metal 2 level
	• SiO_2 etching
	  – to produce the contact holes (vias)
	• Photoresist etching
	  – to remove the photoresist mask
	• Ti/TiN deposition
	  – using the sputtering technique (Sect. 3.1.4)
	  – to produce the diffusion barrier (Sect. 8.5.2)
	• W deposition
	  – using the CVD technique (Sect. 3.10)
	  – to fill the vias (W plugs)
	• W etch-back or CMP
	  – to remove the W film outside the vias
	• Metal 2 deposition
	  – to produce the AlSiCu interconnects in the 2nd metallization level

Table 8.11. (continued)

No.	• Description of each process step • Cross-section through the silicon wafer after the last process step described

14	• Photolithography using mask 11 (metal 2)
	  – to define the 2nd metallization level
	• Metal 2 etching
	  – to produce the AlSiCu interconnects in the 2nd metallization level
	• Photoresist etching
	  – to remove the photoresist mask
	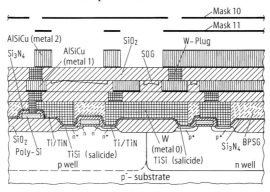

15	• SiO_2 deposition
	  – using the PECVD technique (Sect. 3.1.1)
	  – to produce the passivation layer
	• Si_3N_4 deposition
	  – using the PECVD technique (Sect. 3.1.1)
	  – to produce the passivation layer
	• Photolithography using mask 12 (pad)
	  – to define the pads (connections to package)
	• SiO_2/Si_3N_4 etching
	  – to expose the metal pads
	• Photoresist etching
	  – to remove the photoresist mask
	• Final annealing
	  – to improve the electrical properties of the transistors

Table 8.11. (continued)

No.	• Description of each process step • Cross-section through the silicon wafer after the last process step described
15	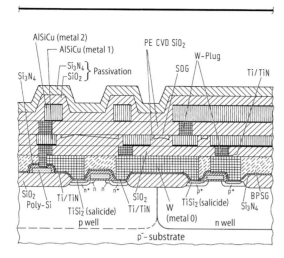

8.6.4
DRAM process

Dynamic semiconductor memory (DRAM – Dynamic Random Access Memory) is still the technological forerunner in microelectronics, as it involves the smallest feature sizes. The highest demands are therefore placed on each of the processing steps used to manufacture these components (see Sect. 8.4.2).

Table 8.14 lists a possible process sequence for manufacturing DRAM Apart from the extremely fine patterns, the outstanding features of this technology include three-dimensional integration of memory cells (trench capacitors), consistent use of planarization and the inclusion of refractory metals [8.20].

Table 8.12. Possible process sequence for the manufacture of analogue/digital BICMOS circuits [8.18]

No.	Process step Cross-section through the silicon wafer after the last process step described	Description of each process step

1	[Cross-section: SiO$_2$ layer on p$^-$ Si]	• Oxidation of the silicon wafers
		  – starting material: p-type silicon
		  – SiO$_2$ film thickness ≈ 500 nm

2	[Cross-section: Arsenic or Antimony implantation through Mask 1 into n$^+$ regions, SiO$_2$ on p$^-$ Si]	• Photolithography using mask 1
		  – to define the buried layer
		  – processing steps:
		    apply photoresist
		    expose using mask 1
		    develop photoresist
		• SiO$_2$ etching
		  – using photoresist mask
		• Photoresist etching
		  – to remove the photoresist mask (photoresist stripping)
		• Oxidation
		  – to produce the screen oxide for subsequent ion implantation
		• Ion implantation
		  – to dope the buried layer
		  – dopant: arsenic or antimony
		  – using SiO$_2$ mask

3	[Cross-section: p Epi-Si layer over n$^+$ buried layers in p$^-$ Si]	• Screen oxide etching
		  – to remove the contaminants in the screen oxide produced by ion implantation
		• Annealing
		  – to drive in the implanted dopant atoms into the silicon crystal
		  – it also repairs the damage caused to the silicon crystal structure by ion implantation
		• SiO$_2$ etching
		  – to remove the oxide film from the whole surface
		• Epitaxy
		  – epitaxial deposition of a monocrystalline p-type silicon layer

Table 8.12. (continued)

No.	• Process step • Cross-section through the silicon wafer after the last process step described	Description of each process step

4	Phosphorus ↓↓↓↓↓ Phosphorus ↓↓↓ — Mask 2
	n p n — Epi-Si
	n⁺ p⁻ n⁺ — Si
		• Oxidation
		  – to produce an SiO_2 masking layer for subsequent ion implantation
		• Photolithography using mask 2
		  – to define the n-wells
		• SiO_2 etching
		  – using photoresist mask
		  – to mask the subsequent ion implantation
		• Photoresist etching
		  – to remove the photoresist mask
		• Oxidation
		  – to produce the screen oxide for subsequent ion implantation
		• Ion implantation of phosphorus
		  – for the n-wells
		  – using oxide mask
		• Screen oxide etching
		  – to remove the screen oxide
		• Annealing
		  – to drive in the phosphorus atoms up to the n^+ buried layer
		• SiO_2 etching
		  – to remove the SiO_2 layer from the whole surface

5	— Mask 3
	SiO_2
	n p n — Epi-Si
	n⁺ p⁻ n⁺ — Si
		• Oxidation
		  – to produce a thin SiO_2 film below the Si_3N_4 layer for local oxidation
		• Si_3N_4 deposition
		  – CVD deposition of an Si_3N_4 layer for local oxidation
		• Photolithography using mask 3
		  – to define the active regions and the thick oxide regions
		• Si_3N_4 etching
		  – using photoresist mask
		  – for local oxidation
		• Photoresist etching
		  – to remove the photoresist mask
		• Local oxidation
		  – Oxidation only in the Si_3N_4 windows, since the Si_3N_4 layer acts as a diffusion barrier to oxygen

Table 8.12. (continued)

No.	• Process step • Cross-section through the silicon wafer after the last process step described	Description of each process step

5		• Si_3N_4 etching
		  – removal of Si_3N_4 layer from the whole surface
		  – The SiO_2 film below the Si_3N_4 layer remains (screen oxide for subsequent ion implantations)

6	Phosphorus, Mask 4, SiO_2, Si	• Photolithography using mask 4
		  – to define the collector connection
		• Ion implantation of phosphorus
		  – to dope the collector connection
		  – using photoresist mask
		• Photoresist etching
		  – to remove the photoresist mask
		• Annealing
		  – to drive in the phosphorus atoms as far as the buried layer
		  – this produces a low resistance collector connection

7	Boron, Mask 5, SiO_2, Si	• Photolithography using mask 5
		  – to define the base region of the bipolar transistors
		• Ion implantation of boron
		  – using photoresist mask
		  – to dope the base region of the bipolar transistors
		• Photoresist etching
		  – to remove the photoresist mask
		• Annealing
		  – to activate the dopant atoms

8	Mask 6, Poly–Si, SiO_2, Si	• Screen oxide etching
		  – to remove the contaminants introduced during ion implantation
		• Oxidation
		  – to produce a high-quality gate oxide for the MOS transistors
		• Poly-Si deposition
		  – for the poly-Si gate of the MOS transistors
		  – The n^+ doping of the poly-Si film is either performed by ion implantation or diffusion (see Fig. 3.6.1, right-hand column)
		• Photolithography using mask 6
		  – to define the poly-Si1 regions

Table 8.12. (continued)

No.	• Process step • Cross-section through the silicon wafer after the last process step described	Description of each process step

8		• Poly-Si etching
		  – to produce the poly-Si1 regions
		• Photoresist etching
		  – to remove the photoresist mask

9	Arsenic, Arsenic, Mask 7, Poly-Si, SiO$_2$, Si	• Photolithography using mask 7
		  – to define the source and drain of the N-channel MOS transistors, and the emitter and collector connection of the bipolar transistors
		• Ion implantation of arsenic
		  – using photoresist mask
		  – to dope the source and drain of the N-channel MOS transistors, and the emitter and collector connection of the bipolar transistors
		• Photoresist etching
		  – to remove the photoresist mask

10	Boron, Poly-Si, Boron, Mask 8, SiO$_2$, Si	• Photolithography using mask 8
		  – to define the source and drain of the P-channel MOS transistors and the base connection of the bipolar transistors
		• Ion implantation of boron or BF$^+_2$
		  – using photoresist mask
		  – to dope the source and drain of the P-channel MOS transistors and the base connection of the bipolar transistors
		• Photoresist etching
		  – to remove the photoresist mask

11		• Screen oxide etching
		• Oxidation
		  – to isolate the poly-Si1 level
		  – to produce the capacitor dielectric
		• Poly-Si2 deposition
		  – using CVD technique
		  – to produce the capacitors and the large-valued resistors
		• Ion implantation of phosphorus
		  – to produce the large-valued resistors (n$^-$ poly-Si)

300   8 Process integration

Table 8.12. (continued)

No.	• Process step • Cross-section through the silicon wafer after the last process step described	Description of each process step

11		• Photolithography using mask 9
		  – to define the low resistance poly-Si2 regions (n⁺ poly-Si2)
		  – to mask the high resistance regions
		• Ion implantation of phosphorus
		  – to produce the low resistance poly-Si2 regions (n⁺ poly-Si2)
		  – using photoresist mask
		• Photoresist etching
		  – to remove the photoresist mask
		• Silicon crystal annealing
		  – annealing of Si wafers to activate the dopant atoms after ion implantation

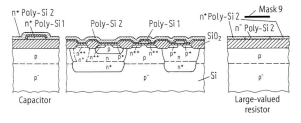

Capacitor   Large-valued resistor

| 12 | 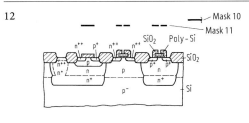 | • Photolithography using mask 10
  – to define the poly-Si2 patterns (large-valued resistors, capacitors and leads)
• Poly-Si2 etching
  – to produce the poly-Si2 patterns
• Photoresist etching
  – to remove the photoresist mask
• SiO₂ deposition
  – CVD deposition of an insulating oxide
• Photolithography using mask 11
  – to define the contact holes
  – SiO₂ etching
  – to produce the contact holes
• Photoresist etching
  – to remove the photoresist mask |

Table 8.12. (continued)

No.	• Process step • Cross-section through the silicon wafer after the last process step described	Description of each process step

13	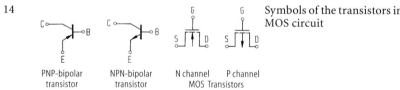	• Metal deposition
		  – sputtering of Ti/TiN/AlSiCu
		• Photolithography using mask 12
		  – to define the interconnects
		• Metal etching
		  – using photoresist mask
		  – to produce the interconnects
		• Photoresist etching
		  – to remove the photoresist mask
		• Deposition of passivation layer
		  – deposition of an SiO_2 layer, Si_3N_4 layer, and possibly a polyimide film (see Sect. 8.5.4)
		  – The passivation layer protects the integrated circuit
		• Photolithography using mask 13
		  – to define the bonding pads
		• Etching of bonding pads
		  – metal areas (pads) are exposed to enable bonding of connection wires to the integrated circuit
		• Photoresist etching
		  – to remove the photoresist mask
		• Final annealing

14	PNP-bipolar transistor NPN-bipolar transistor N channel P channel MOS Transistors	Symbols of the transistors in the BICMOS circuit

Table 8.13. Possible process sequence for the manufacture of bipolar circuits [8.19]

No.	• Description of each process step • Cross-section through the silicon wafer after the last process step described
1	• Oxidation of the silicon wafers – starting material: p-type silicon with $\rho \approx 10\ \Omega$ cm 

Table 8.13. (continued)

No.	• Description of each process step • Cross-section through the silicon wafer after the last process step described

2	• Photolithography using mask 1
	  – to define the buried layer
	  – Processing steps: apply the photoresist
	                              expose using mask 1
	                              develop the photoresist
	• SiO_2 etching
	  – using photoresist mask
	• Photoresist etching
	  – to remove the photoresist mask (resist stripping)
	• Oxidation
	  – to produce the screen oxide for subsequent ion implantation
	• Ion implantation of arsenic
	  – to dope the buried layer
	  – using SiO_2 mask
	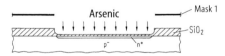

3	• Annealing/oxidation
	  – to produce the buried layer
	  – to produce steps in the Si for alignment purposes
	• SiO_2 etching
	  – to completely remove the oxide
	• Epitaxy
	  – to produce an n-type Si layer approximately 1 µm thick
	• Oxidation
	  – to cover the Si surface with an SiO_2 layer
	• Photolithography using mask 2
	  – to define the isolation regions (channel stoppers) between the transistors
	• Ion implantation of boron
	  – to produce the channel stoppers
	  – using photoresist mask
	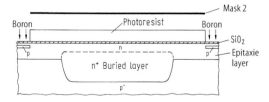

4	• Photoresist etching
	  – to remove the photoresist mask
	• Annealing/oxidation
	  – to produce the p-type insulating surround for lateral isolation of the bipolar transistors

Table 8.13. (continued)

No.	• Description of each process step • Cross-section through the silicon wafer after the last process step described
4	• SiO$_2$ etching – to remove the oxide contaminated by ion implantation • Oxidation – to produce a thin SiO$_2$ film for local oxidation • Polysilicon deposition – to produce a polysilicon film for local oxidation • Si$_3$N$_4$ deposition – to produce an Si$_3$N$_4$ film for local oxidation • Photolithography using mask 3 – to define the thick and thin oxide regions • Si$_3$N$_4$ etching – using photoresist mask for local oxidation 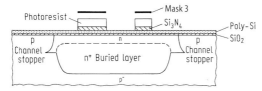
5	• Photoresist etching – to remove the photoresist mask • Local oxidation (field oxide) – to produce thin and thick oxide regions – oxidation only occurs in those areas not covered by Si$_3$N$_4$. Si$_3$N$_4$ acts as a diffusion barrier to oxygen atoms 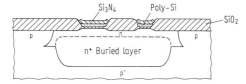
6	• SiO$_2$ etching – to remove the thin oxynitride on the Si$_3$N$_4$ • Si$_3$N$_4$ and polysilicon etching – to remove the LOCOS mask (Local Oxidation of Silicon) • Oxidation – to oxidize off the silicon containing nitrogen in the bird's beak (transition from thin to thick oxide) in order to avoid the "white ribbon" effect (Sect. 3.4.2) • SiO$_2$ etching – to remove the thin oxide • Oxidation – to produce the screen oxide for subsequent ion implantation • Photolithography using mask 4 – to define the collector connection and the back plates of the capacitors (see process step 7)

Table 8.13. (continued)

No.	• Description of each processing step • Cross-section through the silicon wafer after the last processing step described

6	• Ion implantation of phosphorus
	  – to dope the collector connection and the back plates of the capacitors (see process step 7)
	  – using photoresist and thick oxide mask
	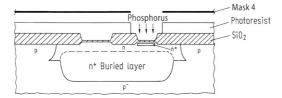

7	• Photoresist etching
	  – to remove the photoresist mask
	• Annealing
	  – to drive in the dopant atoms into the n^+ collector connection and the n^+ back plates of the capacitors (see following steps)
	• SiO$_2$ etching
	  – to remove the contaminated thin oxide
	• Formation of oxide-nitride-oxide (ONO) film (see Sect. 3.7.2)
	  – for capacitors
	• Photolithography using mask 5
	  – to define the capacitors
	• Oxide-nitride-oxide (ONO) etching
	  – to produce the capacitors
	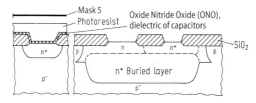

8	• Photoresist etching
	  – to remove the photoresist mask
	• Poly-Si deposition (amorphous, see Sect. 3.8.2)
	  – for large-valued resistors
	• Ion implantation of boron
	  – to set the sheet resistance of the large-valued resistors ($R_S \approx 1$ k$\Omega/\square$)
	• Photolithography using mask 6
	  – to define the large-valued resistors
	• Ion implantation of boron (p^+ polysilicon)
	  – to set the sheet resistance of the small-valued resistors ($R_S \approx 150$ $\Omega/\square$), capacitor electrodes and connecting lines
	  – using photoresist mask

8.6 Detailed process sequence of selected technologies

Table 8.13. (continued)

No.	• Description of each process step • Cross-section through the silicon wafer after the last process step described
8	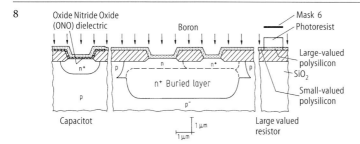

9	• Photoresist etching
	  – to remove the photoresist mask
	• Annealing
	  – to set the required sheet resistance of the poly-Si
	• SiO$_2$ deposition (using TEOS process, Sect. 3.5)
	  – to isolate the base from the emitter, thickness ~ 300 nm
	• Photolithography using mask 7
	  – to define the emitter regions and the poly-Si resistors
	• TEOS-SiO$_2$ etching
	  – to expose the emitter regions and collector connections, and to pattern the poly-Si
	• Polysilicon etching
	  – to expose the emitter regions and the collector connections
	  – to pattern the poly-Si
	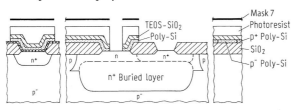

10	• Photoresist etching
	  – to remove the photoresist mask
	• Oxidation
	  – to produce a thin screen oxide for subsequent ion implantation
	• Photolithography using mask 8
	  – to dope the areas under the emitter window
	• Ion implantation of BF$^+_2$
	  – for p-type doping of the base region of the transistor
	• Ion implantation with single- and double-charged phosphorus (P$^+$ and P^{++})
	  – to produce the buried area beneath the platform collector (see Fig. 8.3.8)

Table 8.13. (continued)

No.	• Description of each process step • Cross-section through the silicon wafer after the last process step described
10	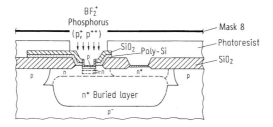

11	• Photoresist etching
	  – to remove the photoresist mask
	• Annealing
	  – to repair the damage caused to the silicon crystal lattice by ion implantation
	• SiO_2 etching
	  – to remove the screen oxide contaminated during ion implantation
	• Thermal oxidation
	  – to produce a thin SiO_2 film than can be etched selectively to silicon
	• Si_3N_4 deposition
	  – to produce a thin stop-layer when etching the overlying poly-Si film
	• Polysilicon deposition
	  – to produce the spacers for emitter doping
	• Polysilicon etching
	  – to produce the spacers
	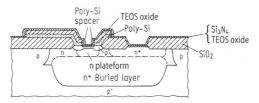

12	• Si_3N_4 etching
	  – to remove the nitride film except in the spacer areas
	• Polysilicon etching
	  – to remove the poly-Si spacers
	• SiO_2 etching
	  – to remove the oxide beneath the Si nitride
	• Polysilicon deposition
	  – for emitter and collector connections
	• Ion implantation of arsenic
	  – for n^+ doping of the poly-Si film

Table 8.13. (continued)

No.	• Description of each process step • Cross-section through the silicon wafer after the last process step described
12	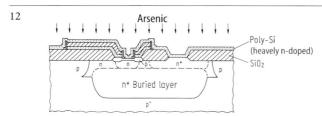

13	• Photolithography using mask 9
	  – to define the emitter and collector connection
	• Polysilicon etching
	  – to produce the emitter and collector connection
	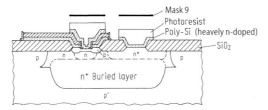

14	• Photoresist etching
	  – to remove the photoresist mask
	• SiO$_2$ deposition (Sect. 3.5)
	  – to produce a diffusion barrier between BPSG and silicon
	• BPSG deposition (Sect. 3.6)
	  – for vertical isolation and smoothing of edges
	• BPSG reflow and emitter drive-in
	  – to round off feature edges
	  – arsenic diffuses out of the poly-Si into the mono-Si to produce the emitter region
	• Photolithography using mask 10
	  – to define the contact holes
	• SiO$_2$/BPSG etching
	  – to produce the contact holes
	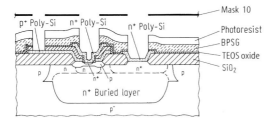

Table 8.13. (continued)

No.	• Description of each process step • Cross-section through the silicon wafer after the last process step described

15	• Photoresist etching
	  – to remove the photoresist mask
	• Ti/TiN deposition
	  – using sputtering technique
	  – to produce a diffusion barrier (Sect. 3.10)
	• W deposition
	  – using the CVD technique (Sect. 3.10)
	  – to fill the contact holes (W plugs)
	• W etch-back or CMP (Sect. 5.1.2)
	  – to remove the W layer outside the contact holes
	• AlSiCu deposition
	  – using sputter technique
	  – to produce the metal 1 interconnects
	• TiN deposition
	  – to produce a barrier layer for the metal 2 level
	• Photolithography using mask 11
	  – to define the interconnects in the metal 1 level
	• Etching of metal 1 interconnects
	  – to produce the interconnects in the metal 1 level
	  – TiN-AlSiCu-TiN-Ti etching
	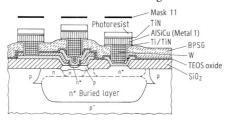

16	• Photoresist etching
	  – to remove the photoresist mask
	• SiO_2 deposition
	  – for isolation between metallization level 1 and 2
	  – process sequence: 1. PECVD deposition of SiO_2
	    2. application of spin-on glass and planarization
	    3. PECVD deposition of SiO_2
	    4. annealing of isolation layer
	• Photolithography using mask 12
	  – to define the via holes (contact holes for connecting metallization 1 and metallization 2 level)
	• Via etching
	  – for connection of the metal 1 and metal 2 level
	• Photoresist etching
	  – to remove the photoresist mask
	• Deposition of the 2nd metallization level
	  – to produce the interconnects in the metal 2 plane
	  – Ti-TiN-AlSiCu-TiN film sequence

8.6 Detailed process sequence of selected technologies 309

Table 8.13. (continued)

No.	• Description of each process step • Cross-section through the silicon wafer after the last process step described

16	• Photolithography using mask 13
	  – to define the interconnects in the metal 2 level
	• Etching of metallization 2 level
	  – to remove the photoresist mask
	• Photoresist etching
	  – to produce interconnects
	• Annealing
	  – to improve the properties of the transistors, interconnects and ohmic contacts
	• Deposition of the passivation layer (Sect. 8.5.4)
	  – to protect the circuit
	  – process sequence: 1. plasma CVD deposition of SiO_2
	    2. plasma CVD deposition of Si_3N_4
	• Photolithography using mask 14
	  – to define the pads (metal areas for connecting to the package pins)
	• Etching of Si_3N_4/SiO_2 passivation layer
	  – to expose the pads
	• Photoresist etching
	  – to remove the photoresist mask
	• Annealing
	  – to improve the bonding capability of the pads
	  – to repair any radiation damage caused during plasma processes

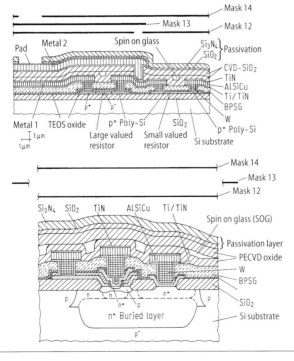

310 8 Process integration

Table 8.14. Possible process sequence in the manufacture of DRAM [8.20]

No.	• Description of each process step • Cross-section through the silicon wafer after the last process step described
1	• Starting material: p-type silicon • Oxidation of the silicon wafers – to produce a screen oxide for subsequent ion implantation

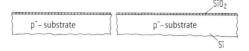

2	• Photolithography using mask 1
	  – to define the alignment marks
	  – processing steps: Apply photoresist
	               Expose using mask 1
	               develop photoresist
	• SiO$_2$/silicon etching
	  – using photoresist mask
	  – to define the alignment marks (not shown)
	• Photoresist etching
	  – to remove the photoresist mask (resist stripping)

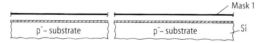

3	• Photolithography using mask 2
	  – to define the n-type memory cell region
	• Ion implantation of phosphorus
	  – to provide all-round isolation of the p-well in the memory cell region
	  – using photoresist mask
	• Photoresist etching
	  – to remove the photoresist
	• Annealing
	  – to drive in the phosphorus

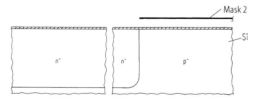

4	• SiO$_2$ etching
	  – to remove the contaminated oxide
	• SiO$_2$/Si$_3$N$_4$/SiO$_2$ deposition
	  – to produce the masking layer for trench etching
	• Photolithography using mask 3
	  – to define the storage capacitors
	• Anisotropic etching of the trench etch masking layer SiO$_2$/Si$_3$N$_4$/SiO$_2$
	  – to produce the etch mask for the trenches

8.6 Detailed process sequence of selected technologies

Table 8.14. (continued)

No.	• Description of each process step • Cross-section through the silicon wafer after the last process step described

4	• Photoresist etching
	  – to remove the photoresist mask
	• Si etching (anisotropic)
	  – using $SiO_2/Si_3N_4/SiO_2$ mask
	  – to produce the storage capacitors

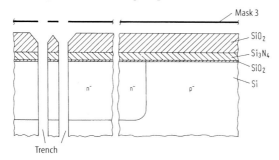

5	• Arsenic glass deposition
	  – to produce the n^+ buried plate
	  – using the arsenic TEOS process (Sect. 3.6.1)
	• SiO_2 deposition
	  – to prevent the As diffusing out of the arsenic glass< – using the TEOS process (Sect. 3.5.1)
	• Photoresist deposition, exposure and development
	  – for partial filling of trenches

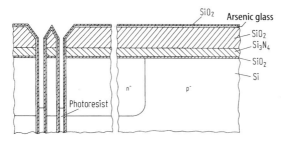

6	• Etching of SiO_2 and arsenic glass
	  – to remove the arsenic glass above the photoresist
	• Photoresist etching
	  – to remove the photoresist in the trenches
	• Oxidation
	  – to produce a thin thermal SiO_2 film on Si
	• SiO_2 deposition
	  – to prevent As diffusing out of the arsenic glass
	  – using the TEOS process (Sect. 3.5)

312   8 Process integration

Table 8.14. (continued)

No.	• Description of each process step • Cross-section through the silicon wafer after the last process step described

6	• Diffusion of arsenic out of the arsenic glass into the silicon
	  – to produce an interconnected n^+ doped buried plate
	• Arsenic glass etching
	  – to remove the arsenic glass layer
	• ONO deposition (Sect. 3.7.3)
	  – to produce the memory capacitor dielectric in the trenches

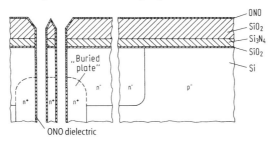

7	• Deposition of n^+ doped poly-Si
	  – to fill the trenches
	• Chemical-mechanical polishing (CMP, Sect. 5.1.2) of poly-Si
	  – so that the poly-Si is only left in the trenches

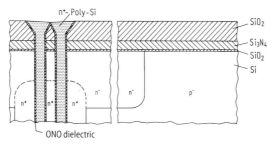

8	• SiO_2 etching
	  – to remove the SiO_2 masking layer
	• Polysilicon etching
	  – up to some 1 µm below the silicon surface
	• Etching of the ONO dielectric
	  – at the exposed side walls of the trenches
	  – to define the collar depths
	• Conformal deposition of an SiO_2 film (see Sect. 3.5.1)
	  – to define the collar thickness
	• Etching of the SiO_2 film (anisotropic)
	  – to produce an SiO_2 spacer (collar) in the upper section of the trenches (etching of SiO_2 lying on poly-Si)

8.6 Detailed process sequence of selected technologies 313

Table 8.14. (continued)

No.	• Description of each process step • Cross-section through the silicon wafer after the last process step described
8	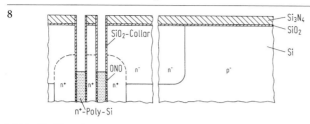

9	• n^+ poly-Si deposition
	  – to fill the storage capacitor trenches in the collar region
	• Poly-Si etching
	  – to prepare for the buried contacts
	• Etching of the SiO_2 collar
	  – to expose the buried contact surfaces
	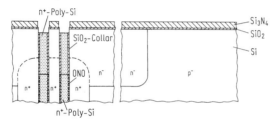

10	• n^+ poly-Si deposition
	  – to complete the buried contacts
	• Chemical-mechanical polishing of poly-Si
	  – with Si_3N_4 film as polish stop
	  – to remove the poly-Si outside the trenches
	• Poly-Si etching
	  – as far as the silicon surface level
	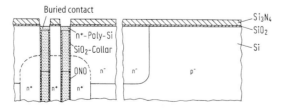

11	• Photolithography using mask 4
	  – to define the isolation regions (trench isolation, Sect. 3.5.4)
	• Si_3N_4 etching
	  – using photoresist mask
	• SiO_2 etching
	  – using photoresist mask

Table 8.14. (continued)

No.	• Description of each process step • Cross-section through the silicon wafer after the last process step described

11	• Silicon etching
	  – using photoresist mask
	  – etch depth equals the depth of the trench isolation
	• Photoresist etching
	  – to remove the photoresist mask
	• Oxidation
	  – to produce a thin thermal SiO_2 film over the silicon
	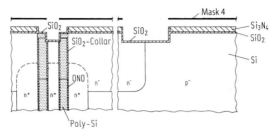

12	• SiO_2 deposition
	  – to fill the trench isolation regions
	  – using the TEOS process (Sect. 3.5)
	• Planarization of the wafer surface
	  – photoresist deposition
	  – etching of photoresist and SiO_2 (selectivity 1:1) to planarize the SiO_2 surface (Sect. 5.3.7)
	  – chemical-mechanical polishing of the wafer surface up to the Si_3N_4 film
	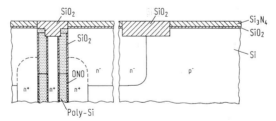

13	• Si_3N_4 etching
	  – to remove the Si_3N_4 film
	• SiO_2 etching
	  – to remove the SiO_2 layer lying below the nitride film
	• Oxidation
	  – to produce a thin screen oxide
	• Photolithography using mask 5
	  – to define the P-channel transistor regions
	• Ion implantation of phosphorus
	  – to produce the n-well
	  – using photoresist mask

8.6 Detailed process sequence of selected technologies

Table 8.14. (continued)

No.	• Description of each process step • Cross-section through the silicon wafer after the last process step described

13	• Ion implantation of arsenic
	  – to set the threshold voltage of the P-channel MOS transistors
	• Photoresist etching
	  – to remove the photoresist mask

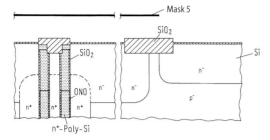

14	• Photolithography using mask 6
	  – to define the N-channel transistor regions
	• Ion implantation of boron
	  – to produce the p-well
	• Ion implantation of boron or BF^+_2
	  – to set the threshold voltage of the N-channel MOS transistors
	• Photoresist etching
	  – to remove the photoresist mask

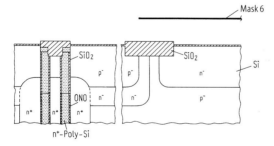

15	• SiO_2 etching
	  – to remove the thin SiO_2 film
	• Oxidation
	  – to produce the gate oxide
	• Polysilicon deposition
	  – to produce the gate electrodes
	• Ion implantation of phosphorus
	  – to dope the polysilicon gate electrodes
	• Metal silicide deposition (e.g. WSi_2)
	  – to reduce the bulk resistance of the polycide interconnects (double layer of polysilicon/silicide)
	• Si_3N_4 deposition
	  – to produce an insulating layer over the polycide for the overlapping contacts

Table 8.14. (continued)

No.	• Description of each process step • Cross-section through the silicon wafer after the last process step described

15	• Photolithography using mask 7
	  – to define the polycide interconnects and the source/drain regions of the MOS transistors
	• Si_3N_4/silicide/polysilicon etching
	  – to produce the polycide interconnects
	• Photoresist etching
	  – to remove the photoresist mask
	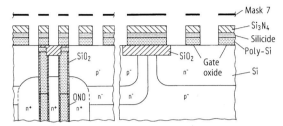

16	• Photolithography using mask 8
	  – to define the LDD zones of the N-channel transistors (Sect. 3.5.3)
	• Ion implantation of arsenic
	  – to define the LDD zones of the N-channel transistors
	• Photoresist etching
	  – to remove the photoresist mask
	• Photolithography using mask 9
	  – to define the LDD zones of the P-channel transistors
	• Ion implantation of BF^+_2 ions
	  – to dope the LDD zones of the P-channel transistors
	• Photoresist etching
	  – to remove the photoresist mask
	• Si_3N_4 deposition
	  – to produce a diffusion barrier
	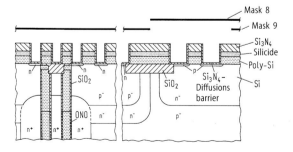

Table 8.14. (continued)

No.	• Description of each process step • Cross-section through the silicon wafer after the last process step described
17	• BPSG deposition (Sect. 3.6.2) 　– to produce the spacers (Sect. 3.5.3) • Photolithography using mask 10 　– to define the N-channel transistor regions outside the cell field • BPSG and Si_3N_4 etching 　– to produce the spacers in the N-channel transistor region outside the cell field • Ion implantation of arsenic 　– for n^+ doping of source and drain in the N-channel transistors outside the cell field • Photoresist etching 　– to remove the photoresist mask • Photolithography using mask 11 　– to define the heavily p-doped source/drain regions of the P-channel transistor • BPSG and Si_3N_4 etching 　– to produce the spacers in the P-channel transistor region • Ion implantation of boron or BF^+_2 　– for p^+ doping of source and drain in the P-channel transistors • Photoresist etching 　– to remove the photoresist mask • Ti deposition 　– to produce a "salicide" film (self aligned silicide Sect. 3.9.1) in those Si regions in the N- and P-channel transistors exposed by masks 10 and 11 • Annealing 　– to form the salicide film • Ti etching 　– to remove the residual titanium layer on the Si_3N_4 and BPSG areas 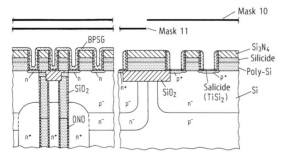
18	• Photolithography using mask 12 　– to remove the BPSG film in the memory cell field • BPSG etching 　– to remove the BPSG film in the memory cell field • Photoresist etching 　– to remove the photoresist mask • Si_3N_4 deposition 　– to produce an etch barrier for subsequent BPSG etching

Table 8.14. (continued)

No.	• Description of each process step • Cross-section through the silicon wafer after the last process step described
18	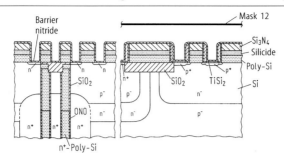

19	• BPSG deposition
	  – using the TEOS process (Sect. 3.6)
	  – for vertical isolation
	• BPSG planarization
	  – e.g. by chemical-mechanical polishing
	• Photolithography using mask 13
	  – to define the overlapping (borderless) contacts to the n-type drain regions of transistors in the memory cell field
	• BPSG etching
	  – to open up the contact holes
	• Photoresist etching
	  – to remove the photoresist mask
	• Si_3N_4 etching
	  – anisotropic etching, during which the Si_3N_4 film remains against the vertical walls of the BPSG contact holes
	  – to expose the Si contact surfaces
	• Ion implantation of arsenic
	  – to reduce the contact resistances
	• n^+ polysilicon deposition
	  – to fill the contact holes
	• Planarization of the polysilicon layer
	  – by chemical-mechanical polishing
	  – planarization stop at BPSG layer
	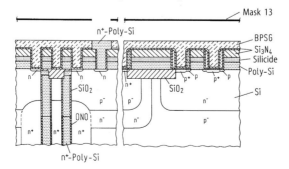

8.6 Detailed process sequence of selected technologies

Table 8.14. (continued)

No.	• Description of each process step • Cross-section through the silicon wafer after the last process step described

20	• Photolithography using mask 14
	  – to define the contact holes outside the memory cell field
	• Two-stage etch process
	  – to etch the BPSG and Si_3N_4 layer
	• Photoresist etching
	  – to remove the photoresist mask

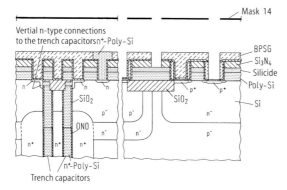

| 21 | • Selective tungsten deposition (Sect. 3.10)
  – only over exposed silicide regions to produce the "W studs"
• Chemical-mechanical polishing of the tungsten to remove the W outside the contact holes
  – to planarize the wafer surface
  – with BPSG layer acting as polish stop
• SiO_2 deposition
  – using the PECVD process (Sect. 3.5)
• Photolithography using mask 15
  – to define the interconnects in the metal 0 level
• SiO_2 etching
  – to remove the SiO_2 layer where the interconnects will later be located (Damascene technique, Sect. 8.5)
• Photoresist etching
  – to remove the photoresist mask
• TiN/W deposition (Sect. 3.10)
  – conformal W deposition to fill the trenches in the SiO_2 layer (W plugs)
• Chemical-mechanical polishing of W
  – to planarize the wafer surface
  – to remove the W outside the trenches |

Table 8.14. (continued)

No.	• Description of each process step • Cross-section through the silicon wafer after the last process step described
21	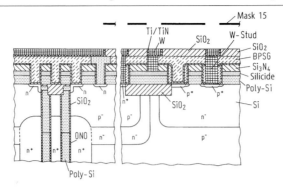

22	• SiO$_2$ deposition
	  – using the PECVD technique (Sect. 3.5)
	• Photolithography using mask 16
	  – to define the contact holes (vias 1) between metal 0 and metal 1 level
	• SiO$_2$ etching
	  – to produce the contact holes between metal 0 and metal 1 level
	• Photoresist etching
	  – to remove the photoresist mask
	• TiN/W deposition
	  – to fill vias 1 (W plugs)
	• Chemical-mechanical polishing of W
	  – to planarize the wafer surface
	  – to remove the W outside the vias 1
	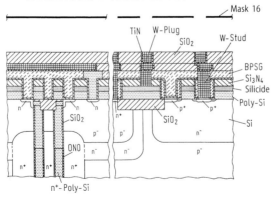

23	• SiO$_2$ deposition
	  – using the PECVD technique
	• Photolithography using mask 17
	  – to define the 1st metallization level

8.6 Detailed process sequence of selected technologies   321

Table 8.14. (continued)

No.	• Description of each process step • Cross-section through the silicon wafer after the last process step described

23	• SiO$_2$ etching
	  – for local removal of the SiO$_2$ layer and subsequent filling with metal (Damascene technique, Sect. 8.5)
	• Photoresist etching
	  – to remove the photoresist mask
	• Metallization (1st metallization level)
	  – 1. Ti deposition
	    2. AlCu deposition
	    TiN deposition to produce a diffusion barrier between Al and W
	    W deposition to fill the trenches in the SiO$_2$ layer
	• Chemical-mechanical polishing of the W/TiN/AlCu/Ti
	  – to planarize the wafer surface
	  – SiO$_2$ acts as polish stop
	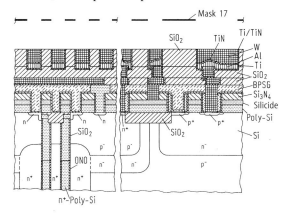

24	• SiO$_2$ deposition
	  – using the PECVD technique
	• Photolithography using mask 18
	  – to define the contact holes between the 1st and 2nd metallization level (vias 2)
	• SiO$_2$ etching
	  – to open up the vias 2
	  – sloping SiO$_2$ sides for better edge coverage (Sect. 8.5.2)
	• Photoresist etching
	  – to remove the photoresist mask
	• Metallization (2nd metallization plane)
	  – 1. Ti deposition
	    2. AlCu deposition
	    3. TiN deposition as diffusion barrier
	• Photolithography using mask 19
	  – to define the interconnects in the 2nd metallization level
	• TiN/Al/Ti etching
	  – to produce the interconnects in the 2nd metallization level

322 **8 Process integration**

Table 8.14. (continued)

No.	• Description of each process step • Cross-section through the silicon wafer after the last process step described
24	• Photoresist etching – to remove the photoresist mask

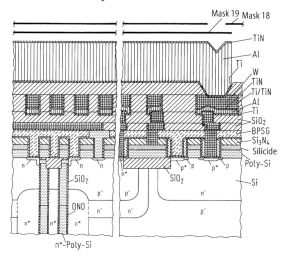

References

Chapter 2

2.1 Horninger, K.: Integrierte MOS-Schaltungen, 2. Aufl. Berlin: Springer 1986
2.2 Rein, H.M.; Ranfft, R.: Integrierte Bipolarschaltungen. Berlin: Springer 1980
2.3 Müller, R.: Bauelemente der Halleiter-Elektronik. Berlin: Springer 1987
2.4 Sze, S.M.: Physics of semiconductor devices. New York: Wiley 1981
2.5. Ruge, I.; Mader H.: Halbleiter-Technologie, 3. Aufl. Berlin: Springer 1991
2.6 Madelung, O.; Schulz, M.; Weiss, H.: Landolt-Börnstein. Neue Serie Bd. 17c, Technologie von Si, Ge und SiC. Berlin: Springer 1984
2.7 Sze, S.M.: VLSI-Technology. New York: McGraw-Hill 1983
2.8 Einspruch, N.G.; Brown, D.M.: VLSI Electronics. New York: Academic Press 1984
2.9 Wolf, S.; Tauber, R.N.: Silicon Processing for the VLSI Era. Subset Beach, California: Lattice Press 1986
2.10 Hacke, H.J.: Montage Integrierter Schaltungen. Berlin: Springer 1987
2.11 Schwabe, U.; Herbst, H.; Jacobs, E.P.; Takacs, D.: IEEE Trans Electron Devices ED-30 (1983) 1339
2.12 Jacobs, E.P.; Takacs, D.; Schwabe, U.: IEDM Techn Dig (1984) 642
2.13 Mader, H.: Microstructuring in semiconductor Technology, Thin Solids Films 175 (1989) 1–16

Chapter 3

3.1 Kern, W.; Schnable, G.L.: IEEE Trans. Electron Devices ED-26 (1979) 674
3.2 Yieh, B.; Nguyen, B.; Tribula, D.: Semiconductor Fabtech. 94 (1994) 205
3.3 Venkatesan, M.; Beinglass, I.: Solid State Technology (March 1993) 49
3.4 Iida, S.: JST News, 2 (1983) 29
3.5 Firmenschrift P5000 der Firma AMAT
3.6 Mathuni, J.: Siemens AG, Private communications
3.7 Chen, J.Y.; Henderson, R.: J. Electrochem. Soc. 131 (1984) 2147
3.8 Grove, A.S.: Physics and Technology of Semiconductor Devices. New York: Wiley 1967, p. 7–34
3.9 Hess, D.W.; Deal, B.: J. Electrochem. Soc. 124 (1977) 735
3.10 Razouk, R.R.; Lie, L.N.; Deal, B.E.: J. Electrochem. Soc. 128 (1981) 2214
3.11 Pawlik, D.: Siemens AG, Private communications
3.12 Widmann, D.: IEEE J. Solid-State-Circuits SC-11 (1976)
3.13 Gossner, H.; Baumgärtner, H.; Hammerl, E.; Wittmann, F.; Eisele, I.; Lorenz, H.: Jpn. J. Appl. Phys. 33 (1994) 2268
3.14 Kasper, E.; Wörner, K.: VLSI Science and Technology (1984) 429
3.15 Maissel, L.: Handbook of Thin Film Technology. New York: McGraw-Hill 1970, Chap. 4
3.16 Deppe, H.R.; Hieke, E.; Sigusch, R.: Semiconductor Silicon (1977) 1082
3.17 Burggraaf, P.: Semiconductor International (Oct. 1982) 37
3.18 Colinge, J.P.: Silicon-On-Insulator Technology. Kluwer Academic Publishers, 1991
3.19 Yallup, K.: Semiconductor Fabtech. (1994) 189
3.20 Sedgwick, T.O.: Semiconductor Silicon (1982) 130
3.21 Lam, H.W.; Tasch, A.F.; Pinzotto, R.F.: VLSI Electronics 4 (1982) 1

3.22 Kolbesen, B.O.; Strunk, H.P.: VLSI Electronics. Microstructure Sci. 12 (1985) 143
3.23 Hoenlein, W.; Siemens AG, Private communications
3.24 Murakami, S. et al.: IEEE J. Solid-State Circuits 26 (1991) 1563
3.25 Borland, J.O.; Schmidt, D.N.; Stivers, A.R.: Extended Abstracts Conf. Solid State Devices and Materials, Tokyo 1986, 53
3.26 Kooi, E.: The Invention of LOCOS, IEEE Case Histories of Achievement in Science and Technology, Vol. 1, 1991
3.27 Chin, K.Y.; Fang, R.; Lin, J.; Moll, J.L.: Technical Digest VLSI Symp., Oiso 1982, p. 28
3.28 Roth, S.S.; Ray, W.; Mazuré, C.; Kirsch, H.C.: IEEE Electron Device Letters 12 (1991) 92
3.29 Hofmann, K.; Weber, W.; Werner, C.; Dorda, G.: Technical Digest IEDM (1984) 104
3.30 Weber, W.; Brox, M.; Künemund, T.; Mühlhoff, H.M.; Schmitt-Landsiedel, D.: IEEE Trans. Electron Devices ED-38 (1991) 1859
3.31 Winnerl, J.; Lill, A.; Schmitt-Landsiedel, D.; Orlowski, M.; Neppl, F.: IEDM Tech. Dig. (1988) 204
3.32 Bergholz, W.; Mohr, W.; Drewes, W.: Materials Science and Engineering 4 (1989) 359
3.33 Kakoschke, R.: Proceedings 3rd International Rapid Thermal Processing Converence, Amsterdam (1995)
3.34 Küsters, K.H. et al.: Digest Symp. VLSI Technology, Karuizawa 1987, 93
3.35 Wieder, A.W.: Siemens Forsch. Entwicklungsber. 13 (1984) 246
3.36 Becker, F.S.; Pawlik, D.; Schäfer, H.; Stuadigl, G.: J. Vac. Sci. Technol. B4,3 (1986) 732
3.37 Adams, A.C.; Capio, C.D.: J. Electrochem. Soc. 132 (1985) 1472
3.38 Ito, T.; Ishikawa, H.; Shinoda, M.: Jpn. J. Appl. Phys. 20-1 Supplement (1981) 33
3.39 Watanabe, T.; Menjoh, A.; Ishikawa, M.; Kumagai, J.: Technical Digest IEDM (1984) 173
3.40 Risch, L.; Krautschneider, W.H.; Hofmann, F.; Schäfer, H.: Extended Abstracts ESSDERC (1995)
3.41 Richardson, W.F. et al.: Technical Digest IEDM (1985) 714
3.42 Sinha, A.K.: J. Vac. Sci. Technol. 19 (1981) 778
3.43 Intemann, A.S.: Siemens AG, Dissertation 1994
3.44 Körner, H.; Erb, H.P.; Melzner, H.: Applied Surface Science (1995)
3.45 Murrmann, H.; Widmann, D.: IEEE Trans. Electron Devices ED-16 (1969) 1022
3.46 Ruge, I.; Mader, H.: Halbleiter-Technologie, 3. Aufl. Berlin (1991) 123–134
3.47 Ahne, H.; Niederle, C.; Rubner, R.: Siemens-Zeitschrift Special, FuE (Frühjahr 1994) 30
3.48 Gerthsen, Ch.; Kneser, H.O.; Vogel, H.: Physik, 14. Aufl. Berlin: Springer 1982, S. 188–197
3.49 Koenig, H.R.; Maissel, L.I.: IBM J. Res. Dev. 14 (1970) 168

Chapter 4

4.1 Dill, F.H.; Neureuther, A.R.; Tuttle, J.A.; Walter, E.J. IEEE Trans. Electron Devices ED-22 (1975) 456
4.2 Oldham, W.G.; Nandgaonkar, S.N.; Neureuther, A.R.; O'Toole, M.M.: IEEE Trans. Electron Devices ED-26 (1979) 717
4.3 Widmann, D.W.; Binder, H.: IEEE Trans. Electron Devices ED-22 (1975) 467
4.4 Mader, L.; Widmann, D.; Oldham, W.G.: Proceedings Microcircuit Engineering (1981) 105
4.5 Sebald, M.; Berthold, J.; Beyer, M.; Leuschner, R.; Nölscher, C.; Scheler, U.; Sezi, R.; Ahne, H.; Birkle, S.: Proc. SPIE 1466 (1991) 227
4.6 Tai, K.L.; Vadimsky, R.G.; Ong, E.: Proc. SPIE 333 (1982) 32
4.7 Hatzakis, M.: Sold State Technol. (Aug. 1981) 74
4.8 Roland, B.; Coopmans, F.: Extended Abstracts 18th Conf. Solid State Devices and Materials, Tokyo 1986, p. 33
4.9 Mac Donald, S.A.; Miller, R.D.; Willson, C.G.: Proc. Kodak Interface 1982, p. 93
4.10 Griffing, B.F.; West, P.R.: Extended Abstracts 16th Conf. Solid State Devices and Materials, Kobe 1984, p. 7
4.11 Herriot, D.R.; Collier, R.J.; Alles, D.S., Stafford, J.W.: IEEE Trans. Electron Devices ED-22 (1975) 385
4.12 Firmenschrift der Firma Canon

4.13 Grassmann, A.; Prein, F.; Zell, T. et al.: Proc. SEMICON Europe, Geneva (April 95)
4.14 Markle, D.A.: Solid State Technology (Sept. 1984)
4.15 Cuthbert, J.D.: Solid State Technology (Aug. 1977) 59
4.16 Mader, L.; Lehner, N.: Proc. SPIE 2440 (1995)
4.17 Horiuchi, T.; Takeuchi, Y.; Matsuo, S.; Harada, K.: IEDM Digest of Techn. Papers (1993) 657
4.18 Shiraishi, N.; Hirukawa, S.; Takeuchi, Y.; Magome, N.: Microlithography World (July/August 1992)
4.19 Ferguson, R.; Ausschnitt, C.; Chang, I.; Farrel, T.; Hashimoto, K.; Liebmann, L.; Martino, R.; Maurer, W.: Symp. on VLSI Technol. (1994) 89
4.20 Jinbo, H.; Yamashita, Y.: IEDM Digest of Techn. Papers (1990) 285
4.21 Lin, B.J.: Solid State Technology (Jan. 1992) 43
4.22 Levenson, M.D.; Viswanathan, N.S.; Simpson R.A.: IEEE Trans. Electron Devices ED-29 (1982) 1828
4.23 Hershel, R.: Proc. SPIE 275 (1981) 23
4.24 Spears, D.L.; Smith, H.I.: Electron. Lett. 8 (1972) 102
4.25 Tischer, P.: From Electronics to Microelectronics. Amsterdam: North-Holland 1980, p. 46
4.26 Taylor, G.N.: Solid State Technol. (June 1984) 124
4.27 Betz, H.; Chen, J.T.; Heuberger, A.; Asmussen, F.; Sotobayashi, H.; Schnabel, W.: J. Electrochem. Soc. 130 (1983) 180
4.28 Heuberger, A.; Betz, H.: Proc. ESSDERC, München 1982, p. 121
4.29 Trinks, U.; Nolden, F.; Jahnke, A.: Nucl. Instrum. Methods 200 (1982) 475
4.30 Heuberger, A.: Tagungsband NTG-Tagung Baden-Baden, März 1983, S. 105
4.31 Doemens, G.: Proc. 11th CIRP Int. Seminar, June 1979
4.32 Roberts, E.: Solid State Technol. (Feb. 1984) 111
4.33 Kyser, D.F.; Viswanathan, N.S.: J. Vac. Sci. Technol. (1975) 1305
4.34 Greeneich, J.S.: Semiconductor Int. (April 1981) 159
4.35 Parikh, M.: J. Vac. Sci. Technol. 14 (1978) 931
4.36 Speth, A.J.; Wilson, A.D.; Kern, A.; Chang, T.H.P.: J. Vac. Sci. Technol. 12 (1975) 1235
4.37 Pfeiffer, H.C.: J. Vac. Sci. Technol. 15 (1978) 887
4.38 Firmenschrift "The MEBES System" der Firma Silicon Valley Group
4.39 Firmenschrift "The AEBLE System" der Firma Silicon Valley Group
4.40 Scott, J.P.: J. Vac. Sci. Technol. 15 (1978) 1016
4.41 Lischke, B. et al.: Proc. Int. Conf. Microlithography, Paris 1977, p. 167
4.42 Friedrich, H.; Zeitler, H.U.; Bierhenke, H.: J. Electrochem. Soc. 124 (1977) 627
4.43 Mader, H.: Lithography. In: Landolt-Börnstein, Neue Serie Bd. 17c, Technologie von Si, Ge und SiC. Berlin: Springer 1984, S. 250–280, 542–555
4.44 Stengl, G.; Löschner, H.; Muray, J.J. Solid State Technol. (Feb. 1986) 119
4.45 Stengl, G.; Löschner, H.; Maurer, W.; Wolf, P.: J. Vac. Sci. Technol. B4, 1 (1986) 194
4.46 Miyauchi, E.; Morita, T.; Takamori, A.; Arimoto, H.; Bamba, H.; Hashimoto, H.: J. Vac. Sci. Technol. B4, 1 (1986) 189
4.47 Bartelt, J.L.: Solid State Technol. (May 1986) 215
4.48 Stengl, G.; Kaitna, R.; Löschner, H.; Rieder, R.; Wolf, P.; Sacher, R.: Proc. Microcircuit Eng. 81, Lausanne 1981, p. 345
4.49 Morimoto, H.; Onoda, H.; Kato, T.: Sasaki, Y.; Saitoh, K.; Kato, T.: J. Vac. Sci. Technol. B4, 1 (1986) 205
4.50 Randall, J.N.; Stern, L.A.; Donnelly, J.P.: J. Vac. Sci. Technol. B4, 1 (1986) 201
4.51 McGillis, D.A.: Lithography. In: Sze, S.M. (Ed.): VLSI Technology, New York: McGraw-Hill 1983, p. 297
4.52 Fichtner, W.: Process Simulation. In: Sze, S.M. (Ed.): VLSI Technology, New York: McGraw-Hill 1983, p. 427
4.53 Brault, R.G.; Miller, L.J.: Polymer Eng. Sci. 20 (1980) 1064
4.54 Karapiperis, K.; Adesida, L.; Lee, S.A.; Wolf, E.D.: J. Vac. Sci. Technol. 19 (1981) 1259
4.55 Ryssel, H.: Proc. Microcircuit Eng., Lausanne, 1981
4.56 Stengl, G.; Kaitna, R.; Löschner, H.; Rieder, R.; Wolf, P.; Sacher, R.: J. Vac. Sci. Technol. 19 (1981) 1164
4.57 Rieder, R.; Löschner, H.; Kaitna, R.; Sacher, R.; Stengl, G.; Wolf, P.: Private Communication 1981

4.58 Ryssel, H.; Glawischnig, H.: Springer Series in Electrophysics; Ion Implantation 11 (1983) 242
4.59 Mohondro, R.: Semiconductor Fabtech (1996) 177
4.60 Csepregi, L.; Iberl, F.; Eichinger, P.: Microcircuit Engineering 80, Amsterdam (1980)
4.61 Ryssel, H.; Prinke, G.; Bernt, H.; Haberger, K.; Hoffmann, K.: Appl. Phys. A27 (1982) 239

Chapter 5

5.1   Mader, H.: Etching processes. In: Landolt-Börnstein. Neue Serie Bd. 17c, Technologie von Si, Ge und SiC. Berlin: Springer 1984, S. 280–305
5.2   Ruge, I.; Mader, H.: Halbleiter-Technologie. 3. Aufl. Berlin: Springer 1991
5.3   Horiike, Y.; Shibaaki, M.: Jpn. J. Appl. Phys. 15 (1976) 13
5.4   Coburn, J.W.; Winters, H.F.: J. Vac. Sci. Techn. 16
5.5   Steinfeld, J.I. et al.: J. Electrochem. Soc. 127 (1980) 514
5.6   Mader, H.: Etching processes. In: Landolt-Börnstein. Neue Serie Bd. 17c, Technologie von Si, Ge und SiC. Berlin: Springer 1984, S. 559–566
5.7   Beinvogl. W.; Mader, H.: Reactive dry etching of very-large-scale integrated circuits. Siemens Forsch. Entwicklungsber. 11 (1982) 181
5.8   Mogab, C.J.: Dry Etching. In: Sze, S.M. (Ed.): VLSI Technology. New York: MacGraw-Hill 1983, p. 303–345
5.9   Beinvogl., W.; Mader, H.: Reaktive Trockenätzverfahren zur Herstellung von hochintegrierten Schaltungen. ntz Arch. 5 (1983) 3–11
5.10 Harshbarger, W.R.; Porter, R.A.; Miller, T.A.; Norton, P.: Appl. Spectrosc. 31 (1977) 201
5.11 Harshbarger, W.R.; Porter, R.A.; Norton, P.: J. Electron. Mater. 7 (1978) 429
5.12 Korman, C.S.: Solid State Technol. 25 (1982) 115
5.13 Poulsen, R.G.; Smith, G.M.: Electrochem. Soc. Meeting, Philadelphia, Pensylvania Abstr. No. 242, May 1978
5.14 Curtis, B.J.; Brunner, H.J.: J. Electrochem. Soc. 127 (1978) 234
5.15 Degenkolb, E.O.; Mogab, C.J.; Goldrick, M.R.; Griffiths, J.R.: Appl. Spectrosc. 30 (1976) 520
5.16 Griffiths, J.E.; Degenkolb, E.O.: Appl. Spectrosc. 31 (1977) 134
5.17 Einspruch, N.G.; Brown, D.M.: VLSI Electronics. Plasma Processing for VLSI. Vol. 8. New York: Academic Press 1984, p. 411–446
5.18 Schwartz, G.C.; Schaible, P.M.: J. Vac. Sci. Technol. 16 (1979) 410
5.19 Adams, A.C.; Capio. C.D.: J. Electrochem. Soc. 128 (1981) 366
5.20 Flamm, D.L.; Wang, D.N.K.; Maydan, D.: J. Electrochem. Soc. 129 (1982) 2755
5.21 Endo, N.; Kurogi, Y.: IEEE Trans. Electron Devices ED-27 (1980) 1346
5.22 Paraszczazak, J.; Hatzakis, H.: J. Vac. Sci. Technol. 19 (1981) 1412
5.23 Ephrath, L.M.; DiMaria, D.J.; Pesavento, F.L.: J. Electrochem. Soc. 128 (1981) 2415
5.24 Gray, R.K.; Lechnaton, J.S.: IBM Techn. Disclosure Bull. 24 (1982) 4725
5.25 Einspruch, N.G.; Brown, D.M.: VLSI Electronics. Plasma Processing for VLSI. Vol. 8. New York: Academic Press 1984, p. 297–339
5.26 Engelhardt, M.; Schwarzl, S.: Personal communications 1987
5.27 Grewal, V.: Persönliche Mitteilungen 1987
5.28 Heath, B.A.; Mayer, T.M.: Reactive ion beam etching. In: Einspruch, N.G.; Brown, D.M. (ed.): VLSI-Electronics. Vol. 8. New York: Academic Press 1984, p. 365–408
5.29 Smith, D.L.: High-Pressure Etching. In: Einspruch, N.G.; Brown, D.M. (Ed.): VLSI Electronics. Vol. 8. New York: Academic Press 1984, p. 253–296
5.30 Gorowitz, B.; Saia, R.J.: Reactive Ion Etching. In: Einspruch, N.G.; Brown, D.M. (Ed.): VLSI Electronics. Vol. 8. New York: Academic Press 1984, p. 297–339
5.31 Horiike, Y.: Emerging Etching Techniques. In: Einspruch, N.G.; Brown, D.M. (Ed.): VLSI Electronics. Vol. 8. New York: Academic Press 1984, p. 447–486
5.32 Ehrlich, D.J.; Tsao, J.Y.: J. Vac. Sci. Technol. B1 4 (1983) 969–984
5.33 Chapmann, B.: Glow discharge processes. New York: Wiley 1980, 326
5.34 Janes, J.; Huth, Ch.: Appl. Phys. Lett. 61 (1992) 261
5.35 Beinvogl, W.; Deppe, H.R.; Stokan, R.; Hasler, B.; IEEE Trans. Electron Devices ED-28 (1981) 1332

5.36 Mathuni, J.: Private communications 1987
5.37 Müller, P.; Heinrich, F., Mader, H.: Microelectronic Engineering Elseview Science Publishers B.V. North Holland (1988)
5.38 Pilz, W.; Sponholz, T.; Pongratz, S.; Mader, H.: Microelectronic Engineering, North Holland 3 (1985) 467
5.39 Howard, B.J.; Steinbrüchel, Ch.: Conference Proc. ULSI-VIIII, Material Research Society (1993) 391–396
5.40 Dry Etching Application Notes. Firmenschrift der Firma ANELVA (1987)
5.41 Betz, H.; Mader, H.; Pelka, J.: Offenlegungsschrift, Deutsches Patent P3600346.8 (1986)
5.42 Mathuni, J.: Private communications 1995
5.43 Erb, H.-P.: Private communications 1995
5.44 ICE-Report No. 48068: Advanced VLSI Fabrication 1995
5.45 Erb, H.-P.; Münch, I.; Irlbacher, W.: Private communications 1993
5.46 Flamm, D.L.; Donelly, V.M.: The Design of Plasma Etchants; Plasma Chemistry and Plasma Processing, Vol. 1, No. 4, 1981
5.47 Frank, E.: Private communications 1995
5.48 Körner, H.: Private communications 1994
5.49 Engelhardt, M.: Private communications 1994
5.50 Schwarzl, S.: Private communications 1994
5.51 Daviet, J.-F.; Peccoud, L.: J. Electrochem. Soc., Vol. 140, No. 11 (1993) 3245–3261
5.52 Field, J.: Solid State Technology, September 1994, 91–98
5.53 Seidel, H.; Csepregi, L.; Heuberger, A.; Baumgärtel, H.: J. Electrochem. Soc. 137 (1990) 3626
5.54 Heuberger, A.: Mikromechanik. Berlin: Springer 1989
5.55 Singer, P.: Semiconductor International, Februar 1994, 48–52
5.56 Klose, R.: Private communications 1991
5.57 Deal, B.E.; Helms, C.R.: Mat. Res. Soc. Symp. 259 (1992) 361
5.58 Singer, P.: Semiconductor International, July 1992, 52–57
5.59 Campbell, G.A.; Chambrier, A. de; Tsukada, T.: SPIE 1803 (1992) 226
5.60 Franz, G.: Oberflächentechnologie mit Niederdruckplasmen. Berlin: Springer 1994
5.61 Janzen, G.: Plasmatechnik. Heidelberg: Hüthig 1992
5.62 Neumann, G.; Scheer, H.-C.: Rev. Sci. Instrum. 63 (1992) 2403
5.63 Heinrich, F.; Hoffmann, P.; Müller, K.P.: Microelectronic Engineering 13 (1991) 433
5.64 Börnig, K.; Janes, J.: Microelectronic Engineering 26 (1995) 217
5.65 Müller, K.P.; Roithner, K.; Timme, H.-J.: Microelectronic Engineering 27 (1995) 457

Chapter 6

6.1 Ruge, I.; Mader, H.: Halbleiter-Technologie, 3. Aufl. Berlin: Springer 1991, S. 82ff.
6.2 Ryssel, H.; Ruge, I.: Ionenimplantation. Stuttgart: Teubner 1978
6.3 Glawischnig, H.; Noack, N.: Ion Implantation. Science and Technology. Orlando, Fla.: Academic 1984, p. 313
6.4 Lindhard, J.; Schwarff, M.; Schiott, H.: Mat. Fys. Med. Dan. Vid. Selsk 33 (1963) 1
6.5 Morgan, D.V.: Channeling: Theory, Observation and Applications. New York: Wiley 1973
6.6 Hofker, W.K.: Philips Res. Rep. Suppl. 8 (1975)
6.7 Tsai, M.Y.; Streetman, B.G.: J. Appl. Phys. 50 (1979) 183
6.8 Runge, H.: Phys. Stat. Sol. (A) 39 (1977) 595
6.9 Hunter, W.R., et al.: IEEE Trans. Electron Devices ED-26 (1979) 353
6.10 Crowder, B.L.: J. Electrochem. Soc. 118 (1971) 943
6.11 Christel, L.A.; Gibbons, J.F.; Mylroie, S.: Nuclear Instrum. Methods 182/183 (1981) 187
6.12 Sze, S.M.: VLSI Technology. New York: McGraw-Hill 1983, p. 169–218
6.13 Antoniadis. D.A.; Hansen, S.E.; Dutton, R.W.: IEEE Trans. Electron Devices ED-26 (1979) 490
6.14 Lorenz, J.; Pelka, J.; Ryssel, H.; Sachs, A.; Seidl, A.; Svoboda, M.: IEEE Trans. Electron Devices ED-32 (1984) 1977
6.15 Bergholz, W.; Zoth, G.; Wendt, H.; Sauter, S.; Asam, G.: Siemens Forsch.- und Entwickl.-Ber 16 (1987) 241

6.16 Gösele, F.W.; Mehrer, U.; Seeger, A.: Diffusion in Crystalline Solids. New York: Academic 1984, p. 64

Chapter 7

7.1   Melzner, H.: Siemens AG, Private communications
7.2   Steinman, A.: Semiconductor Fabtech (1995) 203
7.3   Reichardt, H.: Semiconductor Fabtech (1995) 139
7.4   Kern, W.; Puotinen. D.A.: RCA Review (June 1970) 187
7.5   Schwartzman, S.; Mayer, A.; Kern, W.: RCA Review (March 1985)
7.6   Bitto, F.: Siemens AG, Private communications
7.7   Ohmi, T.: Semiconductor Fabtech (1995) 79
7.8   Rieger, F.: Siemens AG, Private communications
7.9   Schild, R.; Locke, K.; Kozak, M.; Heyns, M.M.: Proc. 2nd Int. Symp. UCPSS, Leuven (Sept. 1994) 31
7.10 Gath, H.C.; Honold, A.; Simon, R.: Semiconductor Fabtech (1994) 51

Chapter 8

8.1   Shimizu, S.; Kusunoki, S.; Kobayashi, M.; Yamaguchi, T.; Kuroi, T.; Fujino, T.; Maeda, H.; Tsutsumi, T.; Hirose, Y.: IEDM Techn. Digest (1994) 67
8.2   Gossner, H.; Eisel, I.; Risch, L.: Jpn. J. Appl. Phys. 33 (1994) 2423
8.3   Preussger, A.; Glenz, E.; Heift, K.; Malek, K.; Schwetlick, W.; Wiesinger, K.; Werner, W.M.: Proc. 3rd Intern. Symp. on Power Semicond. Devices and IC's (1991) 195
8.4   Bauer, F.; Stockmeier, T.; Lendenmann, H.; Dettmer, H.; Fichtner, W.: Elektrotechnik, Heft 3 (1994) 18
8.5   Stoisiek, M.: Siemens AG, Private communications
8.6   Rein, H.M.: Informationstechnik 34 (1992) 209
8.7   Meister, T.F.; Stengl, R.; Weyl, R.; Packan, P.; Schreiter, R.; Popp, J.; Klose, H.; Treitinger, L.: IEDM Techn. Digest (1992) 401
8.8   Rein, H.M.: Proceedings ESSDERC (1995)
8.9   Bertagnolli, E.: Siemens AG, Private communications
8.10 Eimori, T.; Ohno, Y.; Kimura, H.; Matsufusa, J.; Kishimura, S.; Yoshida, A.: IEDM Techn. Digest (1993) 631
8.11 Masuoka, F.: Symp. on VLSI Technology Digest of Techn. Papers (1992) 6
8.12 Heinrich, R.; Heinrigs, W.; Tempel, G.; Winnerl, J.; Zettler, T.: IEDM Techn. Digest (1993) 620
8.13 Onishi, S.; Hamada, K.; Ishihara, K.; Ito, Y.; Yokoyama, S.; Kudo, J.; Sakiyama, K.: IEDM Techn. Digest (1994) 843
8.14 Kiewra, E.; Eckstein, E.; Cote, W.; Hunt, D.; Kocon, W.; Restaino, D.; Wangemann, K.; Feldner, K.; Leslie, T.; Henkel, W.; Roehl, S.; Giammarco, N.; Radens, C.: Proceedings 12th VMIC (1994) 359
8.15 Koburger, C.; Adkisson, J.; Clark, W.; Davari, B.; Geissler, S.; Givens, J.; Hansen, H.; Holmes, S.; Lee, H.K.; Lee, J.; Luce, S.; Martin, D.; Mittl, S.; Nakos, J.; Stiffler, S.: Symp. on VLSI Technology Digest of Techn. Papers (1993) 441
8.17 Arden, W.; Roehl, S.; Sauert, W.: Siemens AG, Private communications
8.18 Müller, K.H.; Poehle, H.; Werner, W.: Siemens AG, Private communications
8.19 Lachner, R.; Werner, W.: Siemens AG, Private communications
8.20 Nesbit, L.; Alsmeier, J.; Chen, B.; DeBrosse, J.; Fahey, P.; Gall, M.; Gambino, J.; Gernhardt, S.; Ishiuchi, H.; Kleinhenz, R.; Mandelman, J.; Mii, T.; Morkado, M.; Nitayama, A.; Parke, S.; Wong, H.; Bronner, G.: IEDM Digest of Techn. Papers (1993) 627
8.21 Mader, H.: AEÜ, Band 42 (1988) 118

Subject Index

Aberration at Electron Lithography 150
abrasive 171
Absorption coefficient
    of X-rays 135
    of photoresist 99
Acceleration voltage at ion implantation 212
Acceptors 219
Activation of dopant atoms 219
active devices 249
active regions 252
Adhesion agent 101
Ag_2S/GeSe-system 113
Air filter 240, 241
alcaline developer 115
Alignment 118, 131, 144, 154
Alignment accuracy of optical equipment 130
Alignment marks detecting 78, 131, 154
Alignment optic 118
Alpha particles 269
Aluminium films 85
    crystal structure 86
    dry etching 189, 203
    Electromigration 87
    for interconnects 285
    Producing 85
    wet etching 172
Aluminium-aluminium contacts 90
Aluminium-silicon contacts 88
    Kirkendahl effect 88
    specific contact resistance 90
    spikes 88
Ammonia gas 114
Amorphization of silicon 216
Amorphous silicon 71
Anisotropy factor 169
Annealing in hydrogen 54
Annealing techniques 35
    forming gas annealing 54
    hydrogen annealing 54
    laser annealing 36
    rapid isothermal annealing 36
    rapid optical annealing 36
    rapid thermal annealing 36
    strip heating 37
Annealing
    of ion implanted layers 209
    of sputtered layers 31
annular illumination 127
Anodically coupled plasma etching 180
Antenna effect 218
Anti-reflective layer 78, 108, 125
ARC (Anti Reflex Coating) 129
Architecture of manufacturing processes 252
Arrhenius relationship 55, 192
Arsenic 187, 216
ASIC (Application Specific Integrated Circuit) 153
Atmospheric pressure CVD reactor 17, 59
Autodoping 44

back scattered electrons 148, 154
back sputtering 31
Backdoor etch 172
background light intensity 126
back-scatter of ions 164
Ballroom design 240
Barrel reactor 176
Barrier layer 234, 260
BCD (Bipolar/CMOS/DMOS) 251
BEOL (Back End Of Line) 252
BESOI technique (Bonded Etched-Back Silicon on Insulator) 34
BESSY 139
Bevelling 278
BIAS sputtering 31
BICMOS process 286
BICMOS technology 249, 251, 255, 265
Bilevel resist technique 112
Bipolar process 286
Bipolar technology 249, 251, 255
Bipolar transistors 250, 251, 264
Bipolar-technology
    comparison to MOS-technology 249
Bird's beak 48

Bird's head   53
Bit line   268, 269, 273
black chromium   121
Bonding wires   285
borderless contacts   282
Boron   187, 207
Boron nitride   284
Bottom resist   111
Box profile   224
BOX technique (Buried Oxide)   62
BPSG (Boron Phosphorus Silicate Glass)   66, 280, 291, 317
Breakdown charge   57, 217
Breakdown field strength   56, 217, 274
Breakdown voltage   57
Bright field method   131
Bromine   186, 187
BST (Barium strontium Titanate)   271
Buried Channel   55, 261
Buried channel transistor   56
Buried contacts   76, 82, 282, 283
Buried layer   255, 296, 302
Buried layer islands   44
Buried plate   77, 311
Buried strap   283
Buried-Strap contact   77
Burn-in   57
butted contacts   282

caesium iodide film   153
CAIBE (Chemically Assisted Ion Beam Etching)   175, 182
Cantilever   26
Capacitance   253
Capacitor dielectric
   multi layer   69
   silicon nitride   69
Capacitors   252, 253
capped contact   281
Carboxylic acid   114
Carbon   187
Caro's acid   102
CDE (Chemical Dry Etching)   175, 176
CEL (Contrast Enhancing Layer)   115
Channel of MOS-transistors
   channel length   256
   channel width   256
Channel stopper   302
Channeling   209, 212
CHE (Channel Hot Electron)   274
CHE cell   274
Chemical dry etching   170, 175, 177
Chemical etching reactions   186
Chemical-physical wet etching   170

Chip area   271
Chip assembly   10
Chip cards   251
Chip yield   98
Chlorine   186, 187
Chromium area on a reticle   126
Chromium deposition, laser-induced   133
Chromium masks   121
circulating air   240
clean materials   242
clean processing   244
clean room   239
clean room classification   239
cleaning procedures   236
CMOS (Complementary MOS)
   analogue process   254
   basic process   252, 254
   digital process   286
   Inverter   43
   Technology   5, 249, 251
   Transistors   251
CMP (Chemical Mechanical Polishing)   66, 171, 279, 281, 312, 321
COG (Chromium On Glass)   127
Collar   77, 312
Components in Integrated Circuits
   active components   249
   passive components   252
contact hole   276
   etching   202
   cleaning   247
   resistance
      Al/monocristaline silicon   81, 90
      Al/polsilicon   81
      Al/polycide   81
Contact printing   117
Contacts   281
   buried   282
   butted   282
   nested capped   282
   non-capped   282
   non-nested   282
   self-aligned   61, 263, 282
   straps   282
   vias   280
Contacts in integrated circuits   281
contaminants   235
   in silicon   230, 235
   in SiO_2 53, 235
contamination   171, 217, 242
Contrast detection method   144
Contrast of photoresist   102, 115
control gate line   273
Cool wall reactor   17, 80

Subject Index 331

cool water supply 242
COP resist 137, 146, 158
Copper
    Diffusion 231
    Dry etching 190
    Interconnections 203, 285
Corrosion of aluminium 204
CO-sputtering 32
Cost of manufacture DRAM 271
COSY (Compact synchrotron) 141
critical pattern planes 259
Crystal damage due to ion implantation 217
Crystallographic plane of silicon 38
CVD (Chemical Vapour Deposition) 13
    diffusion-controlled 14, 16
    Epitaxy 43
    reaction-controlled 14, 16
    Reactors 17
CVD processes 13
    LP CVD (Low Pressure CVD) 15
    MO CVD (Metal Organic CVD) 84
    PE CVD (Plasma Enhanced CVD) 18
Czochralski technique 39
CZ-process 39

Damascene technique 279, 281
Dark field method 131
Dark light transition 123
de-Broglie wavelength 164
deep collector diffusion 265
deep collector implantation 304
Deep UV 125
Defect density 133, 174, 238
defective resist patter 97
Defects 96, 133
Defocusing 121, 123, 276
Degradation 57, 260
degree of contamination 242
deionized water 242
denudes zone 41
Dep./Etch process 20, 175, 279
Depletion transistors 268, 272
Depletion zone 236
Developer 3
Developing concentration 100
Developing temperature 100
Developing time 103
Development of photoresist 99
Diazonaphtoquinone 98
Diffraction contrast method 130
Diffraction orders 121
Diffusion barrier 227, 234, 260
Diffusion constants 14, 221
Diffusion equation 221

Diffusion length 221
Diffusion of doping atoms 14, 43, 207, 219
    at high concentration of doping atoms 223
    at interfaces 225
    at the edge of doped regions 230
    from a dopant layer 207
    from the gas phase 207
    in layers 227
    into epitaxial layer 43
    intrinsic 220
    Oxidation enhanced 224
Diffusion of non-doping materials 231
Diffusion profile 224
Diffusion rate 227
Digital CMOS process 286, 287
Diodes 252, 253
Disilicides 79
dislocations by ion implantation 216, 220
distortion of the wafer 133
DI-water (deionized water) 242
DMOS circuits 10
DMOS transistors 61, 250, 262
Donors 219
Doping dependance of etch rate 172, 198
Doping technique 3, 4, 207
    Ion implantation 209
    thermal doping 208
Doping
    of polysilicon 73
    of Si-wafers 39
Dose curve 103
Dose for resist exposure 103
Drain Source punchthrough 98
DRAM process 295, 310
DRAM
    cell area 271
    memory cell 269, 270
    Process architecture 254
    technology 251, 310
Drive In 288, 296
Dry etching 170, 174
Dry etching processes 196
    of aluminium 203
    of metal silicides 200
    of monocristalline silicon 199
    of polymers 205
    of polysilicon 197
    of refractory metals 200
    of silicon dioxide 201
    of silicon nitride 197
Dummy features 279
Duty factor 55
DUV Lithography 126, 129

332 Subject Index

E^2PROM 75, 272
    Process architecture 254
    technology 249, 251
early gate-drain breakdown 216
ECR etching (Electron Cyclotron Resonance) 182
Edge bevelling 171
Edge bevelling techniques 278
Edge contrast method 130
Edge placement error 96
edge transistion 276
Electromigration 87, 260, 276
Electron back scattering 148
Electron beam direct write machine 117, 145
Electron beam pattern generator 148
    direct write machine 145, 148
        pattern generator 145
        projection printer 145, 153
        proximity printer 145
        wafer stepper 145
Electron beam vaporization 175
Electron cyclotron resonance (ECR) 182
Electron induced chemical etching 175
Electron lithography 143
    alignment techniques 154
    radiation damage 154
    resolution capability 146
Electron resists 146, 158
Electron scattering 147
Electron shower 217
Electron source 149
Electrostatic Charging 155, 241
Electrostatic wafer clambing 186
Ellipsometry 196
Emergency showers 241
Emission spectroscopy 194
Emitter push effect 224
encapsulation 10
Endpoint detection 174, 193
    Emission spectroscopy 194
    Laser interferometry 194
Endpoint detection 193
Epitaxial layers 44
Epitaxy 14, 41
    selective 42, 261
EPROM 75, 273
Etch mask 169, 179
Etch process optimization 188
Etch processes
    Chemical dry etching 170, 176
    Dry etching 170
    Physical-chemical etching 170
    Physical dry etching 170

Wet chemical etching 170
Wet etching 170
Etch profiles 169
    of anisotropic etch processes 169, 179
    of isotropic etch processes 169
Etch rate 169
Etch residues 192
Etching gases 188
Etching solutions 172
etching techniques uses sources 182
Etching technology 3, 4, 169
eutectic point of Al/Si 62
Exposure dose 99, 100, 103
Exposure equipment 117
Exposure latitude 125
Exposure wavelength 105

Far UV 125
fatal defect density 238
FBM resist 137
feature density 272
Feature size 2
    growing 1
    world market 122
FEOL (Front End od Line) 252, 276
Ferroelectrics 275
ferroelectric memory 273, 275
FIB (Focussed Ion Beam) 158
Field doping 48
Field ionization source 159
Field oxidation 303
Field oxide 48, 252, 256
Field oxide transistor 48
Fill factor 122
Fill mask 279
film deposition
    conformal 15
    selective 279
film production 13
film technology 3, 4, 13
Finesonic 247
Fire alarm 241
Flash E^2PROM 251
Flat 38
Floating gate transistors 250, 251, 273
Floating-gate 58, 273
Float-zone silicon 39
Float-zone technique 39
Flood Gun 218
FLOTOX (Floating Gate Tunneling Oxide) 273
Flow-glass 66
Fluorine 187

Subject Index

Fly´s eye lens 127
FN (Fowler-Nordheim-Tunneln) 273
FN flash cell 274
Focal depth 161
Focal plane 123, 165
Focus latitude 125
Forming gas 31
Forming gas annealing 31, 54
Frequency dependance in dry etching 192
Fresh air treatment 240
Fresnel diffraction 120
Fresnel zone method 130
Fuses 272
fuses 36, 272
FZ process 39

Gallium 187
Gas flow
    at CVD 14
    at dry etching 188
Gas pressure at dry etching 179, 191
Gas sensors 241
Gas supply 242
Gate of MOS transistors
    gate oxide thickness 260
    gate length 256
    gate oxide 48, 256
Gaussian profile of doping atoms
Generation processes 188, 193
Getter centres 40
Gettering 40, 234
g-line 125
global alignment 132
Glow discharge 178
Gold 231
Grain boundaries in polysilicon 73
Grain structure
    of aluminium 86
    of polysilicon 72

Hall sensors 250
Hardening of photoresists 101
HBT (Heterojunction Bipolar Transistor) 266
HCl concentration at oxidation 25
HDP (High Density Plasma) 182
HE (Hot-Electron)- degradation 55, 56, 57, 260
heavy metal atoms 216, 231
heavy metal contamination 231, 236
Helicon source 182, 184
Helium nuclei 269
High pressure oxidation 24
High-temperature nitride 68

High-value resistors 75, 253, 286, 300, 304
Hillocks 86
HMDS (Hexamethyldisilazan) 101, 111
Honeycomb floor 240
hot Al 279
hot electrons 54
hot plate 100
Hot wall reactor 17
HTO (High Temperature oxide) 59
Huang cleaning 244

IBE (Ion Beam Etching) 175
IBIM (Ion Beam Induced Mixing) 34
ICP (Inductively coupled Plasma) 182
IGBT (Insulated Gate Bipolar Transistor) 263
i-line 125
Image reversal technique 115
IMD (Intermetal Dielectric) 63
in situ control of dry etching 193
inductively coupled plasma etching 180
inorganic resists 112
Integrated circuit technologies
    applications 251
    properties 251
Integrated circuits
    Fabrication costs 2
    Feature size 2
    growing 1
    Packing density 2
    Productivity 2
    world market 2
Intensity change in photoresists 123
Intensity gradient 122
Intensity profile in photoresists 106, 109
Interconnection 284
    aluminium 203
    metals 284
Interconnection plane 63
interconnections 258
Interface properties
    Interface charge 53
    Si/SiO_2.Interface 53
    Substrate/Epitaxial Layer 44
Interface states 53, 54
Interference effects 109
Intermetal dielectric layer 276
interstitial atoms 224
intrinsic diffusion 220
intrinsic gettering 40
Iodine 186, 187
Ion beam etching 157
Ion beam projection 160, 162
Ion beam writing 158

Subject Index

Ion current density 162
Ion etching 175
Ion exchanger 242
Ion implantation 157, 207, 209
    direction of ion bombardment 213
    doping 209
    film production 34
    machines 209
    mask 215
    oblique implantation 77
Ion implantation machines 209
    high-current 209
    low-current 209
    medium-current 209
Ion lithography 156
    exposure dose 162, 164
    exposure time 162
    pattern generation techniques 157
    resolution capability 162, 165
Ion mask 162
Ion resists 156
Ion sources 182, 209
Iron 231
Isolation of neighbouring transistors 252
isotropic etch profile 169
ITM (Implantation Through Metal) 80

Kirkendahl effect 88
Knock on implantation 217
Koehler illumination 127

Landau decay 185
Lanthanium hexaboride tip 148
Laser annealing 37
Laser interferometer 118, 154, 194
Laser vaporization 36, 75, 175
Latch up effect 43, 258
Lateral scattering of ions in the resist 162
Lattice sites 219
LDD (Lightly Doped Drain) 54, 260, 261, 290, 316
leakage current 236, 260
lens aberration in ion lithography 165
Lens aperture 121
lens optic 118
Life time killer 231
Life time of a MOS-transistor 55
Life time of reactive species 191, 193
Lift off technique 28
Light diffraction 102
Light intensity in resist 102, 105
Light reflexion 108, 276
Line width variations 78, 96, 107, 276
Liquid gallium source 160

Lithography 3, 4, 95
    Electron lithography 156
    Ion lithography 156
    Photo lithography 98
    X-ray lithography 134
Loading effects 193, 198
Local Interconnects 261, 283, 292
local oxidation 5
LOCOS-nitride 47, 197
LOCOS-technique 47, 197, 257, 288
Low alpha moulding material 285
Low pressure plasma 178
Low temperature plasma 178
LP CVD (Low Pressure CVD) 15, 172
LSS theory 211
LTO (Low Temperature Oxide) 59
LTV (Local Thickness Variation) 39

Magnetically enhanced reactive ion etching 180
Magnetron sputtering 32
Magnetron sputtering system 32
Mandrel technique 279, 281
Maragoni drying 247
Mask 4
    Gatemask 257
    Ion implantation 215
    Isolationmask 257
Mask defects 133
Mask distortion 143
Mask pattern edge intensity 120
Mask repairing 133, 159
Mask stepper 117
Mass memory 271
Mass separator 159
Mass spectroscopy 196
Matching 192
MBE (Molecular Beam Epitaxy) 29
MCT (MOS Controlled Thyristor) 263
Mean free paths 191
MEBES (Mask Electron Beam Exposure System) 151
Medium-current ion implantation 210
Megasonic treatment 247
memory cells 267
    dynamic memory 267
    non-volatile memory 272
    static memory 267
MERIE (Magnetically Enhanced Reactive Ion Etching) 180
Metal deposition 27, 29
Metal silicide
    dry etching 190, 200
    fabrication 79

properties 78
wet etching 173
Metallization pattern 174
Metallization planes 63, 321
MFA resist 137
MFC (Mass Flow Control) 241
MIBL (Masked Ion Beam Lithography) 160
Micro mechanics 171
Miller capacitance 230
miniaturization of circuits 258, 261
Mini-environment concept 241
minimum feature size 271
MNOS transistor 69, 273
MO CVD (Metal Organic CVD) 84
Molecular beam epitaxy 29, 42
Molybdenium 83, 187
Monochromacy 153
monochromatic exposure 107
Monocrystalline silicon wafer 38
    Diameter 38
    Flat 38
    Geometry and crystallography 38
Monocrystalline silicon
    dry etching 189, 199
    fabrication 14, 41
    wet etching 172, 173
Monte Carlo Simulation 147, 163
MORI (Mode M=0 Resonant Induction) 182
MOS technologies 249
MOS transistors 250
    fabrication 4
    set up 250, 256
Mosaic target 32
MOS-technology
    comparison to Bipolar-technology 249
Moulding material 285
MTF (Mean Time to Failure) of Aluminium interconnects 87
Muli-level metallization 62
Multi-beam write system 153
Multilayer metallization 62
Multi-wafer system 17

Negative resist 98
nested contacts 281
Neutralization 183
Newton´s interference 107, 125
Nitridation 58
non-capped-contacts 261, 282
non-nested contacts 282
Notching 109
Novolack resin 98
numerical aperture 122, 164
NVM (Non Volatile Memory) 251, 272

OAI (Off Axis Illumination) 127
oblique implantation 77, 261
Off-axis exposure 127
ONO (Oxide/Nitride/Oxide) 270, 312
Optical exposure techniques 116
Optical exposure techniques 116
    Projection exposure 117, 120
Optimization of etch processes 188
Organic films 91
Organic residues 235
Out-diffusion 228
outgoing air 240
Oxidation barrier, Silicon nitride 68
Oxidation constant
    Dependance on HCl-concentration 25
    linear 21
    parabolic 22
Oxidation tube furnace 26
Oxidation
    Doping dependance 24
    high pressure oxidation 24
    RTO (Rapid Thermal Oxidation) 26
    thermal 20
Oxide breakdown 170
Oxide stability 235
Oxinitride 69
Oxygen 187
Oxygen clusters 40

Packing density 259
Pad oxide 287
Pads 171, 285, 294, 301
Parallel plate reactor 178
Parasitic thyristors 257, 258
Parasitic transistors 257, 258
particle contamination 237, 239, 245
Passivation layers 10, 70, 197, 234, 285
Path difference for waves 105
Pattern generator 117
pattern planes 277
pattern transfer 169, 174
Patterngenerator 117
PBL (Poly Buffered LOCOS) 48
PBS resist 137, 146
PE CVD (Plasma Enhanced CVD) 19, 59, 277
    reactors 19
Pedestal collector 266
Pellicle technique 134
PELOX technique (Polysilicon Encapsulated Local Oxidation) 50
P-etch 172
PH_3 source 208
Phase contrast method 131
Phase shifting masks 128

Phosphin 66
Phosphorus 187, 208
Phosphorus doping of polysilicon 71
Phosphorus glass films 64, 209
　　deposited phosphorus glass 65
　　Flow glass 66
　　producing 64
　　thermal phosphorus glass 65, 67
Photo-cathode mask 153
Photoelectrons 136
Photolithography 98
Photon induced chemical etching 175
Photoresist 3
　　stripping 5
Photoresist films 98
　　Absorptions coefficient 99
　　Contrast 100
　　Development 98
　　Exposure 98, 109
　　Exposure dose 99
　　Hardening 101
　　Negative resist 98
　　Positive resist 98
　　Sensitivity 99
Photoresist pattern, formation 102
Photoresist techniques 110
　　Bilevel resist technique 112
　　Single level resist technique 113
　　Trilevel resist technique 111
photosensitive component 99
Physical dry etching 170, 174
Physical-chemical dry etching 170, 178
pile-down 226
pile-up 226
Placement error 96, 261
　　edge placement error 97
　　overlay error 97
　　statistical distribution 97
Planar etch 172
Planar technology 3
Planarization 261, 276, 279
　　of intermediate oxides and intermetal dielectrics 173, 206
　　of metal plugs 173
　　of trench fillings 173
Planarization techniques 276, 279
Plasma diagnostic 196
Plasma etching 175
　　in Barrel reactor 175
　　in parallel plate reactor 178
Plasma nitride 68, 234, 285
Plasma oxide 64, 285
Plasma processes 19, 54, 175
Plasma sources 182

Plasma
　　Low pressure plasma 178
　　Low temperature plasma 178
PMMA resist 146, 163
$POCl_3$-source 208
point defects 219
point-of-use filter 242
poisoned via 280
Polarization of ferroelectric materials 275
Polishing particles, abrasive 171
Polishing slurry 171
Poly load 75, 268
Polycide films 81, 200
Polycrystalline silicon 4, 70
Polyimid films 92, 285
Polymeres
　　Applications 205
　　Dry etching 205
Polysilicon 70
　　Base 75, 286
　　Emitter 75, 286
Polysilicon films 70
　　Applications 74
　　Conductivity 72
　　dry etching 189, 197
　　Producing 70
　　wet etching 172, 173
Positive resist 98
Post exposure bake 105
Postbake 101
Postbake of Photoresist 101
Power delay product 259
Power switch 249
Prebake 101
prebake temperature 100
precipitation of oxygen inside silicon wafer 41
Price formula 238
Process architecture 252, 255
Process integration 249
processing modules (blocks) 5, 252, 254, 255
　　productivity 2
Projection exposure 107
Projection image field 118
Projection scanner 117
PROM 272
Proton beam sensitivity of resists 158
Proximity effect 148, 150
　　Proximity exposure 117, 120
　　Wafer stepper 118, 120
Proximity printer 117
Proximity printing 119
　　in Photolithography 119
　　in X-ray Lithography 135

PSG (Phosphorous Silicate Glass)   63, 172
PSM (Phase Shifting Mask)   127
    alternating   128
    attenuated   128
    halftone PSM   129
    Levenson PSM   129
    rim PSM   129
puddle development   101
punch-through   47
PVD (Physical Vapor Deposition)   29
PVDF   242

Q_{bd} (Charge to breakdown)   56
quadrant sensor   132
quadrupole illumination   127
quartz boat   26

radiation damage   27, 31
    in electron lithography   154
    in X-ray lithography   144
radicals   54, 177
ramping   26
range
    of electrons   155
    of ions   212
Rapid annealing   220
rapid isothermal annealing   37
rapid optical annealing   37
raster scan   151
Rayleigh criterion   124
Rayleigh depth   124
RCA cleaning   346
    RE CVD (Radiation Enhanced CVD)   20
    SA CVD (Sub-Atmospheric Pressure CVD)   15, 42, 63
Reaction constant in dry etching   193
Reaction gas   21, 175, 188
reaction product   175, 176, 187
Reaction rate in CVD   14
Reaction-controlled process   14
Reactive ion etching (RIE)   175, 180, 290
recess etching   77
recessed-LOCOS technique   51
recoil effect   34
recoil implantation (knock-on)   217
recombination processes   188
recrystallization   43
RECVD (Radiation Enhanced CVD)   20
Redeposition   174
redundant memory cells   271
reflectance of wafers   126
Reflow technique   66, 203
Refractive Index
    of photoresist   105
    of substrate   105

refractory metal films
    molybdenum   83
    tantalum   83
    titanium   83
    tungsten   83
refractory metals
    fabrication   83
    dry etching   190, 200
Reliability of CMOS circuits   260, 276
repair of crystal defects   219
residence time of the reactive species   191
Resist   95
    contrast   100, 103
    deposition   102
    electron resist   145
    hardening   101
    implantation   157
    ion resist   158
    photoresist   98
    profile   103
    sensitivity   99
    stripping   287
    technique   110, 113
    thickness   103
    X-ray resist   136
resistivity
    of monocrystalline silicon   39
    of polysilicon   73
Resistor   252
Resolution capability
    of electron lithography   146
    of ion lithography   162, 165
    of optical exposure techniques   119, 125, 127
    of X-ray lithography   135
Reticle   118, 133
Reticle reflexions   126
retrograde wells   258, 261
reverse image   114
reverse osmosis process   242
rf-Power at dry etching   192
RIBE (Reactive Ion Beam Etching)   175, 182
RIE (Reactive Ion Etching)   175, 180, 290
rim   129
RIPE (Resonant Inductive Plasma Etching)   182
RISE (Reactive Ion Stream Etching)   182
ROM (Read Only Memory)   75, 272
RSE (Reactive Sputter Etching)   180
RTA (Rapid Thermal Annealing)   37
RTN (Rapid Thermal Nitridation)   58
RTO (Rapid Thermal Oxidation)   26, 58

SA CVD (Sub-Atmospheric-Pressure CVD)   16, 42

Subject Index

safety precaution 205
Salicide 80
Salicide (self-aligned silicide) 80, 230, 261, 291, 317
scaling factor 259
scaling principle 259
scanning wafer stepper 119
scatter of implanted atoms 215
scattering club 147, 155
Schottky diode 253
screen oxide 211, 216
screen oxide 298
screen oxide films 216
secondary electrons 154
segregation 23, 72, 225
segregation coefficient 225
selectivity 169
sensors 249, 251
    for magnetic field 249
    for pressure 249
    for radiation 249
    for temperature 249
shaped beam 150
sheet resistance
    of monosilicon 90, 228
    of polysilicon 73
    shunt 230
shrink 271
SIC (Selective Implanted Collector) 267
Si-gate MOS process 75
SiGe epitaxy 267
SiGe hetero-epitaxy 42
Silane 70
Silane oxide technique 59
Silication
    of source/drain regions 83
    selective 83
Silicide films
    dry etching 19, 200
    properties 78
    producing 79
    wet etching 173
Silicon epitaxy
    selective 16
Silicon nitride films
    applications 69
    dry etching 174, 197
    high-temperature nitride layers 68
    plasma nitride 68
    producing 68
    wet etching 172
silylation 112
SIMOX process (Separation by Implantation of Oxygen) 34

SIMS (Secondary Ion Mass Spectroscopy) 227
SiO_2 films
    deposited 59, 279
    applications 46, 60
    breakdown characteristic 23
    degradation 57
    dry etching 174, 201
    properties 53
    thermal 46
    wet etching 172
Site-by-site alignment 132
site-by-site alignment 132
size of silicon wafers 38, 271
slurry 171
small-valued resistor 304
Smart-sensor technology 249, 251
SMIF (Standard Mechanical Interface) 242
snowplough effect 225
Sodium 231
Sodium contaminants 235
SOG (Spin-On Glass technique) 33, 64, 91, 279, 293
SOI (Silicon On Insulator) 34, 261, 264, 267
Solid phase epitaxy 220
solubility of oxygen in silicon 40
solvent for photoresist 101
Source line 273
Source/Drain implantation 214
spacer technique 49, 60, 203, 266, 289
spikes in silicon 88
spiking 260
spin coating 33, 91
spin-drying 247
Spin-on glass technique 279
spray cleaner 246
spray development 101
sprinkler jets 241
SPT (Smart-Power Technology) 10, 249, 251, 255
sputter cleaning 31
sputter coating 276
sputter etching 175
sputtering 29
    BIAS sputtering 31
    CO-sputtering 32
    Magnetron sputtering 32
sputtering system 30, 32
SRAM 268
    memory cell 268
    technology 251
stacked capacitor 78, 270
    crown stacked capacitor 270
    hemi-spherical grained silicon stacked capacitor 270

stacking faults 24, 231
standard deviation of ions 211
Standard MOS transistor 250, 256
standing wave effect 105
Step covering 16, 31, 33
STI (Shallow Trench Isolation) 62
storage capacitor 269
storage mode 236
storage node 236
storage ring 139
strap contacts 83, 282
stress relief 285
stress relief buffer 285
striations 107
stringers 276
strip heating 38
stripping photoresist 102, 176
substrate bias voltage 257
substrate doping 39
sub-threshold current 98, 236
supply voltage 259
surface channel 55, 261
Surface contaminants 235
surface strap 283
suszeptor 17
Synchrotron 136, 139

Tantalum 83, 187
target 29, 32
   mosaic target 32
   sintered target 32
TCP (Transmission Coupled Plasma) 180
TDDB (Time Dependent Dielectric Breakdown) 55
temperature relationships of dry etching 192
TEOS (Tetra-Ethyl-Ortho-Silicate) 60, 289
Test of integrated circuits 10
TFT (Thin Film Transistor) 43, 268
thermal doping 208
thermal phosphorus glass 67
thermal SiO_2 layers 46, 172
Ti/TiN-layer 84
TiN-layer 84
Titanium 83, 186, 203
TOC (total organic carbon) 243
top resist 111
topography 277
transistor transconductance 286
transistors
   in integrated circuits 256
   parasitic transistors 257
transit frequency 265
traps 216

Trench capacitor 77, 200, 269
Trench capacitor 77, 200, 269, 311
   buried plate 270
   poly plate 270
trench effect 174
Trench isolation 51, 63, 261, 314
Trench memory cell 77
TRIE (Triode Reactive Ion Etching) 180
Trilevel resist technique 91, 110
TSI (Top Surface Imaging) 110, 125
Tungsten 83, 187, 285
Tungsten plugs 284, 293
tunnel current 56, 273
tunnel oxide 56, 273
tunnel reactor 176

ULPA filters 240
Ultra-filtration 242
ultra-pure water 243
undercutting of a layer 169

V trenches 173
vacuum pick-up tools 244
vaporization source 27
Vapour phase deposition 27
   electron beam vaporization 27
Vapour pressure 187
vectorscan design 150
via crowns 236
via etching 236, 247
vias 293
Voids 276
voltage stability 262

wafer bonding 34, 264
wafer chuck 118
wafer clamping 186
wafer cleaning 244
wafer warpage 132
wafer warpage 133
Water contamination 243
wave interference 105
wave length in a vacuum 105
Well contact 256
Well potential 256
Well process module 252, 254
Wells in CMOS circuits 5, 258
   well contacts 258
Wet chemical etching 170
Wet chemical wafer cleaning 246
Wet etching 170
Wet oxidation 24
wetting agents 171
Whistler waves 185

White ribbon effect 47
Word line 268, 269, 273
write field area 152
write frequency 152

X-ray lithography 134
 alignment procedure 144
 radiation damage 144
 resolving power 135
 wavelength region 135

X-ray masks 142
X-ray projection 134, 144
X-ray resists 136
X-ray sources 137
X-ray tube 138
X-rays 134

Yellow-light room 102
Yield 11, 98, 238

Printing: Mercedes-Druck, Berlin
Binding: Stürtz AG, Würzburg